# LOGIC
## Techniques of Formal Reasoning

# LOGIC
# Techniques of Formal Reasoning

Donald Kalish
Richard Montague

UNIVERSITY OF CALIFORNIA, LOS ANGELES

NEW YORK CHICAGO SAN FRANCISCO ATLANTA

Harcourt, Brace & World, Inc.

*Library of Congress Catalog Card Number* 64-10046

PRINTED IN THE UNITED STATES OF AMERICA

# CONTENTS

# PREFACE

THE expressions 'logic', 'formal logic', 'symbolic logic', and 'mathematical logic' are in the just acceptation synonyms. They refer to a discipline created by Aristotle, extended by the Stoics, studied by the Scholastics, developed in the esoteric writings of Leibniz, and sent on its modern career in the late nineteenth century. The discoveries of Leibniz and the Stoics, though extensive, were until recently either ignored or misinterpreted, and thus had to be duplicated in the early twentieth century. Presently logic is cultivated as a branch of both philosophy and mathematics.

This book is an introduction to logic, requiring no prior knowledge of philosophy or mathematics. It does not aim at communicating results about logical systems (as do several excellent but more advanced texts), but instead at imparting a skill—the ability to recognize and construct correct deductions. Metamathematical results are indeed sometimes mentioned, but only incidentally and as an aid to understanding.

The logical apparatus introduced in the first seven chapters is essentially that of the first-order predicate calculus with identity and descriptive phrases. The *theorems* of those chapters (with a few exceptions connected with descriptive phrases) are thus standard and well known. (Indeed, the completeness and universal validity of the system has been proved in the publication listed in the bibliography as Montague and Kalish [1].) Some originality, however, may be claimed for the *rules* of the system, which constitute, we think, a closer approximation to everyday and mathematical reasoning than has previously been achieved by any formal system, and which suggest a number of simple and practical strategies for the discovery of proofs. In these chapters heavy emphasis is placed on applications to ordinary language.

To achieve the objective of mirroring everyday reasoning, we have sometimes had to sacrifice economy, though never precision. We have constructed a graduated but unified development of logic from the sentential calculus through the theory of descriptions, steadfastly resisting the temptation to introduce fragmentary though perhaps attractive procedures appropriate to isolated branches of logic. (For instance, sentential logic is not developed by means of truth tables, though they are introduced in an optional section.)

In the last two chapters first-order logic is extended (for the first time, it seems) so as to comprehend arbitrary variable-binding operators and is employed in a fairly detailed development of a familiar mathematical theory. The general theory of variable-binding operators and their definitions was developed in collaboration with Professor Dana Scott and will appear also in the monograph Montague, Scott, Tarski [1].

Although the book is written so as to be comprehensible without a teacher, it is primarily designed for use in a first course in symbolic logic, taught in a department of either philosophy or mathematics. Such a course may occupy either one or two semesters, and be set at any undergraduate level, from freshman to senior. A one-semester course should treat at least the first three, and at most the first seven, chapters, with chapter V probably omitted. A year course could be restricted to the first seven chapters (again with the possible omission of chapter V), in which case all the standard material of elementary logic would be treated, or preferably could cover all nine chapters. In a mathematics department it might be found desirable to begin the second semester with chapter VIII and to extend the mathematical development beyond the limits of chapter IX but along the lines suggested there. Another alternative is to cover the book in two one-semester courses set at different levels, the second semester beginning with chapter V, VI, or VIII. To facilitate this alternative, summaries suitable for reviewing the preceding material are given as appendices to chapters IV and VII. As an additional pedagogical aid a list of special symbols is given at the end of chapter IX (pp. 331 – 32).

The sections and chapter marked with an asterisk, together with the exposition in fine print, may be regarded as optional and can be omitted, in whole or in part, without loss of continuity to the main development.

For historical and bibliographical information we have relied heavily on two works of Alonzo Church, *A bibliography of symbolic logic* and *Introduction to mathematical logic*. For helpful suggestions and discussions we are indebted to Dr. J. D. Halpern and to Professors David Kaplan, Benson Mates, Ruth Anna Mathers, and Dana Scott.

# LOGIC
## Techniques of Formal Reasoning

# Chapter I

# 'NOT' and 'IF'

LOGIC is concerned with arguments, good and bad. With the docile and the reasonable, arguments are sometimes useful in settling disputes. With the reasonable, this utility attaches only to good arguments. It is the logician's business to serve the reasonable. Therefore, in the realm of arguments, it is he who distinguishes good from bad.

Virtue among arguments is known as validity. An argument is valid if it is impossible for its premises to be true and its conclusion false. But this is to speak loosely, and the reasonable do not countenance looseness.

That intuition is not a reliable judge of validity will appear from a few examples.

(1) Suppose that a student, Alfred, satisfies the following conditions. If he studies, then he receives good grades; if he does not study, then he enjoys college; if he does not receive good grades, then he does not enjoy college. Is it correct to draw any conclusion concerning Alfred's academic performance?

It is correct to conclude that Alfred receives good grades. This example is simpler than most of those we shall encounter, yet it is sufficiently complex to puzzle an unschooled intuition. The situation is worse when intuitively plausible premises lead by way of an intuitively valid argument to an obviously false conclusion, as in the next two examples.

(2) Suppose that Alfred, an inhabitant of Berkeley, shaves all and only those inhabitants of Berkeley who do not shave themselves. Does he shave himself or not? The answer is neither. For suppose he shaves himself. Then, since he shaves only those inhabitants of Berkeley who do not shave themselves, he clearly does not shave himself. Suppose, on the other hand, that he does not shave himself. Then, since he is himself an inhabitant of Berkeley and he shaves all inhabitants of Berkeley who fail to shave themselves, he must shave himself. Thus both possibilities lead to absurdity.

(3) Is the following boxed sentence true or false?

| The boxed sentence is false. |
|---|

Again, the answer is neither. For suppose the boxed sentence is true.

Then it is true that the boxed sentence is false. Hence the boxed sentence is false. On the other hand, suppose the boxed sentence is false. Then it is false that the boxed sentence is false. Hence the boxed sentence is true. Thus, if the sentence is true, it is false; and if it is false, it is true.

In the next example, which we owe to St. Anselm, the conclusion is not obviously false but only controversial. To those who believe it, however, the argument should appear blasphemously short.

(4) Even the atheist, who denies that God exists, must be able to conceive of Him in order to know what he denies. Suppose that God does not exist. Then the atheist can nevertheless conceive of Him as existing and thereby can conceive of something greater than God. But, by definition, God is that than which nothing greater can be conceived. Hence the atheist's supposition leads to contradiction. Therefore God exists.

**1. Symbols and sentences.** We shall analyze validity in steps. At first we shall restrict ourselves to valid arguments of a very special kind—roughly speaking, those arguments whose validity depends only on the phrases 'it is not the case that' and 'if . . . , then'.

Let us adopt the symbol '∼' as an abbreviation for 'it is not the case that'. Thus

(1) ∼ Socrates is bald

is read

(2) It is not the case that Socrates is bald.

The expression (1) is called the *negation* of 'Socrates is bald'.

The phrase 'if . . . , then' is used to combine two sentences, say,

(3) Rex is a dog

and

(4) Rex is carnivorous,

into a new sentence,

(5) If Rex is a dog, then Rex is carnivorous.

For the phrase 'if . . . , then' let us use the symbol '→', accompanied by a pair of parentheses. Thus (5) becomes

(6) (Rex is a dog → Rex is carnivorous) .

(6) is called the *conditional* formed from (3) and (4); (3) is known as the *antecedent*, and (4) as the *consequent*, of (6).

In addition to '∼' and '→', which abbreviate certain phrases, we shall use the capital letters 'P' through 'Z' as abbreviations for English sentences; 'A' through 'O' will be reserved for later purposes. For example, we may let 'P' stand for 'Rex is a dog' and 'Q' for 'Rex is carnivorous'. The sentence (6) will then acquire the concise form

(7) $$(P \rightarrow Q) \quad .$$

The relation between a capital letter and the sentence that it abbreviates, unlike that between the symbols '$\sim$' and '$\rightarrow$' and their English counterparts, is subject to change. By allowing 'P' and 'Q' to represent other sentences, we may, for instance, consider (7) an abbreviation for

If snow is white, then grass is green

or

If Empedocles is hoary, then Dalmatia is icebound.

Thus, in different contexts, we may ascribe different significance to the capital letters 'P' through 'Z'. The ascription will take the form of a *scheme of abbreviation;* such schemes will be discussed more fully in a moment.

Despite the latitude achieved by permitting shifts of abbreviation, it is conceivable that occasions will arise when we shall require more than eleven sentential abbreviations. To prevent embarrassment, therefore, we admit as possible abbreviations for English sentences not only the letters 'P' through 'Z' but also any variant of these letters obtained by adding a subscript, for example,

$$P_0 \quad ,$$

$$Q_{23} \quad .$$

Thus the language we consider initially is English, supplemented by '$\sim$' and '$\rightarrow$', parentheses, and *sentence letters* (that is, capital letters 'P' through 'Z', with or without numerical subscripts). To avoid accusations of provincialism, we should mention that the preferred status of English is a matter only of the authors' convenience; the subsequent treatment would apply as well to French, German, or Coptic.

Our interest will not extend to all sentences of English. We shall arbitrarily avoid questions, commands, and exclamations, leaving their treatment to logicians of the future. We shall be concerned exclusively with declarative sentences, that is, those sentences which are capable of truth or falsehood. The sentences of our language fall, then, into the following categories:

(1) declarative sentences of English, for example:

The text is readable,
Alfred will pass or the text is not readable;

(2) sentence letters, for example:

$$\text{T} \quad ,$$
$$\text{P}3 \quad ;$$

(3) sentences correctly constructed from sentences of categories (1) and (2) by means of '$\sim$', '$\rightarrow$', and parentheses, for example:

(Alfred will pass or the text is not readable $\rightarrow$ the text is readable) ,
(T $\rightarrow$ Alfred will pass or the text is not readable) ,
$\sim$(T $\rightarrow$ $\sim$P) .

To be more explicit, the class of *sentences* can be exhaustively characterized as follows:

(*1*) *All declarative sentences of English are sentences.*
(*2*) *Sentence letters are sentences.*
(*3*) *The negation of a sentence is a sentence.*
(*4*) *The conditional formed from two sentences is a sentence.*

Clauses (3) and (4) admit of the following alternative formulations:

(*3′*) *If $\phi$ is a sentence, then so is the result of writing '$\sim$' followed by $\phi$.*
(*4′*) *If $\phi$ and $\psi$ are sentences, then so is the result of writing '(' followed by $\phi$ followed by '$\rightarrow$' followed by $\psi$ followed by ')'.*

We shall use typographical displays in such a way that (3′) and (4′) will be respectively synonymous with the following more succinct formulations:

(*3″*) *If $\phi$ is a sentence, then so is*

$$\sim\phi \quad .$$

(*4″*) *If $\phi$ and $\psi$ are sentences, then so is*

$$(\phi \rightarrow \psi) \quad .$$

It is evident from the preceding discussion that no sentence of our language will contain Greek letters; although '$\sim$P' is a sentence, '$\sim\phi$' is not. We reserve Greek letters for the purpose of making general statements *about* sentences and, later, other expressions. Thus (3′), for example, is to be regarded as a generalization having the following assertion among its special cases: if '(P $\rightarrow$ Q)' is a sentence, then the result of writing '$\sim$' followed by '(P $\rightarrow$ Q)' (that is, '$\sim$(P $\rightarrow$ Q)') is again a sentence.

The *symbolic part* of our language comprises those sentences which are constructed exclusively from sentence letters, parentheses, '$\sim$', and '$\rightarrow$'. Accordingly, the class of *symbolic sentences* can be exhaustively characterized as follows:

(1) *Sentence letters are symbolic sentences.*
(2) *If $\phi$ and $\psi$ are symbolic sentences, then so are*

$$\sim\phi$$

*and*

$$(\phi \rightarrow \psi) \quad .$$

In (6) and in the characterizations of a sentence and a symbolic sentence, it may seem curious that parentheses accompany '$\rightarrow$'. Their function is like that of punctuation in written English and becomes conspicuous in the case of complex sentences. For example,

(8) $$(\sim\phi \rightarrow \psi)$$

is a conditional whose antecedent is

$$\sim\phi \quad ;$$

on the other hand,

$$\sim(\phi \rightarrow \psi)$$

is the negation of a conditional. Similarly, it is important to distinguish between

(9) $$((\phi \rightarrow \psi) \rightarrow \chi)$$

and

(10) $$(\phi \rightarrow (\psi \rightarrow \chi)) \quad .$$

To drop the inner parentheses of (9) and (10) would obliterate the distinction.

Although parentheses are generally required to prevent ambiguity, no confusion will arise if we omit the outermost parentheses of a sentence, and this we shall do frequently. Thus, for example, (8) and (9) may alternatively be written

$$\sim\phi \rightarrow \psi$$

and

$$(\phi \rightarrow \psi) \rightarrow \chi$$

respectively. Further, when parentheses lie within parentheses, some pairs may for perspicuity be replaced by pairs of brackets. For example,

$$((\phi \rightarrow \psi) \rightarrow \psi) \rightarrow \phi$$

may also be written

$$([\phi \rightarrow \psi] \rightarrow \psi) \rightarrow \phi \quad .$$

**EXERCISES**

State whether each of the following expressions is a nonsymbolic sentence, a symbolic sentence, or neither. Exercises 1 and 2 are solved for illustration.

1. $(\sim P \rightarrow (Q \rightarrow P))$

According to clauses (2) and (3) of the characterization of the class of sentences (p. 6), '$\sim P$' is a sentence; according to clauses (2) and (4), '$(Q \rightarrow P)$' is a sentence; thus, according to clause (4) again, No. 1 is a sentence. Further, it is a symbolic sentence, for it is constructed exclusively from sentence letters, parentheses, '$\sim$' and '$\rightarrow$'.

2. $(P \rightarrow Q \rightarrow R)$

Because the characterization of the class of sentences on page 6 is stipulated to be exhaustive, No. 2, if a sentence, must result from one of the four clauses of that characterization. But clause (1) is clearly inapplicable, and neither of the clauses (2) or (3) can generate an expression beginning with a parenthesis. Thus clause (4) must yield No. 2; and for this to happen, either 'P' and '$Q \rightarrow R$', or '$P \rightarrow Q$' and 'R', must be sentences. But in view of considerations like the foregoing, neither '$Q \rightarrow R$' nor '$P \rightarrow Q$' is, strictly speaking, a sentence; each fails to be a conditional by a pair of peripheral parentheses. Hence No. 2 is not a sentence.

3. If if if Alfred concentrates, then he will pass, then he will secure employment, then he will marry.
4. ($\sim$Alfred will pass $\rightarrow$ (the text is not readable $\rightarrow$ P))
5. $(((P \rightarrow P) \rightarrow P) \rightarrow P)$
6. If the text is readable $\rightarrow$ Alfred concentrates, then he passes.

**2. From symbols to English and back.** Frequently it will be desirable to *translate* a symbolic sentence into English and perform the reverse process of *symbolizing* an English sentence. We consider first the passage from symbols to English.

As was mentioned earlier, the correlation between sentence letters and the English sentences which they abbreviate is quite arbitrary. Thus the passage from a symbolic sentence to a sentence of English must proceed on the basis of a *scheme of abbreviation*, which will establish such a correlation.

More explicitly, let us understand by an *abbreviation* an ordered pair of sentences, the first of which is a sentence letter and the second an

English sentence. A *scheme of abbreviation* is a collection of abbreviations such that no two abbreviations in the collection have the same first member.

For example, the collections

(1)
Q : the lectures are dull
T : the text is readable
P : Alfred will pass

and

S : the lectures are dull
Q : the lectures are dull
T : the text is readable
P : Alfred will pass

are schemes of abbreviation; but the collection

Q : the lectures are dull
Q : the text is readable
P : Alfred will pass

is not. (The two abbreviations whose first member is 'Q' are at fault.)

The process of *literal translation into English on the basis of a given scheme of abbreviation* begins with a symbolic sentence and if successful ends with a sentence of English. The process consists of the following steps:

(*i*) *Restore any parentheses that may have disappeared as a result of applying the informal conventions of the last section.*

(*ii*) *Replace sentence letters by English sentences in accordance with the given scheme of abbreviation; that is, each sentence letter is to be replaced by the English sentence with which it is paired in the scheme.*

(*iii*) *Replace all occurrences of*

$$\sim\phi \quad ,$$

*where $\phi$ is a sentence, by*

*it is not the case that* $\phi$ .

(*iv*) *Replace all occurrences of*

$$(\phi \rightarrow \psi) \quad ,$$

*where $\phi$ and $\psi$ are sentences, by*

(*if* $\phi$, *then* $\psi$) .

For example, under the scheme of abbreviation (1) the sentence

(2) $$Q \rightarrow (\sim T \rightarrow \sim P)$$

becomes in step (i)

$$(Q \rightarrow (\sim T \rightarrow \sim P)) \quad ,$$

in step (ii)

(The lectures are dull → (∼ the text is readable → ∼ Alfred will pass)) ,

in step (iii)

(The lectures are dull → (it is not the case that the text is readable → it is not the case that Alfred will pass)) ,

and in step (iv)

(3) (If the lectures are dull, then (if it is not the case that the text is readable, then it is not the case that Alfred will pass)) .

(The parentheses in sentences such as (3), as in symbolic sentences, serve as marks of punctuation.)

We shall generally wish to be more liberal in translating from symbols to English than the notion of a literal translation will permit. For example, we should like to consider the sentence

(4) Assuming that the lectures are dull, if the text is not readable, then Alfred will not pass

as a translation of (2) on the basis of the scheme (1). Accordingly, we say that an English sentence is a *translation* (or, when a distinction is to be drawn, a *free translation*) of a symbolic sentence $\phi$ on the basis of a given scheme of abbreviation if it is a stylistic variant of the literal translation of $\phi$ into English on the basis of that scheme.

Because (4) differs only in style from (3), the former as well as the latter qualifies as an English translation of (2) on the basis of scheme (1).

In the realm of free translations, we countenance looseness. Specifically, we attempt no precise description of *stylistic variance;* in this connection, intuition (here identified with linguistic insight) rather than exact rules must guide the reader. To remove this source of looseness would require systematic exploration of the English language, indeed of what might be called the 'logic of ordinary English', and would be either extremely laborious or impossible. In any case, the authors of the present book would not find it rewarding.

Although no exact definition of stylistic variance will be offered, we shall not leave the reader entirely to his own devices. Two stylistic variants of

It is not the case that Alfred concentrated

are

Alfred did not concentrate

and

Alfred failed to concentrate.

English idiom provides a number of stylistic variants for

(5) (If Rex is a dog, then Rex is carnivorous) ,

for example,

(6) Rex is carnivorous if Rex is a dog,
(7) Rex is carnivorous provided that Rex is a dog,
(8) Rex is a dog only if Rex is carnivorous,
(9) Only if Rex is carnivorous is Rex a dog.

Between (5) and (6) the only difference (apart from the use of parentheses) is in word order. (7) comes from (6) upon replacement of 'if' by its intuitive equivalent 'provided that'. To see the intuitive equivalence between (5) and (8), the following consideration should be of assistance: to assert that Rex is a dog only if carnivorous is to deny that Rex is a dog and not carnivorous, which is to assert that if Rex is a dog, he is also carnivorous; but this amounts to (5). (9) comes from (8) by inversion of word order.

If $\phi$ and $\psi$ are any two sentences, a partial list of stylistic variants of

(if $\phi$, then $\psi$)

is the following:

if $\phi$, then $\psi$ ,
$\psi$ if $\phi$ ,
$\psi$ provided that $\phi$ ,
$\phi$ only if $\psi$ ,
only if $\psi$ $\phi$ ,
given that $\phi$, $\psi$ ,
$\psi$ in case $\phi$ ,
$\psi$ assuming that $\phi$ ,
$\psi$ on the condition that $\phi$ .

Further instances of stylistic variance may be obtained by introducing pronouns in place of nouns and by altering word order. Examples of these and other sorts of stylistic variance will be found among the exercises of this and later chapters.

We shall also be interested in the passage from English to symbols. Accordingly, we say that $\phi$ is a *symbolization* of an English sentence $\psi$ on the basis of a given scheme of abbreviation if and only if $\phi$ is a symbolic sentence that has $\psi$ as a translation on the basis of that scheme, in other words, if and only if $\psi$ is a stylistic variant of the literal translation of $\phi$ on the basis of the scheme.

To find a symbolization of a given English sentence on the basis of a given scheme of abbreviation, the reader will find it useful to proceed roughly as follows:

(1) *Introduce 'it is not the case that' and '(if . . . , then)' in place of their respective stylistic variants.*

(2) *Reverse the steps leading from a symbolic sentence to a literal English translation; that is,*

(2*a*) *replace all parts of the form*

$$(\textit{if } \phi, \textit{ then } \psi) \quad ,$$

*where $\phi$ and $\psi$ are sentences, by*

$$(\phi \rightarrow \psi) \quad ;$$

(2*b*) *replace all parts of the form*

$$\textit{it is not the case that } \phi \quad ,$$

*where $\phi$ is a sentence, by*

$$\sim\phi \quad ;$$

(2*c*) *replace English components by sentence letters in accordance with the scheme of abbreviation; that is, replace each English component by a sentence letter with which it is paired in the scheme of abbreviation;*

(2*d*) *omit peripheral parentheses and replace parentheses by brackets in accordance with the informal conventions of the preceding section.*

## EXERCISES

On the basis of the scheme of abbreviation

| | | |
|---|---|---|
| P | : | logic is difficult |
| Q | : | Alfred will pass |
| R | : | Alfred concentrates |
| S | : | the text is readable |
| T | : | Alfred will secure employment |
| U | : | Alfred will marry |
| V | : | the lectures are dull , |

translate the following symbolic sentences into idiomatic English.

1. $P \rightarrow (Q \rightarrow R)$
2. $S \rightarrow (R \rightarrow [\sim P \rightarrow Q])$
3. $(R \rightarrow P) \rightarrow \sim Q$

On the basis of the scheme of abbreviation above, symbolize the following English sentences. Exercise 4 is solved for illustration.

4. Only if Alfred concentrates will he pass.

Sentence No. 4 becomes in step (1) of the procedure given on page 12

(If Alfred will pass, then Alfred concentrates) ;

in taking this step we first made minor stylistic changes and replaced 'will he pass' by 'Alfred will pass', and then, in conformity with the discussion on page 11, we inverted the order of the component sentences. In step (2a) the sentence becomes

$$(\text{Alfred will pass} \rightarrow \text{Alfred concentrates}) \quad ,$$

step (2b) is inapplicable, step (2c) leads to

$$(Q \rightarrow R) \quad ,$$

and step (2d) to

$$Q \rightarrow R \quad .$$

5. If logic is difficult, Alfred will pass only if he concentrates.
6. Alfred will pass on the condition that if he will pass only if he concentrates then he will pass.
7. If if if Alfred concentrates, then he will pass, then he will secure employment, then he will marry.
8. It is not the case that if Alfred will secure employment provided that the text is readable, then he will marry only if he concentrates.

The following sentences are ambiguous, in the sense that the placement of parentheses in their symbolizations is not uniquely determined. Give all plausible symbolizations of these sentences on the basis of the scheme of abbreviation that appears above.

9. Alfred will pass only if he concentrates provided that the text is not readable.
10. It is not the case that Alfred concentrates if the lectures are dull.
11. Alfred will not secure employment if he fails to concentrate on the condition that the lectures are dull.

**3. Derivability and validity.** An *argument*, as we shall understand it, consists of two parts—first, a sequence of sentences called its *premises*, and secondly, an additional sentence called its *conclusion*. In the case of a valid argument the premises constitute conclusive evidence for the conclusion. Ordinarily we shall present an argument by listing first the premises, and then the conclusion, set off by the sign '∴' or the word 'therefore'. Three examples follow.

(1) $P \rightarrow Q \quad . \quad Q \rightarrow R \quad \therefore P \rightarrow R$

(2) ∴ Alfred will pass or he will not pass.

(3) If Socrates did not die of old age, then the Athenians condemned him to death. The Athenians did not condemn Socrates to death. ∴ Socrates died of old age.

It is possible for an argument to have an empty sequence of premises (that is, to have no premises at all), as in example (2). It is also possible for a valid argument to have a false conclusion, as in example (3). (When, at the end of this section, we define validity, it will be seen that argument (3) is indeed valid.) In the case of a valid argument we may be sure only that *if* all the premises are true, *then* the conclusion will be true.

Let us for a moment confine our attention to *symbolic arguments*, that is, arguments whose premises and conclusions are symbolic sentences. To establish the validity of a symbolic argument with the sentence $\phi$ as its conclusion, we construct a *derivation* of $\phi$ from the premises of the argument—that is, a sequence of steps, each justified in some way, which lead from the premises to the establishing of $\phi$.

Before presenting the rules for constructing a derivation, let us consider an example. Suppose that from the premise

(4) $$P$$

we wish to derive the conclusion

(5) $$(P \to Q) \to Q \quad .$$

An appropriate derivation may be constructed as follows. We write first the conclusion, (5), together with an indication that it is to be established:

*Show* $(P \to Q) \to Q$

Now (5) is a conditional, and we may establish a conditional by assuming its antecedent and deriving its consequent. Accordingly we add, as an assumption, the antecedent of (5):

*Show* $(P \to Q) \to Q$
$P \to Q$

Now we may add the premise, (4):

*Show* $(P \to Q) \to Q$
$P \to Q$
$P$

From the second and third lines of the derivation we may infer 'Q' by means of an *inference rule*. (A list of inference rules will be given shortly.) The rule used here is *modus ponens;* by *modus ponens* a symbolic sentence $\psi$ may be inferred from symbolic sentences

$$(\phi \to \psi)$$

and $\phi$. Thus we obtain:

*Show* (P → Q) → Q
P → Q
P
Q

We have now succeeded in deriving the consequent of (5). Hence (5) is established. To indicate this fact we *cancel* the occurrence of '*Show*' that precedes '(P → Q) → Q'. Further, we *box* the second, third, and fourth lines of the derivation to indicate that they have served their purpose. Thus we finally obtain:

~~*Show*~~ (P → Q) → Q

| |
|---|
| P → Q |
| P |
| Q |

The derivation of (5) from (4) is now complete.

The *inference rules* we shall employ are the following:

*Modus ponens* (MP): $$\frac{\begin{array}{c}(\phi \rightarrow \psi)\\ \phi\end{array}}{\psi}$$

*Modus tollens* (MT): $$\frac{\begin{array}{c}(\phi \rightarrow \psi)\\ \sim\psi\end{array}}{\sim\phi}$$

Double negation (DN), in two forms: $$\frac{\sim\sim\phi}{\phi} \qquad \frac{\phi}{\sim\sim\phi}$$

Repetition (R): $$\frac{\phi}{\phi}$$

That is, a symbolic sentence $\psi$ is said to follow by *modus ponens* from two other symbolic sentences if and only if these sentences have the forms

$$(\phi \rightarrow \psi)$$

and $\phi$; a symbolic sentence follows by *modus tollens* from two other symbolic sentences if and only if it has the form

$$\sim\phi$$

and the other sentences have the forms

$$(\phi \rightarrow \psi)$$

and

$$\sim\psi \quad ;$$

a symbolic sentence follows from another by *double negation* if and only if the two symbolic sentences have the forms $\phi$ and

$$\sim\sim\phi \quad ;$$

and a sentence $\phi$ follows by *repetition* from a symbolic sentence $\psi$ if and only if $\phi$ and $\psi$ are the same sentence. MP and MT correspond to familiar forms of reasoning; DN is the principle that a double negative amounts to an affirmative; and the function of the trivial rule R will become clear later.

For example, in the arguments

$$(P \to \sim Q) \quad . \quad P \quad \therefore \sim Q \quad ,$$
$$(P \to \sim Q) \quad . \quad \sim\sim Q \quad \therefore \sim P \quad ,$$
$$\sim\sim(P \to Q) \quad \therefore (P \to Q) \quad ,$$
$$(P \to Q) \quad \therefore \sim\sim(P \to Q) \quad ,$$
$$(P \to Q) \quad \therefore (P \to Q) \quad ,$$

the conclusion follows from the premises by the respective rules MP, MT, DN, DN, and R.

Suppose now that we have certain symbolic premises and that we wish to derive as a conclusion the symbolic sentence $\phi$. We begin by writing

*Show* $\phi$ .

We may continue in one of three ways, each of which has numerous intuitive counterparts in the derivations of mathematics and in the reasonings of law courts and everyday life:

(i) By *direct derivation*. We write next a line that can be established independently (for instance, a premise or a sentence accompanied by a subsidiary derivation) and proceed by inference rules, subsidiary derivations, and citing of premises until we secure $\phi$. A direct derivation of $\phi$, then, will have the form

(6) *Show* $\phi$
.
.
.
$\phi$ .

(ii) By *conditional derivation*, in case $\phi$ is of the form

$$(\psi \to \chi) \quad ,$$

where $\psi$ and $\chi$ are symbolic sentences. In this case we write next, as an assumption, the sentence $\psi$, and proceed by inference rules, subsidiary derivations, and citing of premises until we secure $\chi$. A conditional derivation of

$$(\psi \rightarrow \chi) \quad ,$$

then, will have the form

(7) *Show* $(\psi \rightarrow \chi)$
$\psi$ (Assumption)
.
.
.
$\chi$ .

(iii) By *indirect derivation*. In this case we write next, as an assumption, the sentence

$$\sim \phi$$

and proceed by inference rules, subsidiary derivations, and citing of premises until we secure a symbolic sentence $\chi$ and its negation,

$$\sim \chi \quad .$$

If $\phi$ is itself a negation, say

$$\sim \psi \quad ,$$

we may assume $\psi$ instead of

$$\sim \sim \psi$$

and proceed as above. Thus an indirect derivation of $\phi$ will have the form

(8) *Show* $\phi$
$\sim \phi$ (Assumption)
.
.
.
$\chi$
.
.
.
$\sim \chi$

or, in case $\phi$ is the negation of a sentence $\psi$,

(9) *Show* $\sim\psi$

$\psi$ (Assumption)

.

.

.

$\chi$

.

.

.

$\sim\chi$ .

Indirect derivation is also known as derivation by *reductio ad absurdum*, and depends for its cogency on the following consideration. To show that an assertion $\phi$ holds, it is sufficient to assume that it does not, and from this assumption to derive a contradiction (that is, a pair of sentences, one of which is the negation of the other); for then our assumption must be mistaken, and $\phi$ must hold.

When the derivation of $\phi$, accomplished by one of these methods, is complete, we indicate its completion by cancelling the occurrence of '*Show*' in the first line and boxing the remaining lines. (6) will then become

~~Show~~ $\phi$ (Direct derivation)

.

.

.

$\phi$ ,

(7) will become

~~Show~~ $(\psi \rightarrow \chi)$ (Conditional derivation)

$\psi$

.

.

.

$\chi$ ,

(8) will become

~~Show~~ $\phi$ (Indirect derivation)

$\sim\phi$

.

.

.

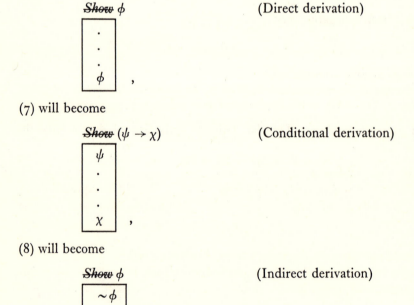

$\chi$
.
.
.
$\sim\chi$ ,

and (9) will become

~~*Show*~~ $\sim\psi$ (Indirect derivation)

$\psi$
.
.
.
$\chi$
.
.
.
$\sim\chi$ .

An example will clarify the use of subsidiary derivations. Suppose that we wish to derive 'P → ([Q → R] → R)' from 'P → Q'. The sentence we wish to derive is a conditional; we therefore start a conditional derivation:

1. *Show* P → ([Q → R] → R)
2. P (Assumption for conditional derivation)

To complete the conditional derivation, we must obtain the consequent of line 1, '[Q → R] → R'. One way of proceeding is to begin with a subsidiary derivation; and since the sentence we wish to establish is a conditional, it is natural to start a conditional derivation:

1. *Show* P → ([Q → R] → R)
2. P
3. *Show* [Q → R] → R
4. Q → R (Assumption for conditional derivation)

We continue the subsidiary derivation, employing inference rules and the premise (in a way that will be explained more fully later), until we reach the consequent of line 3.

1. *Show* P → ([Q → R] → R)
2. P
3. *Show* [Q → R] → R

4. $Q \to R$
5. $P \to Q$ (Premise)
6. $Q$ (2, 5, MP)
7. $R$ (4, 6, MP)

The subsidiary derivation establishing the assertion in line 3 is complete. Thus we may box and cancel to obtain:

1. *Show* $P \to ([Q \to R] \to R)$
2. $P$
3. ~~*Show*~~ $[Q \to R] \to R$ (Conditional derivation)
4. $Q \to R$
5. $P \to Q$
6. $Q$
7. $R$

Now the main derivation is complete. Hence we obtain:

1. ~~*Show*~~ $P \to ([Q \to R] \to R)$ (Conditional derivation)
2. $P$
3. ~~*Show*~~ $[Q \to R] \to R$
4. $Q \to R$
5. $P \to Q$
6. $Q$
7. $R$

The foregoing remarks on derivations constitute only an informal introduction. The following is an explicit set of directions for constructing a *derivation* from given symbolic premises:

(*1*) *If $\phi$ is any symbolic sentence, then*

*Show* $\phi$

*may occur as a line.* (*Such lines may be accompanied by the annotation* '*Assertion*'.)

(*2*) *Any one of the premises may occur as a line.* (*Annotation:* '*Premise*'.)

(*3*) *If $\phi$, $\psi$ are symbolic sentences such that*

*Show* $(\phi \to \psi)$

*occurs as a line, then $\phi$ may occur as the next line.* (*Annotation:* '*Assumption for conditional derivation*' *or simply* '*Assumption*'.)

(*4*) *If $\phi$ is a symbolic sentence such that*

*Show* $\phi$

*occurs as a line, then*

$$\sim\phi$$

*may occur as the next line; if* $\phi$ *is a symbolic sentence such that*

$$\textit{Show} \sim\phi$$

*occurs as a line, then* $\phi$ *may occur as the next line.* (*Annotation: 'Assumption for indirect derivation' or simply 'Assumption'.*)

(5) *A symbolic sentence may occur as a line if it follows by an inference rule from* antecedent lines, *that is, preceding lines which neither are boxed nor contain uncancelled 'Show'.* (*The annotation should refer to the inference rule employed and the numbers of the preceding lines involved.*)

(6) *When the following arrangement of lines has appeared:*

*Show* $\phi$
$\chi_1$
.
.
.
$\chi_m$ ,

*where none of* $\chi_1$ *through* $\chi_m$ *contains uncancelled 'Show' and either*

(*i*) $\phi$ *occurs unboxed among* $\chi_1$ *through* $\chi_m$,

(*ii*) $\phi$ *is of the form*

$$(\psi_1 \to \psi_2)$$

*and* $\psi_2$ *occurs unboxed among* $\chi_1$ *through* $\chi_m$, *or*

(*iii*) *for some sentence* $\chi$, *both* $\chi$ *and its negation occur unboxed among* $\chi_1$ *through* $\chi_m$,

*then one may simultaneously cancel the displayed occurrence of 'Show' and box all subsequent lines.* (*When we say that a sentence* $\phi$ occurs among *certain lines of a derivation, we mean that either* $\phi$ *or* $\phi$ *preceded by '~~Show~~' is one of those lines.*)

Applications of clauses (1) through (4) are quite straightforward. A derivation of '$(\sim Q \to \sim P) \to Q$' from the premise 'P' will illustrate applications of clauses (5) and (6).

| | | |
|---|---|---|
| 1. | *Show* $(\sim Q \to \sim P) \to Q$ | Assertion |
| 2. | $\sim Q \to \sim P$ | Assumption for conditional derivation |
| 3. | *Show* Q | Assertion |
| 4. | $\sim Q$ | Assumption for indirect derivation |
| 5. | P | Premise |
| 6. | $\sim P$ | 2, 4, MP |

The subjunction of line 6 constitutes an application of clause (5), for line 6 follows from the antecedent lines 2 and 4. The derivation can be continued by an application of clause (6), part (iii); for lines 3 – 6 have the pattern

*Show* $\phi$

$\chi_1$

.

.

.

$\chi_m$ ,

where none of $\chi_1$ through $\chi_m$ contains uncancelled '*Show*' and both 'P' and its negation occur unboxed among $\chi_1$ through $\chi_m$. Thus we may simultaneously cancel the last unboxed occurrence of '*Show*' and box all subsequent lines, to obtain:

1. *Show* $(\sim Q \rightarrow \sim P) \rightarrow Q$
2. $\sim Q \rightarrow \sim P$
3. ~~*Show*~~ Q
4. $\sim Q$
5. P
6. $\sim P$

Now an application of clause (6), part (ii), is possible, for lines 1 – 6 have the pattern

*Show* $(\sim Q \rightarrow \sim P) \rightarrow Q$

$\chi_1$

.

.

.

$\chi_m$ ,

where none of $\chi_1$ through $\chi_m$ contains uncancelled '*Show*' and 'Q' occurs unboxed among $\chi_1$ through $\chi_m$. Thus we may again simultaneously cancel the last unboxed occurrence of '*Show*' and box all subsequent lines, to obtain:

1. ~~*Show*~~ $(\sim Q \rightarrow \sim P) \rightarrow Q$
2. $\sim Q \rightarrow \sim P$
3. ~~*Show*~~ Q
4. $\sim Q$
5. P
6. $\sim P$

To ensure correct application of clauses (1) – (6), any parentheses that may have been dropped must be mentally restored. For example, we might be tempted to begin an indirect derivation of '$P \rightarrow Q$' as follows:

*Show* $P \rightarrow Q$
$\sim P \rightarrow Q$ (Assumption for indirect derivation)

But when missing parentheses are restored, this becomes:

*Show* $(P \rightarrow Q)$
$(\sim P \rightarrow Q)$ ,

in which the second line is clearly not a negation, and hence cannot be obtained correctly by clause (4). A correct indirect derivation would begin as follows:

*Show* $P \rightarrow Q$
$\sim(P \rightarrow Q)$ (Clause (4))

Similarly, given

(10) $P \rightarrow Q$

as a line, we might be tempted to infer by DN the sentence

$\sim\sim P \rightarrow Q$ .

But this inference is not permitted by our present rules. Instead, recalling the missing parentheses of (10), we obtain

$\sim\sim(P \rightarrow Q)$ .

A derivation is said to be *complete* if each of its lines either is boxed or contains cancelled '*Show*'.

A symbolic sentence $\phi$ is said to be *derivable* from given symbolic premises if, by using only clauses (1)–(6), a complete derivation from those premises can be constructed in which

~~*Show*~~ $\phi$

occurs as an unboxed line.

For example, the derivation

1. *Show* $P \rightarrow ([Q \rightarrow R] \rightarrow R)$
2. $P$
3. *Show* $[Q \rightarrow R] \rightarrow R$
4. $Q \rightarrow R$
5. $P \rightarrow Q$ (Premise)
6. $Q$
7. $R$

is not complete, but the derivation

| | | |
|---|---|---|
| 1. | ~~*Show*~~ $P \rightarrow ([Q \rightarrow R] \rightarrow R)$ | |
| 2. | $P$ | |
| 3. | ~~*Show*~~ $[Q \rightarrow R] \rightarrow R$ | |
| 4. | $Q \rightarrow R$ | |
| 5. | $P \rightarrow Q$ | |
| 6. | $Q$ | |
| 7. | $R$ | |

*is* complete; and in view of the latter derivation, '$P \rightarrow ([Q \rightarrow R] \rightarrow R)$' is derivable from '$P \rightarrow Q$'.

A very important feature of the notion given above of a derivation, and one that will be shared by all analogous notions to be introduced subsequently, is that there is a purely automatic procedure for checking the correctness of an alleged derivation from a finite class of premises; more generally, whenever a class of premises is such that one can automatically determine of any sentence whether it belongs to that class, there will be an automatic procedure for determining whether any alleged derivation is indeed a correct derivation from that class of premises. Thus the correctness of derivations, unlike the cogency of everyday reasonings, is removed from the realm of controversy.

A *symbolic argument* is said to be *valid* if its conclusion is derivable from its premises. Argument (1) of page 13 happens to be valid in view of the following derivation.

| | | |
|---|---|---|
| 1. | ~~*Show*~~ $P \rightarrow R$ | Assertion |
| 2. | $P$ | Assumption |
| 3. | $P \rightarrow Q$ | 1st premise |
| 4. | $Q$ | 2, 3, MP |
| 5. | $Q \rightarrow R$ | 2nd premise |
| 6. | $R$ | 4, 5, MP |

By an *English argument* is understood an argument whose premises and conclusion are sentences of English. A *symbolization of an English argument on the basis of a given scheme of abbreviation* is a symbolic argument whose premises and conclusion are respectively symbolizations, on the basis of that scheme, of the premises and conclusion of the English argument. A symbolic argument is called simply a *symbolization* of an English argument if there is some scheme of abbreviation on the basis of which it is a symbolization of the English argument. Consider, as an example, the argument:

(11) Free love is justified only if the sex drive is primary. Freud's theory is correct provided that the sex drive is primary.

It is not the case that if psychoanalysis is therapeutic, Freud's theory is correct. ∴ Free love is not justified.

On the basis of a natural scheme of abbreviation, the argument may be symbolized as follows:

$$(12) \qquad Q \rightarrow S \quad . \quad S \rightarrow T \quad . \quad \sim(P \rightarrow T) \quad \therefore \sim Q$$

We confine the application of deductive procedures (that is, the procedures whereby derivations are constructed) to symbolic sentences, for otherwise difficulties of analysis would arise which we prefer either not to treat at all or to treat only in connection with stylistic variance. Thus, in order to define validity for English arguments, we must proceed indirectly, as follows.

An *English argument* is said to be *valid* (in the branch of logic presently under consideration) if and only if it has a symbolization whose conclusion is derivable from its premises. (Other branches of logic, and wider notions of validity, will be considered later.) Argument (11) is valid, in view of the symbolization (12) and the derivation

| | | |
|---|---|---|
| 1. | ~~*Show*~~ $\sim Q$ | Assertion |
| 2. | $Q$ | Assumption |
| 3. | $Q \rightarrow S$ | Premise |
| 4. | $S$ | 2, 3, MP |
| 5. | $S \rightarrow T$ | Premise |
| 6. | ~~*Show*~~ $P \rightarrow T$ | Assertion |
| 7. | $P$ | Assumption |
| 8. | $T$ | 4, 5, MP |
| 9. | $\sim(P \rightarrow T)$ | Premise |

Annotations do not, strictly speaking, form part of a derivation. Their inclusion, however, is often convenient for permitting a quick decision as to the correctness of a proposed derivation. For example, that '$\sim P$' is derivable from the premises '$\sim(Q \rightarrow R)$' and '$P \rightarrow \sim\sim R$' is settled by a brief inspection of the following lines.

| | | |
|---|---|---|
| 1. | ~~*Show*~~ $\sim P$ | Assertion |
| 2. | $P$ | Assumption |
| 3. | $P \rightarrow \sim\sim R$ | Premise |
| 4. | $\sim\sim R$ | 2, 3, MP |
| 5. | ~~*Show*~~ $Q \rightarrow R$ | Assertion |
| 6. | $Q$ | Assumption |
| 7. | $R$ | 4, DN |
| 8. | $\sim(Q \rightarrow R)$ | Premise |

## EXERCISES, GROUP I

Show by constructing derivations that the following arguments are valid.

12. $(P \to Q) \to Q \quad . \quad Q \to P \quad \therefore P$
13. $P \to (Q \to R) \quad . \quad P \to (R \to S) \quad \therefore P \to (Q \to S)$
14. $\therefore ([P \to Q] \to P) \to P$
15. $\sim P \to Q \quad . \quad P \to Q \quad \therefore Q$
16. $Q \to \sim R \quad . \quad \sim P \to R \quad \therefore \sim P \to \sim Q$
17. $\sim (R \to Q) \quad . \quad Q \quad \therefore P$
18. $P \to Q \quad . \quad P \to \sim Q \quad \therefore \sim P$

In solving exercises 12 – 18 (as well as later exercises) the reader will find the following suggestions helpful but not infallible. They are intended merely as informal advice and have not the same status as the official directions for constructing a derivation.

(*1*) *To derive a conditional, use conditional derivation.*

(*2*) *To derive anything else, use indirect derivation unless another procedure is immediately obvious.*

(*3*) *Enter all premises as lines, but not until one or another form of derivation has been commenced.*

(*4*) *Whenever a sentence follows from antecedent lines by MP or MT, enter that sentence as a line.*

(*5*) *When using indirect derivation, determine whether any of the antecedent lines is the negation of a conditional; if so, attempt to derive that conditional.*

We solve exercise 12 for illustration. According to suggestion (2), we begin an indirect derivation:

| | |
|---|---|
| 1. *Show* $P$ | Assertion |
| 2. $\sim P$ | Assumption |

Next, in accordance with suggestion (3), we enter the premises

| | |
|---|---|
| 1. *Show* $P$ | Assertion |
| 2. $\sim P$ | Assumption |
| 3. $(P \to Q) \to Q$ | Premise |
| 4. $Q \to P$ | Premise |

We then notice that 2 and 4 permit an application of MT, and we follow suggestion (4):

| | |
|---|---|
| 1. *Show* $P$ | Assertion |
| 2. $\sim P$ | Assumption |
| 3. $(P \to Q) \to Q$ | Premise |
| 4. $Q \to P$ | Premise |
| 5. $\sim Q$ | 2, 4, MT |

Again we apply MT:

| | | |
|---|---|---|
| 1. | *Show* P | Assertion |
| 2. | $\sim$P | Assumption |
| 3. | (P $\rightarrow$ Q) $\rightarrow$ Q | Premise |
| 4. | Q $\rightarrow$ P | Premise |
| 5. | $\sim$Q | 2, 4, MT |
| 6. | $\sim$(P $\rightarrow$ Q) | 3, 5, MT |

We have now obtained, in our indirect derivation, the negation of a conditional. Thus, following suggestion (5), we should attempt to derive 'P $\rightarrow$ Q', and and this we do, following suggestion (1), by conditional derivation:

| | | |
|---|---|---|
| 1. | *Show* P | Assertion |
| 2. | $\sim$P | Assumption |
| 3. | (P $\rightarrow$ Q) $\rightarrow$ Q | Premise |
| 4. | Q $\rightarrow$ P | Premise |
| 5. | $\sim$Q | 2, 4, MT |
| 6. | $\sim$(P $\rightarrow$ Q) | 3, 5, MT |
| 7. | *Show* P $\rightarrow$ Q | Assertion |
| 8. | P | Assumption |

To complete the subsidiary conditional derivation, we must derive 'Q'; thus, by suggestion (2), we try indirect derivation:

| | | |
|---|---|---|
| 1. | *Show* P | Assertion |
| 2. | $\sim$P | Assumption |
| 3. | (P $\rightarrow$ Q) $\rightarrow$ Q | Premise |
| 4. | Q $\rightarrow$ P | Premise |
| 5. | $\sim$Q | 2, 4, MT |
| 6. | $\sim$(P $\rightarrow$ Q) | 3, 5, MT |
| 7. | *Show* P $\rightarrow$ Q | Assertion |
| 8. | P | Assumption |
| 9. | *Show* Q | Assertion |
| 10. | $\sim$Q | Assumption |

It is now obvious how to complete the subsidiary indirect derivation:

| | | |
|---|---|---|
| 1. | *Show* P | Assertion |
| 2. | $\sim$P | Assumption |
| 3. | (P $\rightarrow$ Q) $\rightarrow$ Q | Premise |
| 4. | Q $\rightarrow$ P | Premise |
| 5. | $\sim$Q | 2, 4, MT |
| 6. | $\sim$(P $\rightarrow$ Q) | 3, 5, MT |
| 7. | *Show* P $\rightarrow$ Q | Assertion |
| 8. | P | Assumption |

| | | |
|---|---|---|
| 9. | ~~*Show*~~ Q | Assertion |
| 10. | $\sim Q$ | Assumption |
| 11. | P | 8, R |
| 12. | $\sim P$ | 2, R |

But this completes the subsidiary conditional derivation, whose completion in turn completes the main derivation:

| | | |
|---|---|---|
| 1. | ~~*Show*~~ P | Assertion |
| 2. | $\sim P$ | Assumption |
| 3. | $(P \to Q) \to Q$ | Premise |
| 4. | $Q \to P$ | Premise |
| 5. | $\sim Q$ | 2, 4, MT |
| 6. | $\sim (P \to Q)$ | 3, 5, MT |
| 7. | ~~*Show*~~ $P \to Q$ | Assertion |
| 8. | P | Assumption |
| 9. | ~~*Show*~~ Q | Assertion |
| 10. | $\sim Q$ | Assumption |
| 11. | P | 8, R |
| 12. | $\sim P$ | 2, R |

Actually, the derivation to which our suggestions led us is unnecessarily complicated. An insight after we reached line 8 would have allowed us to complete the derivation in one additional line, as follows:

| | | |
|---|---|---|
| 1. | ~~*Show*~~ P | Assertion |
| 2. | $\sim P$ | Assumption |
| 3. | $(P \to Q) \to Q$ | Premise |
| 4. | $Q \to P$ | Premise |
| 5. | $\sim Q$ | 2, 4, MT |
| 6. | $\sim (P \to Q)$ | 3, 5, MT |
| 7. | ~~*Show*~~ $P \to Q$ | Assertion |
| 8. | P | Assumption |
| 9. | $\sim P$ | 2, R |

Here the subsidiary derivation of line 7 is accomplished in an unusual way. The assumption, line 8, is made in accordance with conditional derivation, but the boxing and cancelling are done in accordance with indirect derivation. This procedure is legitimate in view of the directions for constructing a derivation; see clauses (3) and (6, iii), pp. 20 – 21. Other mixed forms of derivation are possible and indeed convenient, but in each case the mixture could be avoided at the expense of a few extra lines.

## EXERCISES, GROUP II

Show the following arguments valid by constructing symbolizations and deriving the conclusions of the symbolizations from their premises. Indicate the scheme of abbreviation used. Exercise 19 is solved for illustration.

19. Argument (3) of page 14.

On the basis of the scheme

P : Socrates died of old age
Q : the Athenians condemned Socrates to death ,

we obtain as a symbolization of the argument (3) the symbolic argument

$$\sim P \rightarrow Q \quad . \quad \sim Q \quad \therefore P \quad ;$$

its validity, and hence that of (3), is established by the following derivation:

| | | |
|---|---|---|
| 1. | ~~*Show*~~ P | |
| 2. | $\sim P \rightarrow Q$ | Premise |
| 3. | $\sim Q$ | Premise |
| 4. | $\sim\sim P$ | 2, 3, MT |
| 5. | P | 4, DN |

The reader should note that (3) has other symbolizations; for example, on the basis of the scheme

P : if Socrates did not die of old age, then the Athenians condemned him to death
Q : the Athenians did not condemn Socrates to death
R : Socrates died of old age

the argument

$$P \quad . \quad Q \quad \therefore R$$

is a symbolization of (3); and on the basis of the scheme

P : Socrates did not die of old age
Q : Socrates died of old age
R : the Athenians condemned Socrates to death ,

so is the argument

$$P \rightarrow R \quad . \quad \sim R \quad \therefore Q \quad .$$

But neither of the last two symbolizations is valid (as we shall be able to show in the next chapter). In general, the longer the symbolization, the more likely it is to be valid; and among symbolizations of equal length, the likelihood of validity increases as the number of distinct sentence letters decreases.

20. If Alfred studies, then he receives good grades. If he does not study, then he enjoys college. If he does not receive good grades, then he does not enjoy college. ∴ Alfred receives good grades. (Compare (1), p. 3.)

21. If Herbert can take the apartment only if he divorces his wife, then he should think twice. If Herbert keeps Fido, then he cannot take the apartment. Herbert's wife insists on keeping Fido. If Herbert does not keep Fido, then he will divorce his wife provided that she insists on keeping Fido. ∴ Herbert should think twice.

22. If Herbert grows rich, then he can take the apartment. If he divorces his wife, then he will not receive his inheritance. Herbert will grow rich if he receives his inheritance. Herbert can take the apartment only if he divorces his wife. ∴ If Herbert receives his inheritance, then Fido does not matter.

23. If God exists, then He is omnipotent. If God exists, then He is omniscient. If God exists, then He is benevolent. If God can prevent evil, then if He knows that evil exists, then He is not benevolent if He does not prevent it. If God is omnipotent, then He can prevent evil. If God is omniscient, then He knows that evil exists if it does indeed exist. Evil does not exist if God prevents it. Evil exists. ∴ God does not exist.

24. If business will flourish on the condition that taxes will not increase, then the standard of living will improve. It is not the case that the standard of living will improve. If business does not flourish, then taxes will increase. ∴ Unemployment will be a problem.

25. The standard of living will improve provided that taxes will increase. Business will flourish only if unemployment is not a problem. Business will flourish if the standard of living will improve. ∴ On the condition that taxes will increase, unemployment is not a problem.

**4.* Fallacies.** In the directions for constructing a derivation, a number of restrictions appear whose significance is perhaps not immediately obvious. However, if we define a *fallacy* as a procedure that permits the validation of a *false English argument*, that is, an argument whose premises are true sentences of English and whose conclusion is a false sentence of English, then the neglect of any but one of the restrictions would lead to fallacies.

The fallacies involved here are of two kinds—those of begging the question and those of unwarranted assumptions. These kinds of fallacies can be characterized in a rough, informal way, as follows.

When, in deriving something, one uses the assertion that is to be derived, one is said to be guilty of *begging the question*.

Assumptions are useful tools in logical derivations, but they must be employed with caution. Only on special occasions may they be made or inferences drawn from them. The use of assumptions at other junctures is the fallacy of *unwarranted assumptions*. (The reader should observe the difference between the use of *premises*, which is justified by clause (2), and the use of *assumptions*. A premise may be stated at any point in the course of a derivation; there are no restrictions here whose neglect would lead to fallacies.)

In clauses (3) and (4) of the directions for constructing a derivation, it is specified that an assumption may be made only in connection with a line containing uncancelled '*Show*'. Let us examine the consequences of ignoring this restriction, in connection first with clause (3). Consider the argument

> If Schopenhauer was married, then he had a wife.
> ∴ Schopenhauer had a wife.

Here the premise is obviously true, and the conclusion is, as a matter of history, false; hence the argument is false. But by an incorrect use of clause (3), the validity of this argument could be established by means of the following symbolization and accompanying derivation.

(1) $P \rightarrow Q \quad \therefore Q$

| | | |
|---|---|---|
| 1. | ~~*Show*~~ Q | Assertion |
| 2. | $P \rightarrow Q$ | Premise |
| 3. | P | Unwarranted assumption |
| 4. | Q | 2, 3, MP |

Now let us employ clause (4) incorrectly, again making an assumption in connection with a line not beginning with uncancelled '*Show*'. Consider the false argument

> Snow is white. ∴ Grass is red.

and the following symbolization and derivation:

(2) $P \quad \therefore Q$

| | | |
|---|---|---|
| 1. | ~~*Show*~~ Q | Assertion |
| 2. | P | Premise |
| 3. | $\sim P$ | Unwarranted assumption |

In both (1) and (2), then, we have validated false arguments by committing one form of the fallacy of unwarranted assumptions; in neither case was the assumption made in connection with a line containing uncancelled '*Show*'.

The other restriction involved in clauses (3) and (4) is that the assumption be made immediately after the line with which it is connected. It is not yet possible to illustrate the necessity of this restriction; in fact, its neglect will produce no fallacies until our logical apparatus is enlarged by the introduction, in chapter III, of a new form of derivation.

Clause (5) permits us to draw inferences only from *antecedent lines*. This restriction can be violated in two ways—by applying an inference rule to a line containing uncancelled '*Show*' or by applying a rule to a boxed line. Let us illustrate the first violation.

Argument:

Snow is white. ∴ If snow is white, then grass is red.

Symbolization:

$$P \quad \therefore P \rightarrow Q$$

Derivation:

| | | |
|---|---|---|
| 1. | ~~*Show*~~ P → Q | Assertion |
| 2. | P | Premise |
| 3. | Q | 1, 2, incorrect application of MP |

The application of MP is incorrect because at the time it occurred the '*Show*' in line 1 had not been cancelled. Intuitively, we used the conclusion that was to be derived in order to obtain one of the lines in its derivation. Thus we begged the question.

Let us now illustrate the consequences of applying an inference rule to a boxed line.

Argument:

Snow is white. ∴ Grass is red.

Symbolization:

$$P \quad \therefore Q$$

We consider several stages in constructing a derivation corresponding to this symbolization.

(3)

| | | |
|---|---|---|
| 1. | *Show* Q | Assertion |
| 2. | *Show* Q → P | Assertion |
| 3. | Q | Assumption |
| 4. | P | Premise |

(4)

| | |
|---|---|
| 1. | *Show* Q |
| 2. | ~~*Show*~~ Q → P |
| 3. | Q |
| 4. | P |

(5)

| | | |
|---|---|---|
| 1. | ~~*Show*~~ Q | |
| 2. | ~~*Show*~~ Q → P | |
| 3. | Q | |
| 4. | P | |
| 5. | Q | 3, incorrect application of R |

The partial derivations (3) and (4) are correctly constructed, but (5) involves an unwarranted use of assumptions. Although the assumption made in line 3 is legitimate, an illegitimate consequence has been drawn from it; for line 3 is boxed in stage (4) and hence no longer available for further inferences.

The derivation (5) can be obtained by another fallacious procedure. From (3) we can proceed legitimately not only to (4), but also to the following partial derivation:

(4′)

1. *Show* Q
2. *Show* Q → P
3. Q
4. P
5. Q    3, R

Then we might apply clause (6) of the directions for constructing derivations, ignoring the injunction that when an occurrence of '*Show*' is cancelled, *all* subsequent lines must be boxed:

(4″)

1. *Show* Q
2. ~~*Show*~~ Q → P
3. | Q |
4. | P |
5. Q    3, R

From this (incorrect) partial derivation, we might then proceed legitimately to (5). Since line 5 depends on an assumption, it is legitimately available only as long as the assumption is; thus the failure to box line 5 along with lines 3 and 4 in passing from (4′) to (4″) can be construed, like the previous fallacy, as an unwarranted use of an assumption.

In boxing and cancelling by clause (6), the lines to be boxed must be free of uncancelled '*Show*'. Let us consider two examples in which this restriction is violated.

Argument:

Snow is white. ∴ Grass is red.

Symbolization:

P ∴ Q

Derivation:

(6)

1. ~~*Show*~~ Q    Assertion
2. | P |    Premise
3. | *Show* Q |    Assertion

Derivation:

| | | | |
|---|---|---|---|
| (7) | 1. | ~~*Show*~~ Q | Assertion |
| | 2. | *Show* Q → P | Assertion |
| | 3. | Q | Assumption |

In the derivation (6) we have begged the question, and in (7) we have made unwarranted use of an assumption; in both cases the lines boxed contain uncancelled '*Show*'.

There is another restriction imposed in clause (6). In each of our three forms of derivation certain lines are crucial, and these must occur unboxed. In direct derivation, for instance, the assertion that is to be established must occur in some unboxed line following its initial statement. Let us consider a case in which the crucial line is boxed.

Argument:

Snow is white. ∴ Grass is red.

Symbolization:

P ∴ Q

We consider two stages in constructing a corresponding derivation.

| | | | |
|---|---|---|---|
| (8) | 1. | *Show* Q | Assertion |
| | 2. | ~~*Show*~~ Q → P | Assertion |
| | 3. | Q | Assumption |
| | 4. | P | Premise |

| | | |
|---|---|---|
| (9) | 1. | ~~*Show*~~ Q |
| | 2. | ~~*Show*~~ Q → P |
| | 3. | Q |
| | 4. | P |

The incomplete derivation (8) is correctly constructed. The passage from (8) to (9), however, fails to satisfy the requirements of clause (6), for the occurrence of 'Q' in line 3 of (8) is boxed. This passage is, moreover, intuitively unsatisfactory; it involves an unwarranted use of an assumption.

We leave to the reader illustrations of the same sort in connection with conditional and indirect derivation.

**5. Theorems.** As we stated earlier, it is possible for an argument to have an empty sequence of premises; an example is (2) on page 13. If such an argument is valid and is furthermore symbolic, its conclusion is called a *theorem;* for example, the sentence in exercise 14 (p. 26) is a theorem. A derivation corresponding to such an argument will contain no line justified by clause (2) (the clause that admits premises), and will be called a *proof*. The following theorems, some of which are accompanied by proofs, will be useful in subsequent chapters.

| | | | |
|---|---|---|---|
| T1 | 1. | ~~*Show*~~ $P \to P$ | Assertion |
| | 2. | $P$ | Assumption |
| T2 | 1. | ~~*Show*~~ $Q \to (P \to Q)$ | Assertion |
| | 2. | $Q$ | Assumption |
| | 3. | ~~*Show*~~ $P \to Q$ | Assertion |
| | 4. | $Q$ | 2, R |
| T3 | 1. | ~~*Show*~~ $P \to ([P \to Q] \to Q)$ | Assertion |
| | 2. | $P$ | Assumption |
| | 3. | ~~*Show*~~ $[P \to Q] \to Q$ | Assertion |
| | 4. | $P \to Q$ | Assumption |
| | 5. | $Q$ | 2, 4, MP |

T4 and T5 are known as principles of *syllogism*. Their derivations are left to the reader.

T4 $(P \to Q) \to ([Q \to R] \to [P \to R])$

T5 $(Q \to R) \to ([P \to Q] \to [P \to R])$

T6 and T7 are called principles of *distribution* of '→' over '→'.

| | | | |
|---|---|---|---|
| T6 | 1. | ~~*Show*~~ $(P \to [Q \to R]) \to ([P \to Q] \to [P \to R])$ | Assertion |
| | 2. | $P \to [Q \to R]$ | Assumption |
| | 3. | ~~*Show*~~ $[P \to Q] \to [P \to R]$ | Assertion |
| | 4. | $P \to Q$ | Assumption |
| | 5. | ~~*Show*~~ $P \to R$ | Assertion |
| | 6. | $P$ | Assumption |
| | 7. | $Q$ | 4, 6, MP |
| | 8. | $Q \to R$ | 2, 6, MP |
| | 9. | $R$ | 7, 8, MP |

T7 $([P \to Q] \to [P \to R]) \to (P \to [Q \to R])$

The principle of *commutation:*

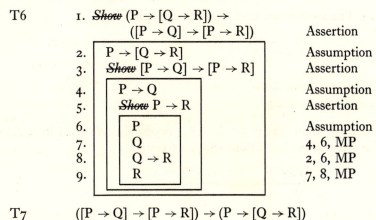

| | | | |
|---|---|---|---|
| T8 | 1. | ~~*Show*~~ $(P \to [Q \to R]) \to (Q \to [P \to R])$ | Assertion |

| | | |
|---|---|---|
| 2. | $P \to [Q \to R]$ | Assumption |
| 3. | ~~*Show*~~ $Q \to [P \to R]$ | Assertion |
| 4. | $Q$ | Assumption |
| 5. | ~~*Show*~~ $P \to R$ | Assertion |
| 6. | $P$ | Assumption |
| 7. | $Q \to R$ | 2, 6, MP |
| 8. | $R$ | 4, 7, MP |

T9 $(P \to [P \to Q]) \to (P \to Q)$

T10

| | | |
|---|---|---|
| 1. | ~~*Show*~~ $([P \to Q] \to Q) \to ([Q \to P] \to P)$ | Assertion |
| 2. | $[P \to Q] \to Q$ | Assumption |
| 3. | ~~*Show*~~ $[Q \to P] \to P$ | Assertion |
| 4. | $Q \to P$ | Assumption |
| 5. | ~~*Show*~~ $P$ | Assertion |
| 6. | $\sim P$ | Assumption |
| 7. | $\sim Q$ | 4, 6, MT |
| 8. | $\sim [P \to Q]$ | 2, 7, MT |
| 9. | ~~*Show*~~ $P \to Q$ | Assertion |
| 10. | $P$ | Assumption |
| 11. | $\sim P$ | 6, R |

T11 and T12 are the two laws of *double negation.*

T11

| | | |
|---|---|---|
| 1. | ~~*Show*~~ $\sim\sim P \to P$ | Assertion |
| 2. | $\sim\sim P$ | Assumption |
| 3. | $P$ | 2, DN |

T12 $P \to \sim\sim P$

T13 – T16 are known as principles of *transposition.* We shall henceforth omit the annotations 'Assertion' and 'Assumption'.

T13

| | | |
|---|---|---|
| 1. | ~~*Show*~~ $(P \to Q) \to (\sim Q \to \sim P)$ | |
| 2. | $P \to Q$ | |
| 3. | ~~*Show*~~ $\sim Q \to \sim P$ | |
| 4. | $\sim Q$ | |
| 5. | $\sim P$ | 2, 4, MT |

T14 $(P \to \sim Q) \to (Q \to \sim P)$

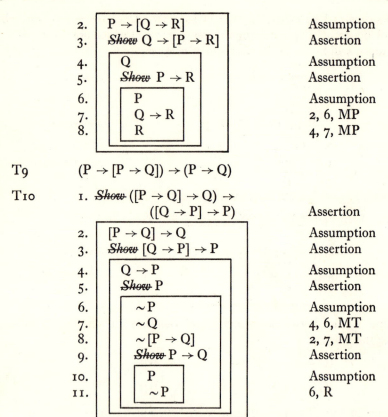

T15 $(\sim P \to Q) \to (\sim Q \to P)$

T16 $(\sim P \to \sim Q) \to (Q \to P)$

T17
1. ~~Show~~ $P \to (\sim P \to Q)$
2. $P$
3. ~~Show~~ $\sim P \to Q$
4. $\sim P$
5. $P$ 2, R

T18 $\sim P \to (P \to Q)$

T19 and T20 are known as laws of *reductio ad absurdum.*

T19
1. ~~Show~~ $(\sim P \to P) \to P$
2. $\sim P \to P$
3. ~~Show~~ $P$
4. $\sim P$
5. $P$ 2, 4, MP

T20 $(P \to \sim P) \to \sim P$

According to T21 and T22, the denial of a conditional leads to the affirmation of its antecedent and the denial of its consequent.

T21 $\sim (P \to Q) \to P$

T22 $\sim (P \to Q) \to \sim Q$

T23 is known as *Peirce's law* (after the nineteenth-century American philosopher C. S. Peirce) and is already familiar to the reader who has solved exercise 14 (p. 26).

T23 $([P \to Q] \to P) \to P$

## EXERCISES

26. Prove T4, T7, T15.
27. Prove T18, T20, T21, T22.

In solving exercises 26 and 27, the reader will again find useful the suggestions made on page 26.

**6. * Historical remarks.** The logic of '$\sim$' and '$\to$' was first studied by the Stoics in the fourth and third centuries B.C. (see Łukasiewicz [2] and Mates [1]). Its first complete formalization (in a technical sense) occurs in Frege [1]. An extensive investigation into this branch of logic is reported in Łukasiewicz and Tarski [1].

The treatment set forth in the present chapter differs in a marked way from that of Frege and Łukasiewicz. The earlier systems relied on only one form of derivation—direct derivation. The simplicity thus obtained was secured at the expense of postulating certain theorems as axioms. The idea of dispensing with axioms in favor of conditional derivation stems from Gentzen [1] and Jaśkowski [1] and depends on a result obtained independently by Herbrand and Tarski—the so-called *deduction theorem* (see Herbrand [1], [2], and Tarski [1]). Systems of the later variety are said to employ *natural deduction* and, as this designation indicates, are intended to reflect intuitive forms of reasoning.

Other systems of natural deduction may be found in Quine [3], Copi [1], Suppes [1], and Hilbert and Ackermann [4].

Our system could be simplified at the expense of making certain derivations unintuitive. In particular, indirect derivation is theoretically superfluous; further, we could replace our four rules of inference by the following two:

*Modus ponens:*

$$\frac{\begin{array}{l}(\phi \rightarrow \psi)\\ \phi\end{array}}{\psi}$$

A variant of *modus tollens:*

$$\frac{\begin{array}{l}(\sim\phi \rightarrow \sim\psi)\\ \psi\end{array}}{\phi}$$

# Chapter II

# 'AND', 'OR', 'IF AND ONLY IF'

**1. Symbols and sentences.** Let us add to our language three new symbols, '$\wedge$', '$\vee$', and '$\leftrightarrow$', symbolizing respectively 'and', 'or', and 'if and only if'. Each of them is used, like '$\rightarrow$', to combine two sentences, say,

(1) Socrates is snub-nosed

and

(2) Socrates is bald,

into a new sentence. Thus we obtain

(3) (Socrates is snub-nosed $\wedge$ Socrates is bald) ,

(4) (Socrates is snub-nosed $\vee$ Socrates is bald) ,

and

(5) (Socrates is snub-nosed $\leftrightarrow$ Socrates is bald) ,

which are read

Socrates is snub-nosed and Socrates is bald,
Socrates is snub-nosed or Socrates is bald,

and

Socrates is snub-nosed if and only if Socrates is bald.

(3) is called the *conjunction* of (1) and (2), which are its *conjuncts*. (4) is the *disjunction* of (1) and (2), which are its *disjuncts*. (5) is the *biconditional* formed from (1) and (2), which are its *constituents*.

The symbols '$\sim$', '$\rightarrow$', '$\wedge$','$\vee$', and '$\leftrightarrow$' are called *sentential connectives*. The corresponding phrases 'it is not the case that', 'if . . . , then', 'and', 'or', and 'if and only if' are called *phrases of connection*.

The *language* we now consider is that of chapter I, supplemented by '$\wedge$', '$\vee$', and '$\leftrightarrow$'. Thus the sentences of our language fall into the following categories:

(1) declarative sentences of English,
(2) sentence letters,
(3) sentences correctly constructed from sentences of categories (1) and (2) by means of sentential connectives and parentheses.

To be more explicit, the class of *sentences* can be exhaustively characterized as follows:

(1) *All declarative sentences of English are sentences.*
(2) *Sentence letters are sentences.*
(3) *If* $\phi$ *and* $\psi$ *are sentences, then so are*

$$\sim\phi \quad ,$$
$$(\phi \rightarrow \psi) \quad ,$$
$$(\phi \wedge \psi) \quad ,$$
$$(\phi \vee \psi) \quad ,$$

*and*

$$(\phi \leftrightarrow \psi) \quad .$$

The *symbolic sentences* of our language are those sentences which are constructed exclusively from sentence letters, parentheses, and sentential connectives. More precisely, the class of *symbolic sentences* is exhaustively characterized by clauses (2) and (3) above, reading everywhere 'symbolic sentences' for 'sentences'.

Notice that '∧', '∨', and '↔', like '→', bring with them a pair of parentheses. The ambiguity which would otherwise result is illustrated by the invitation

Bring your wife or come stag and have a good time.

As in chapter I, we shall usually omit the outermost parentheses of a sentence and sometimes replace parentheses by brackets. In addition, '→' and '↔' will be regarded as marking a greater break than the other connectives. Consequently, the parentheses accompanying '∧' and '∨' may be omitted in certain contexts, as in the following examples. The sentence

$$(P \wedge Q) \rightarrow (R \vee S)$$

becomes

$$P \wedge Q \rightarrow R \vee S \quad ,$$

and

$$(P \vee Q) \leftrightarrow (R \wedge S)$$

becomes

$$P \vee Q \leftrightarrow R \wedge S \quad .$$

Thus we arrive informally at a larger class of sentences than that provided by clauses (1) – (3). Sentences of the smaller class may, in those few cases when it is necessary to draw a distinction, be called *sentences in the official sense.*

## EXERCISES

Which of the following are sentences in the official sense? Exercises 1 and 2 are solved for illustration.

1. $\sim(\sim P \rightarrow (Q \vee R))$

According to clauses (2) and (3) of the characterization of sentences (p. 40), '$(Q \vee R)$' is a sentence, and '$\sim P$' is a sentence. By clause (3), then, '$(\sim P \rightarrow (Q \vee R))$' is a sentence; thus, by clause (3) again, No. 1 is a sentence in the official sense.

2. $(P \vee (Q \rightarrow R) \wedge \sim P)$

Because the characterization of sentences given by clauses (1) – (3) is stipulated to be exhaustive, and clauses (1) and (2), together with the first and fifth parts of clause (3), are inapplicable, No. 2, if a sentence, must be either a disjunction, a conditional, or a conjunction of two sentences. Thus either

$$P$$

and

(1) $$(Q \rightarrow R) \wedge \sim P$$

are sentences, or

(2) $$P \vee (Q$$

and

(3) $$R) \wedge \sim P$$

are sentences, or

(4) $$P \vee (Q \rightarrow R)$$

and

$$\sim P$$

are sentences. But (1), (2), and (4) (and, incidentally, (3)) are not sentences in the official sense, as the reader can easily verify. Thus No. 2 is not a sentence.

3. $((P \leftrightarrow Q) \leftrightarrow ((P \wedge Q) \vee (\sim P \wedge \sim Q)))$
4. $\sim((P \vee Q \rightarrow R) \rightarrow ((P \rightarrow R) \wedge (Q \rightarrow R)))$
5. $\sim(\sim(\sim \sim P \vee Q) \vee R) \rightarrow (P \leftrightarrow (Q \leftrightarrow R))$

From each of the following sentences obtain a sentence in the official sense by restoring parentheses omitted by informal conventions. Exercise 6 is solved for illustration.

6. $(P \land Q \rightarrow P \lor Q) \leftrightarrow P \lor (Q \land R)$

This sentence becomes

$$(((P \land Q) \rightarrow (P \lor Q)) \leftrightarrow (P \lor (Q \land R))) \quad .$$

7. $Q \lor R \rightarrow (P \land R \rightarrow (Q \leftrightarrow R \lor P))$
8. $(P \land Q) \lor (\sim P \land \sim Q) \rightarrow ((P \leftrightarrow Q) \leftrightarrow (Q \leftrightarrow P))$
9. $(P \rightarrow Q \lor R) \lor (Q \lor R \rightarrow P) \leftrightarrow (Q \lor R \leftrightarrow P)$
10. $\sim(P \lor (Q \land R) \rightarrow ((P \lor Q) \land (P \lor R) \leftrightarrow P \land Q))$

**2. Translation and symbolization.** The process of *literal translation into English on the basis of a given scheme of abbreviation* begins as before with a symbolic sentence and if successful ends with a sentence of English. The process consists of the following steps:

(*i*) *Restore any parentheses that may have disappeared as a result of the informal conventions of the last section.*
(*ii*) *Replace sentence letters by English sentences in accordance with the scheme of abbreviation.*
(*iii*) *Eliminate sentential connectives in favor of the corresponding phrases of connection, preserving all parentheses.*

As before, we call an English sentence a *free translation* (or simply a *translation*) of a symbolic sentence $\phi$ on the basis of a given scheme of abbreviation if it is a stylistic variant of the literal English translation of $\phi$ based on that scheme.

Let us consider, for example, the scheme of abbreviation

P : Alfred dances with Alice
Q : Alfred dances with Mary
R : Alfred will enjoy the party

and the symbolic sentence

(1) $$P \land Q \rightarrow R \quad .$$

In step (i) of the process of literal translation into English, this sentence becomes

$$((P \land Q) \rightarrow R) \quad ,$$

in step (ii)

((Alfred dances with Alice ∧ Alfred dances with Mary) → Alfred will enjoy the party) , and in step (iii)

(2) (If (Alfred dances with Alice and Alfred dances with Mary), then Alfred will enjoy the party) .

Thus (2) is the literal translation of (1) into English on the basis of the scheme above, and the more idiomatic sentence,

(3) Alfred will enjoy the party if he dances with both Alice and Mary,

being a stylistic variant of (2), qualifies as a free translation of (1) on the basis of the same scheme.

As regards stylistic variance, we persist in our loose practices, giving no exact definition but only a few examples. 'But', 'although', 'even though', 'both . . . and', and sometimes the relative pronouns 'who', 'which', 'that', serve as stylistic variants of 'and'; 'unless', 'either . . . or', as stylistic variants of 'or'; and 'exactly on condition that', and sometimes 'just in case', as stylistic variants of 'if and only if'. Thus each of

Alfred slept through class, but he passed,
Even though Alfred slept through class, he passed,
Alfred, who slept through class, passed

is regarded as a stylistic variant of

Alfred slept through class and Alfred passed;

The needle on the ammeter will deflect unless the battery is dead

as a stylistic variant of

The needle on the ammeter will deflect or the battery is dead;

and

Alfred will be elected exactly on condition that he stand for office,

Just in case Alfred stands for office will he be elected

as stylistic variants of

Alfred will be elected if and only if Alfred stands for office.

As a further illustration of stylistic variance, observe that the phrase 'neither . . . nor' is expressible by means of phrases of connection.

Neither $\phi$ nor $\psi$ ,

where $\phi$ and $\psi$ are sentences, has the stylistic variants

(It is not the case that $\phi$ and it is not the case that $\psi$)

and

It is not the case that ($\phi$ or $\psi$) .

It should be emphasized that the *binary* sentential connectives, that is, '$\rightarrow$', '$\wedge$', '$\vee$', '$\leftrightarrow$', may stand only between sentences (or, in later chapters, between formulas). English usage, however, provides other contexts for the corresponding phrases of connection, as in the sentence (3) above,

(4) Socrates is either snub-nosed or bald,

and

(5) Arcadia lies between Laconia and Achaea.

(3) and (4) have *canonical* stylistic variants, that is, stylistic variants in which phrases of connection operate only on sentences. We have seen this already in the case of (3), and (4) can be expanded into

(Socrates is snub-nosed or Socrates is bald) .

But (5) cannot be similarly treated; it is clearly not synonymous with

Arcadia lies between Laconia and Arcadia lies between Achaea.

As in chapter I, we say that $\phi$ is a *symbolization* of an English sentence $\psi$ on the basis of a given scheme of abbreviation just in case $\phi$ is a symbolic sentence which has $\psi$ as a translation on the basis of that scheme.

To find a symbolization of a given English sentence on the basis of a given scheme of abbreviation, the reader will find it useful to proceed roughly as follows:

(*1*) *Introduce phrases of connection, accompanied by parentheses and occurring canonically* (*that is, standing only before or between sentences*), *in place of their stylistic variants.*

(*2*) *Reverse the steps leading from a symbolic sentence to a literal English translation; that is,*

(*2a*) *replace all parts having one of the forms*

*it is not the case that* $\phi$ ,
(*if* $\phi$, *then* $\psi$) ,
($\phi$ *and* $\psi$) ,
($\phi$ *or* $\psi$) ,
($\phi$ *if and only if* $\psi$) ,

*where* $\phi$ *and* $\psi$ *are sentences, by*

$$\sim\phi \quad ,$$
$$(\phi \rightarrow \psi) \quad ,$$
$$(\phi \wedge \psi) \quad ,$$
$$(\phi \vee \psi) \quad ,$$
$$(\phi \leftrightarrow \psi)$$

*respectively;*

(2*b*) *replace English components by sentence letters in accordance with the scheme of abbreviation;*

(2*c*) *omit parentheses and insert brackets in accordance with the informal conventions of the preceding section.*

## EXERCISES

11. On the basis of the scheme of abbreviation

P : Alice will dance with Alfred
Q : Mary will dance with Alfred
R : Alfred will improve his deportment

translate the following symbolic sentence into idiomatic English:

$$([\sim P \wedge \sim Q] \vee R) \wedge (R \leftrightarrow P \wedge Q) \quad .$$

Symbolize each of the following sentences on the basis of the scheme of abbreviation that accompanies it. Exercise 12 is solved for illustration.

12. Errors will decrease in the subject's performance just in case neither motivation is absent nor learning has not occurred. (P : errors will decrease in the subject's performance; Q : motivation is absent; R : learning has occurred)

In step (1) of the informal procedure for symbolizing English sentences (pp. 44 - 45), we may transform No. 12 into

> (Errors will decrease in the subject's performance if and only if it is not the case that (motivation is absent or it is not the case that learning has occurred)) .

In step (2a) the sentence becomes

> (Errors will decrease in the subject's performance $\leftrightarrow$ $\sim$(motivation is absent $\vee$ $\sim$learning has occurred)) ;

in step (2b),

$$(P \leftrightarrow \sim(Q \vee \sim R)) \quad ;$$

and in step (2c),

$$P \leftrightarrow \sim(Q \vee \sim R) \quad .$$

13. Assuming either that logic is difficult or that the text is not readable, Alfred will pass only if he concentrates. (P : logic is difficult; Q : the text is readable; R : Alfred will pass; S : Alfred concentrates)

14. Unless logic is difficult, Alfred will pass if he concentrates. (P : logic is difficult; R : Alfred will pass; S : Alfred concentrates)

15. Mary will arrive at 10:30 A.M. unless the plane is late. (P : Mary will arrive at 10:30 A.M.; Q : the plane will be late)

16. Assuming that the professor is a Communist, he will sign the loyalty oath; but if he is an idealist, he will neither sign the loyalty oath nor speak to those who do. (P : the professor is a Communist; Q : the professor will sign the loyalty oath; R : the professor is an idealist; S : the professor will speak to those who sign the loyalty oath)

17. If Alfred and Mary are playing dice together, it is the first throw of the game, and Mary is throwing the dice, then she wins the game on the first throw if and only if she throws 7 or 11. (P : Alfred is playing dice; Q : Mary is playing dice; R : Alfred and Mary are playing dice together; S : it is the first throw of the game; T : Mary is throwing the dice; U : Mary wins on the first throw; V : Mary throws 7 or 11; W : Mary throws 7; X : Mary throws 11)

18. If the world is a progressively realized community of interpretation, then either quadruplicity will drink procrastination or, provided that the Nothing negates, boredom will ensue seldom more often than frequently. (P : the world is a progressively realized community of interpretation; Q: quadruplicity will drink procrastination; R : the Nothing negates; S : boredom will ensue seldom more often than frequently)

The following sentences are ambiguous in the sense that the placement of parentheses in their symbolizations is not uniquely determined. Give all plausible symbolizations of each on the basis of the given scheme of abbreviation. Exercise 19 is solved for illustration.

19. Errors will occur in the subject's performance if and only if motivation is absent or learning has not taken place.

Given the scheme of abbreviation

P : errors will occur in the subject's performance
Q : motivation is absent
R : learning has taken place ,

No. 19 becomes in step (1) either

(Errors will occur in the subject's performance if and only if (motivation is absent or it is not the case that learning has taken place))

or

(Errors will occur in the subject's performance if and only if motivation is absent) or it is not the case that learning has taken place) ;

and in step (2) it correspondingly becomes either

$$P \leftrightarrow Q \vee \sim R$$

or

$$(P \leftrightarrow Q) \vee \sim R \quad .$$

20. If either a war or a depression occurs then neither science nor music and literature will flourish unless the government supports research and provides patronage for artists. (P : a war occurs; Q : a depression occurs; R : science will flourish; S : music will flourish; T : literature will flourish; U : the government will support research; V : the government will provide patronage for artists)

21. If Mary belongs to a sorority then she will be graduated from college only if she resists temptation, provided that she is not attractive or intelligent. (P : Mary belongs to a sorority; Q : Mary will be graduated from college; R : Mary resists temptation; S : Mary is attractive; T : Mary is intelligent)

**3. Inference rules; theorems with unabbreviated proofs.** The *sentential calculus* is that branch of logic which essentially involves the sentential connectives. For a complete formulation of the sentential calculus, we must add to the apparatus of chapter I the following inference rules for the new sentential connectives.

Simplification (S), in two forms: $\dfrac{(\phi \wedge \psi)}{\phi}$ $\quad$ $\dfrac{(\psi \wedge \phi)}{\phi}$

Adjunction (Adj): $\dfrac{\begin{matrix}\phi \\ \psi\end{matrix}}{(\phi \wedge \psi)}$

Addition (Add), in two forms: $\dfrac{\phi}{(\phi \vee \psi)}$ $\quad$ $\dfrac{\phi}{(\psi \vee \phi)}$

*Modus tollendo ponens* (MTP), in two forms: $\dfrac{\begin{matrix}(\phi \vee \psi) \\ \sim \phi\end{matrix}}{\psi}$ $\quad$ $\dfrac{\begin{matrix}(\phi \vee \psi) \\ \sim \psi\end{matrix}}{\phi}$

Biconditional-conditional (BC), in two forms: $\dfrac{(\phi \leftrightarrow \psi)}{(\phi \rightarrow \psi)}$ $\quad$ $\dfrac{(\phi \leftrightarrow \psi)}{(\psi \rightarrow \phi)}$

Conditional-biconditional (CB): $\dfrac{\begin{matrix}(\phi \rightarrow \psi) \\ (\psi \rightarrow \phi)\end{matrix}}{(\phi \leftrightarrow \psi)}$

Thus one symbolic sentence is said to follow from another by *simplification* just in case the former is a conjunct of the latter; a symbolic sentence follows by *adjunction* from two others just in case it is their conjunction; and so on. When

$$(\phi \vee \psi)$$

or

$$(\psi \vee \phi)$$

is inferred from $\phi$ by Add, the symbolic sentence $\psi$ is called the *added disjunct*.

The new rules are exemplified by the following arguments.

Simplification:

$$(\sim P \wedge Q) \quad \therefore \sim P$$
$$(P \wedge (Q \rightarrow R)) \quad \therefore (Q \rightarrow R)$$

Adjunction:

$$\sim P \quad . \quad Q \quad \therefore (\sim P \wedge Q)$$

Addition:

P ∴ (P ∨ ~Q) (Here '~Q' is the added disjunct.)
~Q ∴ (P ∨ ~Q) (Here 'P' is the added disjunct.)

*Modus tollendo ponens:*

$$(P \vee \sim Q) \quad . \quad \sim P \quad \therefore \sim Q$$
$$(P \vee \sim Q) \quad . \quad \sim\sim Q \quad \therefore P$$

Biconditional-conditional:

$$(P \leftrightarrow (Q \vee R)) \quad \therefore (P \rightarrow (Q \vee R))$$
$$(P \leftrightarrow (Q \vee R)) \quad \therefore ((Q \vee R) \rightarrow P)$$

Conditional-biconditional:

$$(P \rightarrow (Q \vee R)) \quad . \quad ((Q \vee R) \rightarrow P) \quad \therefore (P \leftrightarrow (Q \vee R))$$

We have at our disposal several conventions for dropping parentheses, but they must be used with caution. In particular, when applying inference rules, we must mentally restore omitted parentheses. For example,

(1) $$\frac{P \wedge Q \rightarrow R}{Q \rightarrow R}$$

might seem to be a case of simplification, and

(2) $$\frac{P}{P \vee Q \rightarrow R}$$

a case of addition. But when parentheses are restored according to our conventions, (1) and (2) become respectively

$$\frac{((P \land Q) \to R)}{(Q \to R)}$$

and

$$\frac{P}{((P \lor Q) \to R)} \quad ,$$

which clearly do not constitute applications of our rules. Indeed, the inferences (1) and (2) *should not* fall under our rules. For (1) is comparable to the inference from

> If San Francisco is larger than New York and New York is larger than Los Angeles, then San Francisco is larger than Los Angeles

to

> If New York is larger than Los Angeles then San Francisco is larger than Los Angeles,

and (2) to the inference from

Socrates is bald

to

> If either Socrates is bald or Socrates is snub-nosed, then Socrates is unmarried.

'Or' has two senses.

$\phi$ or $\psi$

may mean either

either $\phi$ or $\psi$, but not both ,

or

either $\phi$, or $\psi$, or both .

We select the second sense. This choice is reflected in our adoption of the rule Add; for if the first sense were selected, the inference

> Socrates is bald. ∴ Socrates is bald or Socrates is snub-nosed

would lead from truth to falsehood. Though we shall use '∨' in the second sense of 'or', it is clear that the first sense can still be expressed by a combination of '∨', '∼', and '∧'.

'If and only if' is a composite of 'if' and 'only if'.

$$\phi \text{ if and only if } \psi$$

asserts

$$\phi \text{ if } \psi$$

and

$$\phi \text{ only if } \psi \quad .$$

This is reflected in our two rules BC and CB, by which the biconditional

$$\phi \leftrightarrow \psi$$

is related to the conditional

$$\psi \rightarrow \phi$$

and its *converse*,

$$\phi \rightarrow \psi \quad .$$

The definitions of an *argument*, a *symbolic argument*, and an *English argument* are carried over intact from chapter I (pp. 13 and 24). The directions for constructing a *derivation from given symbolic premises* (pp. 20 – 21), and the definitions of a *complete* derivation (p. 23), of *derivability* (p. 23), of a *valid symbolic argument* (p. 24), of a *valid English argument* (p. 25), of a *theorem* (p. 34), and of a *proof* (p. 35) also remain unchanged. The interpretation, however, of the phrase 'an inference rule' (which occurs in clause (5) of the directions for constructing a derivation) is extended so as to include our new inference rules (S, Adj, Add, MTP, BC, and CB) as well as MP, MT, DN, and R.

It is convenient, in developing the full sentential calculus, to depart from the order of presentation of chapter I. We begin, not with derivations involving premises, but with proofs of theorems. The first group of theorems primarily concern the connectives '$\sim$', '$\rightarrow$', and '$\wedge$'. T24 is the *commutative law* for '$\wedge$'.

| T24 | | | |
|---|---|---|---|
| | 1. | ~~Show~~ $P \wedge Q \leftrightarrow Q \wedge P$ | |
| | 2. | ~~Show~~ $P \wedge Q \rightarrow Q \wedge P$ | |
| | 3. | $P \wedge Q$ | |
| | 4. | $Q$ | 3, S |
| | 5. | $P$ | 3, S |
| | 6. | $Q \wedge P$ | 4, 5, Adj |
| | 7. | ~~Show~~ $Q \wedge P \rightarrow P \wedge Q$ | |
| | | Similar | |
| | 8. | $P \wedge Q \leftrightarrow Q \wedge P$ | 2, 7, CB |

T25 is the *associative law* for '$\wedge$'.

T25 $$P \wedge (Q \wedge R) \leftrightarrow (P \wedge Q) \wedge R$$

Here, as in the case of T24, we prove explicitly only one conditional. The converse can be proved in a similar way, and the biconditional will then follow by CB.

| | | |
|---|---|---|
| 1. | ~~*Show*~~ $P \wedge (Q \wedge R) \rightarrow (P \wedge Q) \wedge R$ | |
| 2. | $P \wedge (Q \wedge R)$ | |
| 3. | $P$ | 2, S |
| 4. | $Q \wedge R$ | 2, S |
| 5. | $Q$ | 4, S |
| 6. | $P \wedge Q$ | 3, 5, Adj |
| 7. | $R$ | 4, S |
| 8. | $(P \wedge Q) \wedge R$ | 6, 7, Adj |

A new informal convention for the omission of parentheses will prove convenient. In a repeated conjunction in which all terms are associated to the left, the internal parentheses may be omitted. For instance,

$$(P \wedge Q) \wedge R$$

becomes

$$P \wedge Q \wedge R \quad ,$$

and

$$([P \wedge Q] \wedge R) \wedge S$$

becomes

$$P \wedge Q \wedge R \wedge S$$

However, the parentheses of

$$P \wedge (Q \wedge R)$$

may not be omitted. T25 may now be formulated as

$$P \wedge (Q \wedge R) \leftrightarrow P \wedge Q \wedge R \quad .$$

T26, like T4 and T5, is a principle of *syllogism*.

T26 $$(P \rightarrow Q) \wedge (Q \rightarrow R) \rightarrow (P \rightarrow R)$$

T27 is known as the law of *exportation*.

T27

| | |
|---|---|
| 1. | ~~*Show*~~ $(P \wedge Q \rightarrow R) \leftrightarrow (P \rightarrow [Q \rightarrow R])$ |
| 2. | ~~*Show*~~ $(P \wedge Q \rightarrow R) \rightarrow (P \rightarrow [Q \rightarrow R])$ |

| | | |
|---|---|---|
| 3. | $P \wedge Q \rightarrow R$ | |
| 4. | ~~*Show*~~ $P \rightarrow [Q \rightarrow R]$ | |
| 5. | $P$ | |
| 6. | ~~*Show*~~ $Q \rightarrow R$ | |
| 7. | $Q$ | |
| 8. | $P \wedge Q$ | 5, 7, Adj |
| 9. | $R$ | 3, 8, MP |
| 10. | ~~*Show*~~ $(P \rightarrow [Q \rightarrow R]) \rightarrow (P \wedge Q \rightarrow R)$ | |
| 11. | $P \rightarrow [Q \rightarrow R]$ | |
| 12. | ~~*Show*~~ $P \wedge Q \rightarrow R$ | |
| 13. | $P \wedge Q$ | |
| 14. | $P$ | 13, S |
| 15. | $Q \rightarrow R$ | 11, 14, MP |
| 16. | $Q$ | 13, S |
| 17. | $R$ | 15, 16, MP |
| 18. | $(P \wedge Q \rightarrow R) \leftrightarrow (P \rightarrow [Q \rightarrow R])$ | 2, 10, CB |

T28 $(P \wedge Q \rightarrow R) \leftrightarrow (P \wedge \sim R \rightarrow \sim Q)$

T29 is the principle of *distribution* of '$\rightarrow$' over '$\wedge$'.

T29 $(P \rightarrow Q \wedge R) \leftrightarrow (P \rightarrow Q) \wedge (P \rightarrow R)$

T30 and T31 are *factor* principles.

T30 $(P \rightarrow Q) \rightarrow (R \wedge P \rightarrow R \wedge Q)$

T31 $(P \rightarrow Q) \rightarrow (P \wedge R \rightarrow Q \wedge R)$

T32 is Leibniz' *praeclarum theorema.*

T32

| | | |
|---|---|---|
| 1. | ~~*Show*~~ $(P \rightarrow R) \wedge (Q \rightarrow S) \rightarrow (P \wedge Q \rightarrow R \wedge S)$ | |
| 2. | $(P \rightarrow R) \wedge (Q \rightarrow S)$ | |
| 3. | ~~*Show*~~ $P \wedge Q \rightarrow R \wedge S$ | |
| 4. | $P \wedge Q$ | |
| 5. | $P \rightarrow R$ | 2, S |
| 6. | $P$ | 4, S |
| 7. | $R$ | 5, 6, MP |
| 8. | $Q \rightarrow S$ | 2, S |
| 9. | $Q$ | 4, S |
| 10. | $S$ | 8, 9, MP |
| 11. | $R \wedge S$ | 7, 10, Adj |

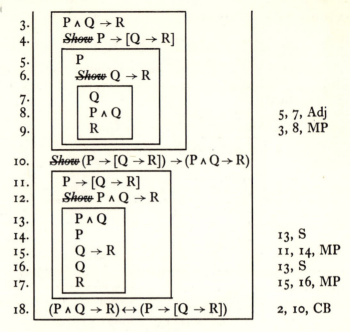

T33 is a principle of *dilemma*.

T33

| | | |
|---|---|---|
| 1. | ~~Show~~ $(P \rightarrow Q) \wedge (\sim P \rightarrow Q) \rightarrow Q$ | |
| 2. | $(P \rightarrow Q) \wedge (\sim P \rightarrow Q)$ | |
| 3. | ~~Show~~ $Q$ | |
| 4. | $\sim Q$ | |
| 5. | $P \rightarrow Q$ | 2, S |
| 6. | $\sim P$ | 4, 5, MT |
| 7. | $\sim P \rightarrow Q$ | 2, S |
| 8. | $\sim\sim P$ | 4, 7, MT |

T34, like T19 and T20, is a law of *reductio ad absurdum*.

T34 $(P \rightarrow Q) \wedge (P \rightarrow \sim Q) \rightarrow \sim P$

T35

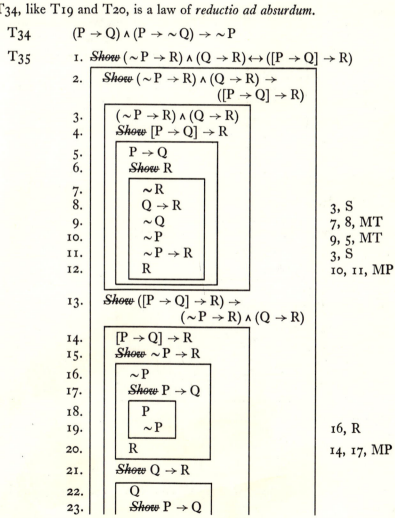

| | | |
|---|---|---|
| 1. | ~~Show~~ $(\sim P \rightarrow R) \wedge (Q \rightarrow R) \leftrightarrow ([P \rightarrow Q] \rightarrow R)$ | |
| 2. | ~~Show~~ $(\sim P \rightarrow R) \wedge (Q \rightarrow R) \rightarrow$ $([P \rightarrow Q] \rightarrow R)$ | |
| 3. | $(\sim P \rightarrow R) \wedge (Q \rightarrow R)$ | |
| 4. | ~~Show~~ $[P \rightarrow Q] \rightarrow R$ | |
| 5. | $P \rightarrow Q$ | |
| 6. | ~~Show~~ $R$ | |
| 7. | $\sim R$ | |
| 8. | $Q \rightarrow R$ | 3, S |
| 9. | $\sim Q$ | 7, 8, MT |
| 10. | $\sim P$ | 9, 5, MT |
| 11. | $\sim P \rightarrow R$ | 3, S |
| 12. | $R$ | 10, 11, MP |
| 13. | ~~Show~~ $([P \rightarrow Q] \rightarrow R) \rightarrow$ $(\sim P \rightarrow R) \wedge (Q \rightarrow R)$ | |
| 14. | $[P \rightarrow Q] \rightarrow R$ | |
| 15. | ~~Show~~ $\sim P \rightarrow R$ | |
| 16. | $\sim P$ | |
| 17. | ~~Show~~ $P \rightarrow Q$ | |
| 18. | $P$ | |
| 19. | $\sim P$ | 16, R |
| 20. | $R$ | 14, 17, MP |
| 21. | ~~Show~~ $Q \rightarrow R$ | |
| 22. | $Q$ | |
| 23. | ~~Show~~ $P \rightarrow Q$ | |

| | | |
|---|---|---|
| 24. | $Q$ | 22, R |
| 25. | $R$ | 14, 23, MP |
| 26. | $(\sim P \rightarrow R) \wedge (Q \rightarrow R)$ | 15, 21, Adj |
| 27. | $(\sim P \rightarrow R) \wedge (Q \rightarrow R) \leftrightarrow ([P \rightarrow Q] \rightarrow R)$ | 2, 13, CB |

T36 is the law of *contradiction.*

T36

| | | |
|---|---|---|
| 1. | ~~*Show*~~ $\sim(P \wedge \sim P)$ | |
| 2. | $P \wedge \sim P$ | |
| 3. | $P$ | 2, S |
| 4. | $\sim P$ | 2, S |

T37 shows how to express '→' by means of '~' and '∧', and T38 how to express '∧' by means of '~' and '→'. T39 and T40 provide alternative expressions for the negation of a conjunction and of a conditional.

T37

| | | |
|---|---|---|
| 1. | ~~*Show*~~ $(P \rightarrow Q) \leftrightarrow \sim(P \wedge \sim Q)$ | |
| 2. | ~~*Show*~~ $(P \rightarrow Q) \rightarrow \sim(P \wedge \sim Q)$ | |
| 3. | $P \rightarrow Q$ | |
| 4. | ~~*Show*~~ $\sim(P \wedge \sim Q)$ | |
| 5. | $P \wedge \sim Q$ | |
| 6. | $P$ | 5, S |
| 7. | $Q$ | 3, 6, MP |
| 8. | $\sim Q$ | 5, S |
| 9. | ~~*Show*~~ $\sim(P \wedge \sim Q) \rightarrow (P \rightarrow Q)$ | |
| 10. | $\sim(P \wedge \sim Q)$ | |
| 11. | ~~*Show*~~ $P \rightarrow Q$ | |
| 12. | $P$ | |
| 13. | ~~*Show*~~ $Q$ | |
| 14. | $\sim Q$ | |
| 15. | $P \wedge \sim Q$ | 12, 14, Adj |
| 16. | $\sim(P \wedge \sim Q)$ | 10, R |
| 17. | $(P \rightarrow Q) \leftrightarrow \sim(P \wedge \sim Q)$ | 2, 9, CB |

T38 $P \wedge Q \leftrightarrow \sim(P \rightarrow \sim Q)$

T39 $\sim(P \wedge Q) \leftrightarrow (P \rightarrow \sim Q)$

T40 $\sim(P \rightarrow Q) \leftrightarrow P \wedge \sim Q$

T41 is the law of *indempotence* for '∧', and T42 – T44 will be cited in chapter V.

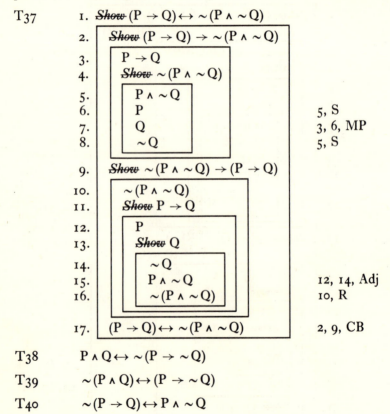

T41 $P \leftrightarrow P \land P$

T42 $P \land \sim Q \rightarrow \sim(P \rightarrow Q)$

T43 $\sim P \rightarrow \sim(P \land Q)$

T44 $\sim Q \rightarrow \sim(P \land Q)$

## EXERCISES

22. Prove T28, T29, T34.
23. Prove T38, T44.

In solving these exercises, which essentially involve only '$\sim$', '$\rightarrow$', and '$\land$', the reader will find the following informal suggestions, some of which appeared in chapter I, helpful (but not infallible).

(*1*) *To derive a conditional, use conditional derivation.*
(*2*) *To derive a conjunction, derive first both conjuncts and then use Adj.*
(*3*) *To derive a biconditional, derive first the two corresponding conditionals and then use CB.*
(*4*) *To derive anything else, use indirect derivation unless another procedure is immediately obvious.*
(*5*) *Whenever a sentence follows from antecedent lines by MP, MT, S, MTP, or BC, enter that sentence as a line.*
(*6*) *When using indirect derivation, determine whether any of the antecedent lines is the negation of a conditional; if so, attempt to derive that conditional.*

T40 is proved for illustration. We begin the proof by writing '*Show*' followed by the sentence to be proved. This sentence is a biconditional; thus, following (3) and (1) of the foregoing suggestions, we begin a conditional derivation of one of the corresponding conditionals.

1. *Show* $\sim(P \rightarrow Q) \leftrightarrow P \land \sim Q$
2. *Show* $\sim(P \rightarrow Q) \rightarrow P \land \sim Q$
3. $\sim(P \rightarrow Q)$

In order to complete the subsidiary conditional derivation we must derive a conjunction; thus, following suggestions (2) and (4), we begin an indirect derivation of one of its conjuncts.

1. *Show* $\sim(P \rightarrow Q) \leftrightarrow P \land \sim Q$
2. *Show* $\sim(P \rightarrow Q) \rightarrow P \land \sim Q$
3. $\sim(P \rightarrow Q)$
4. *Show* P
5. $\sim P$

Consideration of the lines now before us and suggestion (6) lead us to begin next a conditional derivation of the sentence whose negation occurs in line 3, and this can be completed after an application of rule R.

1. *Show* $\sim(P \to Q) \leftrightarrow P \wedge \sim Q$
2. *Show* $\sim(P \to Q) \to P \wedge \sim Q$
3. $\sim(P \to Q)$
4. *Show* P
5. $\sim P$
6. ~~*Show*~~ $P \to Q$
7. P
8. $\sim P$ — 5, R

We now complete the subsidiary derivation of 'P' and, returning to suggestions (2) and (4), begin an indirect derivation of '$\sim Q$', which we can complete by employing again suggestion (6).

1. *Show* $\sim(P \to Q) \leftrightarrow P \wedge \sim Q$
2. *Show* $\sim(P \to Q) \to P \wedge \sim Q$
3. $\sim(P \to Q)$
4. ~~*Show*~~ P
5. $\sim P$
6. ~~*Show*~~ $P \to Q$
7. P
8. $\sim P$ — 5, R
9. $\sim(P \to Q)$ — 3, R
10. ~~*Show*~~ $\sim Q$
11. Q
12. ~~*Show*~~ $P \to Q$
13. Q — 11, R
14. $\sim(P \to Q)$ — 3, R

Now, following through suggestion (2), we employ adjunction in order to complete the subsidiary derivation started in line 2.

1. *Show* $\sim(P \to Q) \leftrightarrow P \wedge \sim Q$
2. ~~*Show*~~ $\sim(P \to Q) \to P \wedge \sim Q$
3. $\sim(P \to Q)$
4. ~~*Show*~~ P
5. $\sim P$
6. ~~*Show*~~ $P \to Q$
7. P
8. $\sim P$ — 5, R
9. $\sim(P \to Q)$ — 3, R

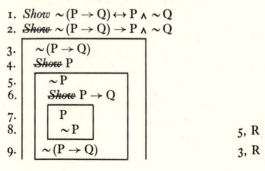

| | | |
|---|---|---|
| 10. | ~~*Show*~~ $\sim Q$ | |
| 11. | $Q$ | |
| 12. | ~~*Show*~~ $P \rightarrow Q$ | |
| 13. | $Q$ | 11, R |
| 14. | $\sim(P \rightarrow Q)$ | 3, R |
| 15. | $P \wedge \sim Q$ | 4, 10, Adj |

To complete the main derivation we derive next the converse of the conditional that occurs in line 2 and then, following through suggestion (3), employ CB.

| | | |
|---|---|---|
| 1. | ~~*Show*~~ $\sim(P \rightarrow Q) \leftrightarrow P \wedge \sim Q$ | |
| 2. | ~~*Show*~~ $\sim(P \rightarrow Q) \rightarrow P \wedge \sim Q$ | |
| 3. | $\sim(P \rightarrow Q)$ | |
| 4. | ~~*Show*~~ $P$ | |
| 5. | $\sim P$ | |
| 6. | ~~*Show*~~ $P \rightarrow Q$ | |
| 7. | $P$ | |
| 8. | $\sim P$ | 5, R |
| 9. | $\sim(P \rightarrow Q)$ | 3, R |
| 10. | ~~*Show*~~ $\sim Q$ | |
| 11. | $Q$ | |
| 12. | ~~*Show*~~ $P \rightarrow Q$ | |
| 13. | $Q$ | 11, R |
| 14. | $\sim(P \rightarrow Q)$ | 3, R |
| 15. | $P \wedge \sim Q$ | 4, 10, Adj |
| 16. | ~~*Show*~~ $P \wedge \sim Q \rightarrow \sim(P \rightarrow Q)$ | |
| 17. | $P \wedge \sim Q$ | |
| 18. | ~~*Show*~~ $\sim(P \rightarrow Q)$ | |
| 19. | $P \rightarrow Q$ | |
| 20. | $P$ | 17, S |
| 21. | $Q$ | 19, 20, MP |
| 22. | $\sim Q$ | 17, S |
| 23. | $\sim(P \rightarrow Q) \leftrightarrow P \wedge \sim Q$ | 2, 16, CB |

**4. Abbreviated derivations.** It will be useful, from time to time, to append to the directions for constructing a derivation what we shall call *abbreviatory clauses;* we shall thus arrive at a more comprehensive class of derivations, called *abbreviated derivations*, which will be constructed on the basis of clauses (1) – (6) along with the new clauses.

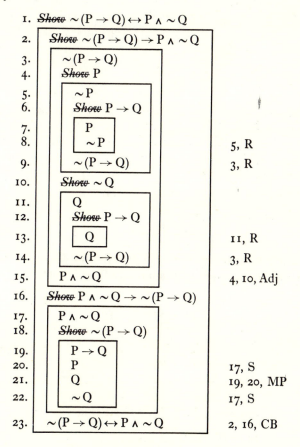

Abbreviatory clauses must satisfy two conditions. In the first place, they must be theoretically dispensable; that is to say, whenever a conclusion can be derived from a class of premises by means of an abbreviated derivation, it must also be derivable from the same premises by means of some *unabbreviated derivation* (a derivation constructed on the basis of clauses (1) – (6) alone). In the second place, abbreviated derivations must share the important characteristic of unabbreviated derivations pointed out earlier: there must be an automatic procedure for checking the correctness of an abbreviated derivation (at least when membership in the class of premises is automatically decidable).

For the sentential calculus we shall adopt only two abbreviatory clauses, (7) and (8), which, when adjoined to clauses (1) – (6), will constitute the directions for constructing an *abbreviated derivation*, or, as we shall henceforth say, a *derivation*. The fact that clauses (7) and (8) satisfy the two conditions mentioned above will be relatively obvious but will not be established in a formal way.

For clause (7) we must introduce a new notion. By an *instance* of a symbolic sentence $\phi$ we shall understand any symbolic sentence obtained from $\phi$ by replacing sentence letters uniformly by symbolic sentences. (The replacement is *uniform* just in case all occurrences of a sentence letter are replaced by the same sentence.)

For example, T2,

$$Q \to (P \to Q) \quad ,$$

has as an instance

$$Q \to ({\sim} R \to Q) \quad .$$

In this case, we have replaced 'P' by '∼R' and 'Q' by 'Q' itself. We may indicate this replacement diagrammatically as follows:

$$\frac{Q \quad P}{Q \quad {\sim}R}$$

Other instances of T2 are

(1) $\quad {\sim}R \to (Q \to {\sim}R) \quad ,$

(2) $\quad P \to ([R \to T] \to P) \quad ,$

(3) $\quad Q \vee P \to ([P \to Q] \to Q \vee P) \quad ,$

(4) $\quad Q \to (P \to Q) \quad ,$

(5) $\quad P \to (Q \to P) \quad ,$

obtained by the respective replacements

(1′) $\quad \dfrac{Q \quad P}{{\sim}R \quad Q} \ ,$

(2′) $\dfrac{Q \qquad P}{P \qquad R \rightarrow T}$ ,

(3′) $\dfrac{Q \qquad P}{Q \vee P \qquad P \rightarrow Q}$ ,

(4′) $\dfrac{Q \quad P}{Q \quad P}$ ,

(5′) $\dfrac{Q \quad P}{P \quad Q}$ .

As a further example, T26,

$$(P \rightarrow Q) \wedge (Q \rightarrow R) \rightarrow (P \rightarrow R) \quad ,$$

has as an instance

$$([S \rightarrow T] \rightarrow T) \wedge (T \rightarrow S) \rightarrow ([S \rightarrow T] \rightarrow S) \quad ,$$

obtained by the replacement

$$\frac{P \qquad Q \qquad R}{S \rightarrow T \qquad T \qquad S} \quad .$$

The relation '$\phi$ is an instance of $\psi$' is intended to apply only to symbolic sentences. (This accords with our previous decision to apply deductive procedures only to symbolic sentences.)

It is because of clause (7) that theorems are important. Having proved a theorem, we may then use it in a later derivation without repeating its proof. Indeed, we may use as well any of its instances, for their proofs could be obtained by a simple procedure of replacement from the original proof.

(7) *If $\phi$ is an instance of a theorem already proved, then $\phi$ may occur as a line.* (*Annotation: the number of the theorem of which $\phi$ is an instance, sometimes together with a diagrammatic indication of the replacement involved.*

For example, clause (7) provides the justification for lines 3 and 5 in the following abbreviated derivation.

| | | |
|---|---|---|
| 1. | ~~Show~~ $([P \rightarrow Q] \rightarrow Q) \wedge (Q \rightarrow P) \rightarrow P$ | |
| 2. | $([P \rightarrow Q] \rightarrow Q) \wedge (Q \rightarrow P)$ | |
| 3. | $([P \rightarrow Q] \rightarrow Q) \wedge (Q \rightarrow P) \rightarrow ([P \rightarrow Q] \rightarrow P)$ | T26$\left(\dfrac{P \qquad Q \qquad R}{P \rightarrow Q \qquad Q \qquad P}\right)$ |
| 4. | $[P \rightarrow Q] \rightarrow P$ | 2, 3, MP |
| 5. | $([P \rightarrow Q] \rightarrow P) \rightarrow P$ | T23 |
| 6. | $P$ | 4, 5, MP |

(The reader should compare this brief proof with the unabbreviated derivation, given on page 26 ff., corresponding to exercise 12 of chapter I.)

It is often convenient to compress several steps into one, omitting some lines that an unabbreviated derivation would require. Such compression will be allowed only when no clauses other than (2) (premises), (5) (inference rules), and (7) (instances of previously proved theorems) are involved, and is legitimized by clause (8).

(8) *A sentence may occur as a line if it follows from antecedent lines by a succession of steps, and each intermediate step can be justified by one of clauses* (2), (5), *or* (7). (*The annotation should determine the succession of steps leading to the line in question. This can be done by indicating, in order of application, the antecedent lines, the premises, the inference rules, and the previously proved theorems employed. In addition, when the rule Add is used, the added disjunct should be indicated; and when an instance of a previously proved theorem is involved, the relevant replacement should be indicated.*)

For example, clause (8) provides the justification for line 5 in the following abbreviated derivation.

| | | |
|---|---|---|
| 1. | ~~Show~~ $(P \rightarrow Q) \rightarrow (R \wedge P \rightarrow R \wedge Q)$ | |
| 2. | $P \rightarrow Q$ | |
| 3. | ~~Show~~ $R \wedge P \rightarrow R \wedge Q$ | |
| 4. | $R \wedge P$ | |
| 5. | $R \wedge Q$ | 4, S, 2, MP, 4, S, Adj |

The omitted lines are, in order:

| | | |
|---|---|---|
| i. | P | (4, S) |
| ii. | Q | (i, 2, MP) |
| iii. | R | (4, S) |

Line 5 follows from (ii) and (iii) by Adj. Clause (8) also provides the justification for lines 3, 4, and 5 in the derivation accompanying the following argument:

$$(Q \rightarrow P) \rightarrow P \quad . \quad P \rightarrow Q \quad \therefore Q$$

| | | |
|---|---|---|
| 1. | ~~Show~~ Q | |
| 2. | $\sim Q$ | |
| 3. | $\sim P$ | 2, 2nd premise, MT |
| 4. | $Q \rightarrow P$ | 2, $T18\left(\frac{P}{Q}\ \frac{Q}{P}\right)$, MP |
| 5. | P | 4, 1st premise, MP |

An unabbreviated derivation of the conclusion from the given premises can be obtained by inserting in the abbreviated derivation the omitted lines, along with a subsidiary derivation of the instance of the theorem employed. The following is an example of such a derivation.

| | | |
|---|---|---|
| 1. | ~~*Show*~~ Q | |
| 2. | $\sim Q$ | |
| 3. | $P \rightarrow Q$ | Premise |
| 4. | $\sim P$ | 2, 3, MT |
| 5. | ~~*Show*~~ $\sim Q \rightarrow (Q \rightarrow P)$ | |
| 6. | $\sim Q$ | |
| 7. | ~~*Show*~~ $Q \rightarrow P$ | |
| 8. | $Q$ | |
| 9. | $\sim Q$ | 6, R |
| 10. | $Q \rightarrow P$ | 2, 5, MP |
| 11. | $(Q \rightarrow P) \rightarrow P$ | Premise |
| 12. | $P$ | 10, 11, MP |

Abbreviated derivations, unlike unabbreviated derivations, require as an essential part their annotations—at least those annotations given in connection with clause (8). For without an indication of omitted steps, it would be impossible, even for the simple logical system now under consideration, to give an automatic procedure for checking the correctness of an abbreviated derivation.

## EXERCISES

Corresponding to the following argument, two abbreviated derivations are given.

(6) $\sim P \rightarrow R \quad . \quad P \rightarrow Q \quad . \quad Q \rightarrow R \quad \therefore R$

(7)

| | | |
|---|---|---|
| 1. | ~~*Show*~~ *R* | |
| 2. | $\sim R$ | |
| 3. | $\sim Q$ | |
| 4. | $\sim P$ | |
| 5. | $R$ | |

(8)

| | | |
|---|---|---|
| 1. | ~~*Show*~~ *R* | |
| 2. | $P \rightarrow R$ | 2nd premise, 3rd premise, Adj, T26, MP |
| 3. | $R$ | 2, 1st premise, Adj, $\mathrm{T33}\left(\frac{P \quad Q}{P \quad R}\right)$, MP |

24. Annotate the derivation (7).

25. Construct an unabbreviated derivation corresponding to (6), using indirect derivation. (Consider (7) and its annotations.)

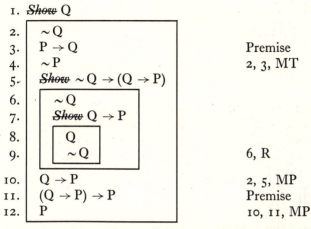

26. Construct an unabbreviated derivation corresponding to (6), using direct derivation. (Consider (8).)

Show, by constructing derivations in which at least one step is justified either by clause (7) or by clause (8), that the following arguments are valid.

27. (P → Q) → R . ∼P ∴ R
28. (P → Q) → R . Q ∴ R

(The reader should attempt two-line derivations in each case.)

**5. Theorems with abbreviated proofs.** Many of the remaining derivations of this chapter are considerably simplified by the addition of clauses (7) and (8) to the directions for constructing a derivation.

Now we consider theorems primarily concerned with 'v' as well as '∼', '→', and '∧'. T45 shows how to express 'v' in terms of '∼' and '→', and T46 shows how to express '→' in terms of '∼' and 'v'. Clause (8) is used in the proof of T45.

| T45 | | | |
|---|---|---|---|
| | 1. | ~~Show~~ P v Q ↔ (∼P → Q) | |
| | 2. | ~~Show~~ P v Q → (∼P → Q) | |
| | 3. | P v Q | |
| | 4. | ~~Show~~ ∼P → Q | |
| | 5. | ∼P | |
| | 6. | Q | 5, 3, MTP |
| | 7. | ~~Show~~ (∼P → Q) → P v Q | |
| | 8. | ∼P → Q | |
| | 9. | ~~Show~~ P v Q | |
| | 10. | ∼(P v Q) | |
| | 11. | ~~Show~~ ∼P | |
| | 12. | P | |
| | 13. | P v Q | 12, Add |
| | 14. | ∼(P v Q) | 10, R |
| | 15. | P v Q | 8, 11, MP, Add |
| | 16. | P v Q ↔ (∼P → Q) | 2, 7, CB |

T46 (P → Q) ↔ ∼P v Q

T47 is the law of *idempotence* for 'v'.

| T47 | | |
|---|---|---|
| | 1. | ~~Show~~ P ↔ P v P |
| | 2. | ~~Show~~ P → P v P |

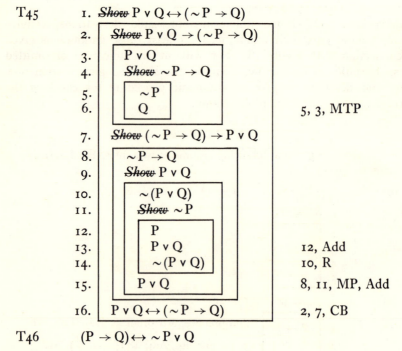

3. $P$
4. $P \lor P$ — 3, Add
5. ~~*Show*~~ $P \lor P \rightarrow P$
6. $P \lor P$
7. ~~*Show*~~ $P$
8. $\sim P$
9. $P$ — 6, 8, MTP
10. $P \leftrightarrow P \lor P$ — 2, 5, CB

T48 and T49, like T33, are principles of *dilemma*.

T48

1. ~~*Show*~~ $(P \lor Q) \land (P \rightarrow R) \land (Q \rightarrow S) \rightarrow R \lor S$
2. $(P \lor Q) \land (P \rightarrow R) \land (Q \rightarrow S)$
3. ~~*Show*~~ $\sim R \rightarrow S$
4. $\sim R$
5. $(P \lor Q) \land (P \rightarrow R)$ — 2, S
6. $Q \rightarrow S$ — 2, S
7. $S$ — 5, S, 4, MT, 5, S, MTP, 6, MP
8. $R \lor S$ — T45$\left(\frac{P \quad Q}{R \quad S}\right)$, BC, 3, MP

T49

1. ~~*Show*~~ $(P \lor Q) \land (P \rightarrow R) \land (Q \rightarrow R) \rightarrow R$
2. $(P \lor Q) \land (P \rightarrow R) \land (Q \rightarrow R)$ — T48$\left(\frac{P\ Q\ R\ S}{P\ Q\ R\ R}\right)$, 2, MP
3. $R$ — T47$\left(\frac{P}{R}\right)$, BC, MP

T50 is a principle of *composition*.

T50

1. ~~*Show*~~ $(P \rightarrow R) \land (Q \rightarrow R) \leftrightarrow (P \lor Q \rightarrow R)$
2. ~~*Show*~~ $(P \rightarrow R) \land (Q \rightarrow R) \rightarrow (P \lor Q \rightarrow R)$
3. $(P \rightarrow R) \land (Q \rightarrow R)$
4. ~~*Show*~~ $P \lor Q \rightarrow R$
5. $P \lor Q$
6. $R$ — 5, 3, Adj, T49, MP
7. ~~*Show*~~ $(P \lor Q \rightarrow R) \rightarrow (P \rightarrow R) \land (Q \rightarrow R)$

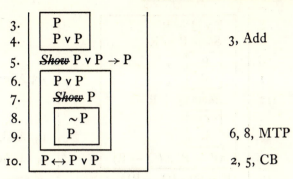

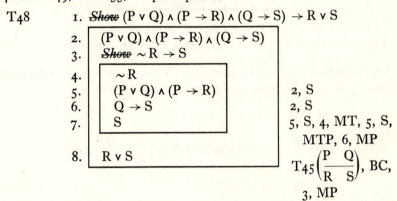

| | | |
|---|---|---|
| 8. | $P \vee Q \to R$ | |
| 9. | ~~*Show*~~ $P \to R$ | |
| 10. | $P$ | |
| 11. | $R$ | 10, Add, 8, MP |
| 12. | ~~*Show*~~ $Q \to R$ | |
| 13. | $Q$ | |
| 14. | $R$ | 13, Add, 8, MP |
| 15. | $(P \to R) \wedge (Q \to R)$ | 9, 12, Adj |
| 16. | $(P \to R) \wedge (Q \to R) \leftrightarrow (P \vee Q \to R)$ | 2, 7, CB |

T51 $(P \vee Q) \wedge (P \to R) \wedge (\sim P \wedge Q \to R) \to R$
T52 $(P \to R) \wedge (\sim P \wedge Q \to R) \leftrightarrow (P \vee Q \to R)$

The following patterns of inference, each corresponding to a theorem proved by now, will appear frequently in what follows:

(1) $\phi \to \psi$
(2) $\sim \phi \to \psi$

$\psi$ (1), (2), Adj, T33 $\left(\frac{P \quad Q}{\phi \quad \psi}\right)$, MP

(3) $\phi \vee \psi$
(4) $\phi \to \chi$
(5) $\psi \to \chi$

$\chi$ (3), (4), Adj, (5), Adj, T49 $\left(\frac{P \quad Q \quad R}{\phi \quad \psi \quad \chi}\right)$, MP

(6) $\phi \to \chi$
(7) $\psi \to \chi$

$\phi \vee \psi \to \chi$ (6), (7), Adj, T50 $\left(\frac{P \quad Q \quad R}{\phi \quad \psi \quad \chi}\right)$, BC, MP

We shall refer jointly to these three patterns as inferences by *the derived rule of separation of cases*, and we shall employ 'SC' to indicate their use. More explicitly, we shall abbreviate the lengthy annotations above by 'SC', together with the numbers of the antecedent lines involved. Thus, for example, the second pattern becomes:

(3) $\phi \vee \psi$
(4) $\phi \to \chi$
(5) $\psi \to \chi$

$\chi$ (3), (4), (5), SC

Illustrations of the first and third forms of SC are provided by the following derivations.

Argument:

$$P \to Q \quad . \quad \sim P \to R \quad . \quad R \to Q \quad \therefore Q$$

Derivation:

1. ~~*Show*~~ Q

| | | |
|---|---|---|
| 2. | $\sim P \to Q$ | 2nd premise, 3rd premise, Adj, T26$\left(\dfrac{P \quad Q \quad R}{\sim P \quad R \quad Q}\right)$, MP |
| 3. | $Q$ | 1st premise, 2, SC |

Argument:

$$\therefore \sim P \vee Q \to (P \to Q)$$

Derivation:

1. ~~*Show*~~ $\sim P \vee Q \to (P \to Q)$

| | | |
|---|---|---|
| 2. | $\sim P \to (P \to Q)$ | T18 |
| 3. | $Q \to (P \to Q)$ | T2 |
| 4. | $\sim P \vee Q \to (P \to Q)$ | 2, 3, SC |

Another useful pattern of inference is called *the derived rule of conditional-disjunction*, and is justified by T45:

(8) $$\frac{\sim \phi \to \psi}{\phi \vee \psi}$$ (8), T45, BC, MP

We shall henceforth replace the annotation 'T45, BC, MP' by the abbreviation 'CD'. An illustration is provided by the following proof.

1. ~~*Show*~~ $(P \to Q) \to \sim P \vee Q$

| | | |
|---|---|---|
| 2. | $P \to Q$ | |
| 3. | ~~*Show*~~ $\sim\sim P \to Q$ | |
| 4. | $\sim\sim P$ | |
| 5. | $Q$ | 4, DN, 2, MP |
| 6. | $\sim P \vee Q$ | 3, CD |

T53 is the *commutative* law for 'v'.

T53 $\quad P \vee Q \leftrightarrow Q \vee P$

The *associative* law for 'v':

T54 $\quad P \vee (Q \vee R) \leftrightarrow (P \vee Q) \vee R$

As in a conjunction, internal parentheses may be omitted from a repeated disjunction in which all terms are associated to the left. Thus

T54 becomes

$$P \vee (Q \vee R) \leftrightarrow P \vee Q \vee R \quad .$$

T55 is the *distributive law* for '→' over 'v'.

T55 $(P \rightarrow Q \vee R) \leftrightarrow (P \rightarrow Q) \vee (P \rightarrow R)$

T56 $(P \rightarrow Q) \rightarrow (R \vee P \rightarrow R \vee Q)$

T57 $(P \rightarrow Q) \rightarrow (P \vee R \rightarrow Q \vee R)$

T58 $(P \rightarrow Q) \vee (Q \rightarrow R)$

The law of the *excluded middle:*

T59 $P \vee {\sim} P$

T60, like T50, is a principle of *composition.*

T60 $(P \rightarrow R) \vee (Q \rightarrow R) \leftrightarrow (P \wedge Q \rightarrow R)$

T61 and T62 are laws of *distribution.*

T61 $P \wedge (Q \vee R) \leftrightarrow (P \wedge Q) \vee (P \wedge R)$

T62 $P \vee (Q \wedge R) \leftrightarrow (P \vee Q) \wedge (P \vee R)$

T63 – T66 are known as *De Morgan's laws*, after the nineteenth-century logician, Augustus De Morgan; T67, a corollary of T66, will play an important role in chapter V.

T63 $P \wedge Q \leftrightarrow {\sim}({\sim} P \vee {\sim} Q)$

T64 $P \vee Q \leftrightarrow {\sim}({\sim} P \wedge {\sim} Q)$

T65 ${\sim}(P \wedge Q) \leftrightarrow {\sim} P \vee {\sim} Q$

T66 ${\sim}(P \vee Q) \leftrightarrow {\sim} P \wedge {\sim} Q$

T67 ${\sim} P \wedge {\sim} Q \rightarrow {\sim}(P \vee Q)$

T68 and T69 provide redundant but useful forms of expression.

T68 $P \leftrightarrow (P \wedge Q) \vee (P \wedge {\sim} Q)$

T69 $P \leftrightarrow (P \vee Q) \wedge (P \vee {\sim} Q)$

We come now to theorems which primarily concern '↔'. All but T78, T80, T94, and T95 (which are included because of their unintuitive character) will be found extremely useful in what follows.

T70 $Q \rightarrow (P \wedge Q \leftrightarrow P)$

T71 ${\sim} Q \rightarrow (P \vee Q \leftrightarrow P)$

T72 $(P \rightarrow Q) \leftrightarrow (P \wedge Q \leftrightarrow P)$

T73 $(P \to Q) \leftrightarrow (P \vee Q \leftrightarrow Q)$

T74 $(P \leftrightarrow Q) \wedge P \to Q$

T75 $(P \leftrightarrow Q) \wedge Q \to P$

T76 $(P \leftrightarrow Q) \wedge {\sim}P \to {\sim}Q$

T77 $(P \leftrightarrow Q) \wedge {\sim}Q \to {\sim}P$

T78 $(P \to [Q \leftrightarrow R]) \leftrightarrow ([P \to Q] \leftrightarrow [P \to R])$

T79 $(P \to [Q \leftrightarrow R]) \leftrightarrow (P \wedge Q \leftrightarrow P \wedge R)$

T80 $(P \leftrightarrow Q) \vee (P \leftrightarrow {\sim}Q)$

T81 $(P \leftrightarrow Q) \leftrightarrow (P \to Q) \wedge (Q \to P)$

T82 $(P \leftrightarrow Q) \leftrightarrow {\sim}([P \to Q] \to {\sim}[Q \to P])$

T83 $(P \leftrightarrow Q) \leftrightarrow (P \wedge Q) \vee ({\sim}P \wedge {\sim}Q)$

T84 $P \wedge Q \to (P \leftrightarrow Q)$

T85 ${\sim}P \wedge {\sim}Q \to (P \leftrightarrow Q)$

T86 $([P \leftrightarrow Q] \to R) \leftrightarrow (P \wedge Q \to R) \wedge ({\sim}P \wedge {\sim}Q \to R)$

T87 ${\sim}(P \leftrightarrow Q) \leftrightarrow (P \wedge {\sim}Q) \vee ({\sim}P \wedge Q)$

T88 $P \wedge {\sim}Q \to {\sim}(P \leftrightarrow Q)$

T89 ${\sim}P \wedge Q \to {\sim}(P \leftrightarrow Q)$

T90 ${\sim}(P \leftrightarrow Q) \leftrightarrow (P \leftrightarrow {\sim}Q)$

T91 $P \leftrightarrow P$

T92 $(P \leftrightarrow Q) \leftrightarrow (Q \leftrightarrow P)$

T93 $(P \leftrightarrow Q) \wedge (Q \leftrightarrow R) \to (P \leftrightarrow R)$

T94 $(P \leftrightarrow [Q \leftrightarrow R]) \leftrightarrow ([P \leftrightarrow Q] \leftrightarrow R)$

T95 $(P \leftrightarrow Q) \leftrightarrow ([P \leftrightarrow R] \leftrightarrow [Q \leftrightarrow R])$

T96 $(P \leftrightarrow Q) \leftrightarrow ({\sim}P \leftrightarrow {\sim}Q)$

T97 $(P \leftrightarrow R) \wedge (Q \leftrightarrow S) \to ([P \to Q] \leftrightarrow [R \to S])$

T98 $(P \leftrightarrow R) \wedge (Q \leftrightarrow S) \to (P \wedge Q \leftrightarrow R \wedge S)$

T99 $(P \leftrightarrow R) \wedge (Q \leftrightarrow S) \to (P \vee Q \leftrightarrow R \vee S)$

T100 $(P \leftrightarrow R) \wedge (Q \leftrightarrow S) \to ([P \leftrightarrow Q] \leftrightarrow [R \leftrightarrow S])$

T101 $(Q \leftrightarrow S) \to ([P \to Q] \leftrightarrow [P \to S]) \wedge ([Q \to P] \leftrightarrow [S \to P])$

| | |
|---|---|
| T102 | $(Q \leftrightarrow S) \rightarrow (P \land Q \leftrightarrow P \land S)$ |
| T103 | $(Q \leftrightarrow S) \rightarrow (P \lor Q \leftrightarrow P \lor S)$ |
| T104 | $(Q \leftrightarrow S) \rightarrow ([P \leftrightarrow Q] \leftrightarrow [P \leftrightarrow S])$ |
| T105 | $P \land (Q \leftrightarrow R) \rightarrow (P \land Q \leftrightarrow R)$ |

Each of the following theorems is an occasionally useful biconditional corresponding to a conditional listed in chapter I. The number of the corresponding conditional (and in some cases that of its converse) is indicated in parentheses.

| | | |
|---|---|---|
| T106 | $(P \rightarrow [Q \rightarrow R]) \leftrightarrow ([P \rightarrow Q] \rightarrow [P \rightarrow R])$ | (T6, T7) |
| T107 | $(P \rightarrow [Q \rightarrow R]) \leftrightarrow (Q \rightarrow [P \rightarrow R])$ | (T8) |
| T108 | $(P \rightarrow [P \rightarrow Q]) \leftrightarrow (P \rightarrow Q)$ | (T9) |
| T109 | $([P \rightarrow Q] \rightarrow Q) \leftrightarrow ([Q \rightarrow P] \rightarrow P)$ | (T10) |
| T110 | $P \leftrightarrow \sim\sim P$ | (T11, T12) |
| T111 | $(P \rightarrow Q) \leftrightarrow (\sim Q \rightarrow \sim P)$ | (T13) |
| T112 | $(P \rightarrow \sim Q) \leftrightarrow (Q \rightarrow \sim P)$ | (T14) |
| T113 | $(\sim P \rightarrow Q) \leftrightarrow (\sim Q \rightarrow P)$ | (T15) |
| T114 | $(\sim P \rightarrow P) \leftrightarrow P$ | (T19) |
| T115 | $(P \rightarrow \sim P) \leftrightarrow \sim P$ | (T20) |

## EXERCISES, GROUP I

29. In the following derivation, which is a proof of T83, insert with annotation all lines omitted by the use of clause (8).

| | | |
|---|---|---|
| 1. | ~~Show~~ $(P \leftrightarrow Q) \leftrightarrow (P \land Q) \lor (\sim P \land \sim Q)$ | |
| 2. | ~~Show~~ $(P \leftrightarrow Q) \rightarrow (P \land Q) \lor (\sim P \land \sim Q)$ | |
| 3. | $P \leftrightarrow Q$ | |
| 4. | ~~Show~~ $\sim(P \land Q) \rightarrow \sim P \land \sim Q$ | |
| 5. | $\sim(P \land Q)$ | |
| 6. | ~~Show~~ $\sim P$ | |
| 7. | $P$ | |
| 8. | $P \land Q$ | 3, BC, 7, MP, 7, Adj |
| 9. | $\sim(P \land Q)$ | 5, R |
| 10. | $\sim P \land \sim Q$ | 3, BC, 6, MT, 6, Adj |
| 11. | $(P \land Q) \lor (\sim P \land \sim Q)$ | 4, CD |

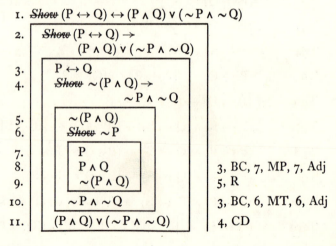

12. ~~Show~~ $P \wedge Q \rightarrow (P \leftrightarrow Q)$
13. $P \wedge Q$
14. $P \leftrightarrow Q$ — 13, S, T2, MP, 13, S, $T2\left(\frac{P\ Q}{Q\ P}\right)$, MP, CB
15. ~~Show~~ $\sim P \wedge \sim Q \rightarrow (P \leftrightarrow Q)$
16. $\sim P \wedge \sim Q$
17. $P \leftrightarrow Q$ — 16, S, T18, MP, 16, S, $T18\left(\frac{P\ Q}{Q\ P}\right)$, MP, CB
18. $(P \leftrightarrow Q) \leftrightarrow (P \wedge Q) \vee (\sim P \wedge \sim Q)$ — 12, 15, SC, 2, CB

Of the preceding theorems, prove

30. T55, T58, T60 – T62;
31. T63, T64, T68, T69;
32. T70 – T73;
33. T78 – T80;
34. T86, T87, T90, T95;
35. T97 – T100.

In solving exercises 30 – 35 (as well as later exercises), the reader will find the following informal suggestions helpful (though again not infallible).

(*1*) *To derive a sentence*

$$\phi \vee \psi \rightarrow \chi \quad ,$$

*derive first*

$$\phi \rightarrow \chi$$

*and*

$$\psi \rightarrow \chi \quad ,$$

*and then use SC.*

(*2*) *To derive*

$$\phi \rightarrow \psi \quad ,$$

*where* $\phi$ *is not a disjunction, use conditional derivation.*

(*3*) *To derive a conjunction, derive first both conjuncts, and then use Adj.*

(*4*) *To derive*

$$\phi \vee \psi \quad ,$$

*derive first*

$$\sim\phi \rightarrow \psi$$

*and then use CD.*

(5) *To derive a biconditional, derive first the two corresponding conditionals, and then use CB.*

(6) *To derive anything else, use either indirect derivation or separation of cases.*

A partial derivation of T94 is given for illustration. After writing the initial assertion line, we begin, according to suggestions (5) and (2), a conditional derivation of one of the relevant conditionals.

1. *Show* $(P \leftrightarrow [Q \leftrightarrow R]) \leftrightarrow ([P \leftrightarrow Q] \leftrightarrow R)$
2. *Show* $(P \leftrightarrow [Q \leftrightarrow R]) \rightarrow ([P \leftrightarrow Q] \leftrightarrow R)$
3. $P \leftrightarrow [Q \leftrightarrow R]$

To complete the conditional derivation started in line 2 we must derive a biconditional neither of whose constituents is a disjunction; thus we again follow suggestions (5) and (2). (Note that it would be superfluous here to enter a line with '*Show*' followed by the biconditional in question.)

1. *Show* $(P \leftrightarrow [Q \leftrightarrow R]) \leftrightarrow ([P \leftrightarrow Q] \leftrightarrow R)$
2. *Show* $(P \leftrightarrow [Q \leftrightarrow R]) \rightarrow ([P \leftrightarrow Q] \leftrightarrow R)$
3. $P \leftrightarrow [Q \leftrightarrow R]$
4. *Show* $[P \leftrightarrow Q] \rightarrow R$
5. $P \leftrightarrow Q$

Since 'R', the sentence to be derived to complete the subsidiary derivation for line 4, is not an obvious consequence of the antecedent lines available (lines 3 and 5), we turn to suggestion (6). To begin an indirect derivation of 'R' would not improve our position, for it is not obvious that a contradiction can be derived from '$\sim$R' along with lines 3 and 5. The most useful strategy is to separate cases. We see clearly that 'R' follows once we assume 'P':

1. *Show* $(P \leftrightarrow [Q \leftrightarrow R]) \leftrightarrow ([P \leftrightarrow Q] \leftrightarrow R)$
2. *Show* $(P \leftrightarrow [Q \leftrightarrow R]) \rightarrow ([P \leftrightarrow Q] \leftrightarrow R)$
3. $P \leftrightarrow [Q \leftrightarrow R]$
4. *Show* $[P \leftrightarrow Q] \rightarrow R$
5. $P \leftrightarrow Q$
6. ~~*Show*~~ $P \rightarrow R$

| | | |
|---|---|---|
| 7. | P | |
| 8. | Q | 5, BC, 7, MP |
| 9. | $Q \leftrightarrow R$ | 3, BC, 7, MP |
| 10. | R | 9, BC, 8, MP |

Now we derive 'R' with $\sim$'P' as an assumption:

1. *Show* $(P \leftrightarrow [Q \leftrightarrow R]) \leftrightarrow ([P \leftrightarrow Q] \leftrightarrow R)$
2. *Show* $(P \leftrightarrow [Q \leftrightarrow R]) \rightarrow ([P \leftrightarrow Q] \leftrightarrow R)$
3. $P \leftrightarrow [Q \leftrightarrow R]$
4. *Show* $[P \leftrightarrow Q] \rightarrow R$

| | | |
|---|---|---|
| 5. | P ↔ Q | |
| 6. | ~~Show~~ P → R | |
| 7. | P | |
| 8. | Q | 5, BC, 7, MP |
| 9. | Q ↔ R | 3, BC, 7, MP |
| 10. | R | 9, BC, 8, MP |
| 11. | ~~Show~~ ∼P → R | |
| 12. | ∼P | |
| 13. | ∼[Q ↔ R] | 3, BC, 12, MT |
| 14. | Q ↔ ∼R | T90$\left(\frac{P \quad Q}{Q \quad R}\right)$, BC, 13, MP |
| 15. | ∼Q | 5, BC, 12, MT |
| 16. | R | 14, BC, 15, MT, DN |

Then, in employing SC, we are able to complete the subsidiary derivation started in line 4.

| | | |
|---|---|---|
| 1. | *Show* (P ↔ [Q ↔ R]) ↔ ([P ↔ Q] ↔ R) | |
| 2. | *Show* (P ↔ [Q ↔ R]) → ([P → Q] → R) | |
| 3. | P ↔ [Q ↔ R] | |
| 4. | ~~Show~~ [P ↔ Q] → R | |
| 5. | P ↔ Q | |
| 6. | ~~Show~~ P → R | |
| 7. | P | |
| 8. | Q | 5, BC, 7, MP |
| 9. | Q ↔ R | 3, BC, 7, MP |
| 10. | R | 9, BC, 8, MP |
| 11. | ~~Show~~ ∼P → R | |
| 12. | ∼P | |
| 13. | ∼[Q ↔ R] | 3, BC, 12, MT |
| 14. | Q ↔ ∼R | T90$\left(\frac{P \quad Q}{Q \quad R}\right)$, BC, 13, MP |
| 15. | ∼Q | 5, BC, 12, MT |
| 16. | R | 14, BC, 15, MT, DN |
| 17. | R | 6, 11, SC |

To complete the subsidiary derivation started in line 2, the converse of the conditional in line 4 must be derived; and then to complete the main derivation, the converse of the conditional in line 2 must be derived. We leave these details to the reader.

## EXERCISES, GROUP II

Show the following arguments valid by constructing symbolizations and deriving the conclusions of the symbolizations from their premises.

Indicate in each case the scheme of abbreviation used. (Before symbolizing Nos. 36 – 39, the reader should review the remarks made in connection with the solution of exercise 19 of chapter I, p. 29.)

36. If Mary joins a sorority and gives in to her inclinations, then her social life will flourish. If her social life flourishes, her academic life will suffer. Fortunately, Mary will give in to her inclinations, but her academic life will not suffer. Therefore she will not join a sorority.

37. Either love is blind and men are not aware of the fact that love is blind, or love is blind and women take advantage of the fact that love is blind. If men are not aware of the fact that love is blind, then love is not blind. Therefore women take advantage of the fact that love is blind.

38. If Alfred is a lover of logic who organizes his time, then he enjoys Mozart in the morning or whiskey at night, but not both. If he enjoys whiskey at night, then either he enjoys Mozart in the morning and organizes his time, or he does not enjoy Mozart in the morning and does not organize his time, or else he is not a lover of logic. Alfred enjoys whiskey at night provided that he both enjoys Mozart in the morning and organizes his time. Therefore Alfred is a lover of logic only if he does not organize his time.

39. Either it is not the case that Alfred pays attention and does not lose track of the argument, or it is not the case that he does not take notes and does not do well in the course. Alfred neither does well in the course nor loses track of the argument. If Alfred studies logic, then he does not do well in the course only if he does not take notes and pays attention. Therefore Alfred does not study logic.

## * EXERCISES, GROUP III

We can generalize the definitions of *conjunction* and *disjunction:* by the *conjunction* or *disjunction* of the sentences $\phi_1, \ldots, \phi_n$ we shall understand the sentence

$$\phi_1 \wedge \ldots \wedge \phi_n$$

or

$$\phi_1 \vee \ldots \vee \phi_n$$

respectively. (In case $n$ is 1, both these sentences reduce to $\phi_1$.)

By a *basic sentence* we understand either a sentence letter or the negation of a sentence letter. We say that a sentence is in *conjunctive normal form* if it is a conjunction of disjunctions of basic sentences, and in *disjunctive normal form* if it is a disjunction of conjunctions of basic sentences. Two symbolic sentences $\phi$ and $\psi$ are said to be *equivalent* if the biconditional

$$\phi \leftrightarrow \psi$$

is a theorem. It happens that every symbolic sentence is equivalent to some sentence in conjunctive normal form, and also to one in disjunctive normal form. The theorems in the next exercise illustrate this point.

40. Prove:

$$(P \rightarrow [Q \leftrightarrow R]) \leftrightarrow (\sim P \vee \sim Q \vee R) \wedge (\sim P \vee \sim R \vee Q)$$

$$(P \rightarrow [Q \leftrightarrow R]) \leftrightarrow \sim P \vee (Q \wedge R) \vee (\sim Q \wedge \sim R)$$

$$\sim [(\sim P \wedge \sim Q) \vee (\sim R \wedge \sim S)] \leftrightarrow (P \vee Q) \wedge (R \vee S)$$

$$\sim [(\sim P \wedge \sim Q) \vee (\sim R \wedge \sim S)] \leftrightarrow (P \wedge R) \vee (P \wedge S) \vee (Q \wedge R) \vee (Q \wedge S)$$

41. For each of the following sentences, find an equivalent sentence in conjunctive normal form and one in disjunctive normal form.

$$P \rightarrow Q \wedge R$$
$$\sim (P \vee Q \rightarrow R)$$
$$P \leftrightarrow (Q \leftrightarrow R)$$
$$([P \rightarrow Q] \rightarrow P) \rightarrow P$$
$$(P \rightarrow Q) \wedge (P \rightarrow \sim Q) \leftrightarrow P$$

42. Formulate a general procedure for transforming any symbolic sentence (of the sentential calculus) into an equivalent sentence in conjunctive normal form.

43. Formulate a general procedure for transforming any sentence in conjunctive normal form into an equivalent sentence in disjunctive normal form.

**6. * Truth-value analysis of sentences.** Inability to construct a derivation corresponding to a given argument is not conclusive evidence that no such derivation exists. The question naturally arises whether there is an automatic method of testing an argument for validity. For symbolic arguments of the sentential calculus there is such a method. Its presentation requires some new notions.

A sentence may have one of two *truth values*, truth ($T$) or falsehood ($F$). We may consider arbitrary *assignments* of truth values to sentence letters; such an assignment correlates with each sentence letter either $T$ or $F$. Given an assignment $A$, we may determine the *truth value of* an arbitrary symbolic sentence $\phi$ *with respect to* $A$ by the following rules:

(1) If $\phi$ is a sentence letter, then the truth value of $\phi$ is the truth value correlated with $\phi$ by $A$.

(2) The truth value of

$$\sim \phi$$

is $T$ if that of $\phi$ is $F$, and $F$ if that of $\phi$ is $T$.

(3) The truth value of

$$\phi \rightarrow \psi$$

is $F$ if that of $\phi$ is $T$ and that of $\psi$ is $F$; otherwise the truth value of

$$\phi \rightarrow \psi$$

is $T$.

(4) The truth value of

$$\phi \wedge \psi$$

is $T$ if the truth values of $\phi$ and $\psi$ are both $T$; otherwise the truth value of

$$\phi \wedge \psi$$

is $F$.

(5) The truth value of

$$\phi \vee \psi$$

is $F$ if the truth values of $\phi$ and $\psi$ are both $F$; otherwise the truth value of

$$\phi \vee \psi$$

is $T$.

(6) The truth value of

$$\phi \leftrightarrow \psi$$

is $T$ if $\phi$ and $\psi$ have the same truth value; otherwise the truth value of

$$\phi \leftrightarrow \psi$$

is $F$.

A *tautology* is a symbolic sentence whose truth value is $T$ with respect to every possible assignment. For instance,

(1) $$\mathrm{P} \rightarrow \mathrm{P}$$

is a tautology. For consider any assignment $A$. It will correlate with 'P' either $T$ or $F$. In neither case can the antecedent of (1) have the value $T$ and the consequent the value $F$. Hence, by rule 3, the value of (1) is always $T$.

It is convenient, in deciding whether a symbolic sentence is a tautology, to construct a *truth table*. The following is a truth table for the sentence '$(\mathrm{P} \rightarrow \mathrm{Q}) \leftrightarrow \sim(\mathrm{P} \wedge \sim \mathrm{Q})$'.

| P | Q | $\mathrm{P} \rightarrow \mathrm{Q}$ | $\sim \mathrm{Q}$ | $\mathrm{P} \wedge \sim \mathrm{Q}$ | $\sim(\mathrm{P} \wedge \sim \mathrm{Q})$ | $(\mathrm{P} \rightarrow \mathrm{Q}) \leftrightarrow \sim(\mathrm{P} \wedge \sim \mathrm{Q})$ |
|---|---|---|---|---|---|---|
| *T* | *T* | *T* | *F* | *F* | *T* | *T* |
| *T* | *F* | *F* | *T* | *T* | *F* | *T* |
| *F* | *T* | *T* | *F* | *F* | *T* | *T* |
| *F* | *F* | *T* | *T* | *F* | *T* | *T* |

The table can be regarded as an abbreviation of the following considerations. Let $A$ be an arbitrary assignment. If $A$ correlates $T$ with both 'P' and 'Q' (first row), then (by rule 3, p. 73) the value of 'P → Q' with respect to $A$ is $T$, that of '∼Q' is (by rule 2) $F$, hence (by rule 4) the value of 'P ∧ ∼Q' is $F$, hence (by rule 2) that of '∼(P ∧ ∼Q)' is $T$, and hence (by rule 6) the value of '(P → Q) ↔ ∼(P ∧ ∼Q)' is $T$. If $A$ correlates $T$ with 'P' and $F$ with 'Q' (second row), then the value of 'P → Q' is $F$, that of '∼Q' is $T$, that of 'P ∧ ∼Q' is $T$, that of '∼(P ∧ ∼Q)' is $F$, and hence that of '(P → Q) ↔ ∼(P ∧ ∼Q)' is $T$. If $A$ correlates $F$ with 'P' and $T$ with 'Q' (third row), then, similarly, the value of '(P → Q) ↔ ∼(P ∧ ∼Q)' is $T$. Finally, if $A$ correlates $F$ with both 'P' and 'Q', the value of '(P → Q) ↔ ∼(P ∧ ∼Q)' is $T$. Since in all cases the value of '(P → Q) ↔ ∼(P ∧ ∼Q)' is $T$, this sentence is a tautology.

In general, to construct a truth table for a given symbolic sentence proceed as follows. Provide a column for each component of the sentence, including the sentence itself. Arrange the components in order of increasing complexity, placing all sentence letters to the left of a double line. To the left of the double line, construct a row for each possible combination of truth values. (If there are $n$ sentence letters, there will be $2^n$ rows.) Proceed to the right in each row, filling in truth values for the sentences heading the columns, on the basis of previous entries and the rules on pages 73 – 74. The sentence is then a tautology just in case its table contains nothing but '$T$' in the final column.

It happens that a sentence is a theorem of the sentential calculus if and only if it is a tautology. (This fact will not be proved here; part of it will be proved in chapter V.) Thus truth tables provide an automatic test for validity in the case of symbolic arguments without premises.

## EXERCISES

Determine by means of truth tables which of the following sentences are tautologies. (In order to show that a sentence is *not* a tautology, it is sufficient to give, rather than a complete truth table, one row whose last entry is '$F$'.) Exercise 44 is solved for illustration.

44. (P ∧ Q → R) → (P → R)

The following single row of a truth table is sufficient to show that No. 44 is not a tautology:

| P | Q | R | P ∧ Q | P → R | P ∧ Q → R | (P ∧ Q → R) → (P → R) |
|---|---|---|---|---|---|---|
| *T* | *F* | *F* | *F* | *F* | *T* | *F* |

45. (P → R) → (P ∨ Q → R)
46. T58, T61, T66, T73, T78, T94.
47. The converse of T32 and the converse of T93.

**7. *Truth-value analysis of arguments.** A symbolic sentence $\phi$ is *tautologically implied* by other symbolic sentences just in case there is no assignment of truth values with respect to which each of the latter sentences has the truth value $T$ while $\phi$ has the truth value $F$. It happens (but will not be proved) that a symbolic argument is valid in the sentential calculus if and only if its conclusion is tautologically implied by its premises.

In the case of symbolic arguments with finitely many premises, truth tables are convenient for determining tautological implication. Consider, for example, the argument

(1) $\qquad \mathrm{P} \rightarrow \mathrm{Q} \quad . \quad \mathrm{P} \rightarrow {\sim}\mathrm{Q} \quad \therefore {\sim}\mathrm{P}$

The following is a truth table for this argument:

| P | Q | ~P | ~Q | P → Q | P → ~ Q |
|---|---|---|---|---|---|
| *T* | *T* | *F* | *F* | *T* | *F* |
| *T* | *F* | *F* | *T* | *F* | *T* |
| *F* | *T* | *T* | *F* | *T* | *T* |
| *F* | *F* | *T* | *T* | *T* | *T* |

The table can be regarded as an abbreviation of the following considerations. Let $A$ be an arbitrary assignment. If $A$ correlates $T$ with both 'P' and 'Q' (first row), then the value of '~P' with respect to $A$ is $F$, and that of '~Q' is $F$, that of 'P → Q' is $T$, and that of 'P → ~Q' is $F$. The other rows have a similar meaning, and the four rows exhaust all possible cases. We observe that in all cases in which the premises of (1) all receive the value $T$ (rows 3 and 4), the conclusion also receives the value $T$. Thus the conclusion of (1) is tautologically implied by its premises, and hence (1) is valid.

In general, to construct a truth table for a given symbolic argument proceed as follows. Provide a column for each component of the premises and the conclusion, including these sentences themselves. Arrange the components in order of increasing complexity, placing all sentence letters to the left of a double line. To the left of the double line, construct a row for each possible combination of truth values. Proceed to the right in each row, filling in truth values for the sentences heading the columns, on the basis of previous entries and the rules on pages 73 – 74. The conclusion of the argument is tautologically implied by its premises just in case there is no row in which the premises all have the value $T$ and the conclusion has the value $F$.

Since a symbolic argument is valid in the sentential calculus just in case its conclusion is tautologically implied by its premises, truth tables provide both an automatic test for validity and a method of showing invalidity in the case of symbolic arguments with finitely many premises.

The truth-table method of showing invalidity applies directly only to symbolic arguments. The reader will recall that an English argument is valid in the sentential calculus just in case it has a symbolization whose conclusion is derivable from its premises (within the sentential calculus). Hence, to show the invalidity of an English argument within this branch of logic, it is necessary to show that no symbolization of it is valid. Using truth tables, we may be able to show that various particular symbolizations are invalid, but without more precise information about the notion of a symbolization (in particular, about the notion of stylistic variance, on which it depends) we shall be unable to establish general assertions about all possible symbolizations of a given argument. Thus, although it would be desirable to develop a test of validity for arguments expressed in idiomatic English, this does not seem possible without a more detailed analysis of the notion of stylistic variance than we care to undertake.

## EXERCISES

Test each of the following arguments for validity, using the method of truth tables. It is sufficient in showing invalidity to exhibit a single row of a truth table. (The invalidity of Nos. 48 and 49 was forecast on page 29.)

48. $P \quad . \quad Q \quad \therefore R$

49. $P \rightarrow R \quad . \quad \sim R \quad \therefore Q$

50. $P \rightarrow Q \vee R \quad . \quad R \rightarrow Q \wedge P \quad . \quad Q \rightarrow R \quad \therefore P \leftrightarrow Q$

51. $\sim S \vee P \vee Q \quad . \quad \sim Q \rightarrow \sim R \quad . \quad P \rightarrow R \wedge S \quad \therefore Q$

52. $\therefore \sim Q \rightarrow (Q \vee [R \rightarrow \sim S] \rightarrow \sim [R \vee S])$

Although we cannot establish in a clear-cut way the invalidity (within the sentential calculus) of arguments expressed in idiomatic English, we can at least, if they are invalid, establish the invalidity of any one of their symbolizations. Together with each of the arguments 53 – 57 below we give an *interesting* symbolization, that is, one which appears to reflect to a maximal degree the logical structure of the argument that it symbolizes. Such a symbolization will be one that is as long as possible and, among those of maximal length, one which uses the smallest possible number of different sentence letters. For each of the symbolizations below, either derive its conclusion from its premises or show by the method of truth tables that it is invalid. In finding derivations, the reader will again find helpful the suggestions made on pages 69 – 70. (Nos. 53 and 54 correspond to an example on page 3.)

53. If Alfred studies, then he receives good grades; if he does not study, then he enjoys college; if he does not receive good grades, then he does not enjoy college. Therefore Alfred studies.

$$P \rightarrow Q \quad . \quad \sim P \rightarrow R \quad . \quad \sim Q \rightarrow \sim R \quad \therefore P$$

54. If Alfred studies, then he receives good grades; if he does not study, then he enjoys college; if he does not receive good grades, then he does not enjoy college. Therefore Alfred enjoys college.

$$P \to Q \quad . \quad \sim P \to R \quad . \quad \sim Q \to \sim R \quad \therefore R$$

55. Alfred will marry either Alice or Mary, but not both. If he marries Alice but not Mary, then he is fortunate. If he does not marry Alice but marries Mary, he is also fortunate. Therefore Alfred is fortunate.

$$(P \vee Q) \wedge \sim(P \wedge Q) \quad . \quad P \wedge \sim Q \to R \quad . \quad \sim P \wedge Q \to R \quad \therefore R$$

56. Alfred passes just in case he is both intelligent and industrious, and he is intelligent. Alfred is industrious if and only if he both is intelligent and does not pass, provided that he does not succumb to temptation. Therefore Alfred succumbs to temptation.

$$(R \leftrightarrow P \wedge Q) \wedge P \quad . \quad \sim S \to (Q \leftrightarrow P \wedge \sim R) \quad \therefore S$$

57. If men prefer chaste women just in case chastity is a virtue, then men prefer virtuous women; but men do not prefer virtuous women. Therefore it is not the case that if men prefer chaste women, then chastity is a virtue.

$$([P \leftrightarrow Q] \to R) \wedge \sim R \quad \therefore \sim(P \to Q)$$

Give an interesting symbolization of each of the following arguments, and for each symbolization either derive its conclusion from its premises or show by the method of truth tables that it is invalid.

58. If God is willing to prevent evil but unable to do so, he is impotent. If God is able to prevent evil but unwilling to do so, he is malevolent. Evil exists if and only if God is either unwilling or unable to prevent it. God exists only if he is neither impotent nor malevolent. Therefore if God exists, evil does not exist.

59. Caesar went to England, and either Pompey went to Spain or Crassus marched against the Parthians. It is not the case that both Caesar went to England and Crassus was not killed by the Parthians. If Caesar went to England and Cicero denounced Catiline, then Pompey did not go to Spain. If Cicero did not denounce Catiline, then Crassus marched against the Parthians. Therefore Caesar was murdered on the Ides of March, 44 B.C.

60. If neither an adequate social life nor a comfortable income can replace Alfred's love of logic, then, provided that he has normal intelligence, he can expect nothing better than an instructorship. An adequate social life can replace Alfred's love of logic just in case a comfortable income can replace his love of logic. Therefore either an adequate social life can replace Alfred's love of logic or he can expect nothing better than an instructorship.

**8.* Historical remarks.** What is now called the sentential calculus has been designated variously the 'calculus of equivalent statements' (MacColl [1]), the 'Aussagenkalkül' (Schröder [1]), the 'propositional calculus' (Russell [2]), and the 'theory of deduction' (Russell [3]).

The sentential calculus was first investigated by the Stoics (see Łukasiewicz [1] and Mates [1]). Its modern development was adumbrated by Leibniz (see Couturat [1] and [2]), in Boole [1] and [2], De Morgan [1], Peirce [1], and MacColl [1], but properly begins with Frege [1]. Its fullest development, based on the ideas of Frege, occurs in Whitehead and Russell [1].

Besides the system of symbols used here, which is due to Tarski, there are two other principal systems. The notation that is most frequent in the literature is that of Whitehead and Russell [1]; its symbols are the following:

'$\sim$' for '$\sim$'
'$\supset$' for '$\rightarrow$'
'$\cdot$' for '$\wedge$'
'$\vee$' for '$\vee$'
'$\equiv$' for '$\leftrightarrow$' .

(To indicate grouping, Whitehead and Russell use, along with parentheses, a complicated system of dots.)

A much more economical notation is that of Łukasiewicz [1]. The essential feature is the placement of sentential connectives. A binary connective is placed *before*, rather than *between*, the two sentences which it connects. Thus, using our symbols '$\sim$' and '$\rightarrow$', we would write

$$\rightarrow \sim P \rightarrow QR$$

for

$$(\sim P \rightarrow (Q \rightarrow R)) \quad ,$$

and

$$\rightarrow\rightarrow \sim PQR$$

for

$$((\sim P \rightarrow Q) \rightarrow R) \quad .$$

The advantage of Łukasiewicz' notation is that it makes unnecessary special symbols for grouping, such as parentheses.

The notion of an instance of a theorem and the principle that an instance of a theorem is again a theorem occur more or less explicitly in Frege [3], Couturat [3], and Russell [2].

The method of truth tables occurs informally in Frege [1], and explicitly in Peirce [2]. The assertion that a symbolic argument is valid in the sentential calculus if and only if its premises tautologically imply its conclusion was first established in Post [1].

Our formulation of the sentential calculus admits of a certain simplification, at the expense of detracting from its intuitive character. We could, as in chapter I, dispense with indirect proof and replace the four inference rules of that chapter by two, *modus ponens* and a variant of *modus tollens;* further, we could replace the primitive inference rules introduced in the present chapter by the following rules, which embody in some sense 'definitions' of '$\wedge$', '$\vee$', '$\leftrightarrow$' in terms of '$\sim$' and '$\rightarrow$':

$$\frac{(\phi \wedge \psi)}{\sim(\phi \rightarrow \sim\psi)} \qquad \frac{\sim(\phi \rightarrow \sim\psi)}{(\phi \wedge \psi)}$$

$$\frac{\phi \vee \psi}{(\sim\phi \rightarrow \psi)} \qquad \frac{(\sim\phi \rightarrow \psi)}{(\phi \vee \psi)}$$

$$\frac{(\phi \leftrightarrow \psi)}{((\phi \rightarrow \psi) \wedge (\psi \rightarrow \phi))} \qquad \frac{((\phi \rightarrow \psi) \wedge (\psi \rightarrow \phi))}{(\phi \leftrightarrow \psi)} .$$

The observation made at the end of section 4, that there is no automatic procedure which, in the absence of annotations, will permit a decision as to the correctness of an abbreviated derivation, holds in the sentential calculus only when infinitely many premises are permitted. It can be shown, however, that even when the class of premises, though allowed to be infinite, is required to be decidable (that is, loosely speaking, such that it can automatically be decided of any given sentence whether it is a member of the class), it is not generally possible to formulate an automatic test for a correct abbreviated derivation from that class of premises without recourse to annotations. In abbreviated derivations of the quantifier calculus (that is, the system of chapters III and IV), annotations are indispensable even when only finitely many premises are involved.

## 9. Appendix: list of theorems of chapters I and II.

| | |
|---|---|
| T1 | $P \rightarrow P$ |
| T2 | $Q \rightarrow (P \rightarrow Q)$ |
| T3 | $P \rightarrow ([P \rightarrow Q] \rightarrow Q)$ |
| T4 | $(P \rightarrow Q) \rightarrow ([Q \rightarrow R] \rightarrow [P \rightarrow R])$ |
| T5 | $(Q \rightarrow R) \rightarrow ([P \rightarrow Q] \rightarrow [P \rightarrow R])$ |
| T6 | $(P \rightarrow [Q \rightarrow R]) \rightarrow ([P \rightarrow Q] \rightarrow [P \rightarrow R])$ |
| T7 | $([P \rightarrow Q] \rightarrow [P \rightarrow R]) \rightarrow (P \rightarrow [Q \rightarrow R])$ |
| T8 | $(P \rightarrow [Q \rightarrow R]) \rightarrow (Q \rightarrow [P \rightarrow R])$ |
| T9 | $(P \rightarrow [P \rightarrow Q]) \rightarrow (P \rightarrow Q)$ |

| | |
|---|---|
| T10 | $([P \to Q] \to Q) \to ([Q \to P] \to P)$ |
| T11 | $\sim\sim P \to P$ |
| T12 | $P \to \sim\sim P$ |
| T13 | $(P \to Q) \to (\sim Q \to \sim P)$ |
| T14 | $(P \to \sim Q) \to (Q \to \sim P)$ |
| T15 | $(\sim P \to Q) \to (\sim Q \to P)$ |
| T16 | $(\sim P \to \sim Q) \to (Q \to P)$ |
| T17 | $P \to (\sim P \to Q)$ |
| T18 | $\sim P \to (P \to Q)$ |
| T19 | $(\sim P \to P) \to P$ |
| T20 | $(P \to \sim P) \to \sim P$ |
| T21 | $\sim(P \to Q) \to P$ |
| T22 | $\sim(P \to Q) \to \sim Q$ |
| T23 | $([P \to Q] \to P) \to P$ |
| T24 | $P \wedge Q \leftrightarrow Q \wedge P$ |
| T25 | $P \wedge (Q \wedge R) \leftrightarrow (P \wedge Q) \wedge R$ |
| T26 | $(P \to Q) \wedge (Q \to R) \to (P \to R)$ |
| T27 | $(P \wedge Q \to R) \leftrightarrow (P \to [Q \to R])$ |
| T28 | $(P \wedge Q \to R) \leftrightarrow (P \wedge \sim R \to \sim Q)$ |
| T29 | $(P \to Q \wedge R) \leftrightarrow (P \to Q) \wedge (P \to R)$ |
| T30 | $(P \to Q) \to (R \wedge P \to R \wedge Q)$ |
| T31 | $(P \to Q) \to (P \wedge R \to Q \wedge R)$ |
| T32 | $(P \to R) \wedge (Q \to S) \to (P \wedge Q \to R \wedge S)$ |
| T33 | $(P \to Q) \wedge (\sim P \to Q) \to Q$ |
| T34 | $(P \to Q) \wedge (P \to \sim Q) \to \sim P$ |
| T35 | $(\sim P \to R) \wedge (Q \to R) \leftrightarrow ([P \to Q] \to R)$ |
| T36 | $\sim(P \wedge \sim P)$ |
| T37 | $(P \to Q) \leftrightarrow \sim(P \wedge \sim Q)$ |
| T38 | $P \wedge Q \leftrightarrow \sim(P \to \sim Q)$ |

| | |
|---|---|
| T39 | $\sim(P \wedge Q) \leftrightarrow (P \rightarrow \sim Q)$ |
| T40 | $\sim(P \rightarrow Q) \leftrightarrow P \wedge \sim Q$ |
| T41 | $P \leftrightarrow P \wedge P$ |
| T42 | $P \wedge \sim Q \rightarrow \sim(P \rightarrow Q)$ |
| T43 | $\sim P \rightarrow \sim(P \wedge Q)$ |
| T44 | $\sim Q \rightarrow \sim(P \wedge Q)$ |
| T45 | $P \vee Q \leftrightarrow (\sim P \rightarrow Q)$ |
| T46 | $(P \rightarrow Q) \leftrightarrow \sim P \vee Q$ |
| T47 | $P \leftrightarrow P \vee P$ |
| T48 | $(P \vee Q) \wedge (P \rightarrow R) \wedge (Q \rightarrow S) \rightarrow R \vee S$ |
| T49 | $(P \vee Q) \wedge (P \rightarrow R) \wedge (Q \rightarrow R) \rightarrow R$ |
| T50 | $(P \rightarrow R) \wedge (Q \rightarrow R) \leftrightarrow (P \vee Q \rightarrow R)$ |
| T51 | $(P \vee Q) \wedge (P \rightarrow R) \wedge (\sim P \wedge Q \rightarrow R) \rightarrow R$ |
| T52 | $(P \rightarrow R) \wedge (\sim P \wedge Q \rightarrow R) \leftrightarrow (P \vee Q \rightarrow R)$ |
| T53 | $P \vee Q \leftrightarrow Q \vee P$ |
| T54 | $P \vee (Q \vee R) \leftrightarrow (P \vee Q) \vee R$ |
| T55 | $(P \rightarrow Q \vee R) \leftrightarrow (P \rightarrow Q) \vee (P \rightarrow R)$ |
| T56 | $(P \rightarrow Q) \rightarrow (R \vee P \rightarrow R \vee Q)$ |
| T57 | $(P \rightarrow Q) \rightarrow (P \vee R \rightarrow Q \vee R)$ |
| T58 | $(P \rightarrow Q) \vee (Q \rightarrow R)$ |
| T59 | $P \vee \sim P$ |
| T60 | $(P \rightarrow R) \vee (Q \rightarrow R) \leftrightarrow (P \wedge Q \rightarrow R)$ |
| T61 | $P \wedge (Q \vee R) \leftrightarrow (P \wedge Q) \vee (P \wedge R)$ |
| T62 | $P \vee (Q \wedge R) \leftrightarrow (P \vee Q) \wedge (P \vee R)$ |
| T63 | $P \wedge Q \leftrightarrow \sim(\sim P \vee \sim Q)$ |
| T64 | $P \vee Q \leftrightarrow \sim(\sim P \wedge \sim Q)$ |
| T65 | $\sim(P \wedge Q) \leftrightarrow \sim P \vee \sim Q$ |
| T66 | $\sim(P \vee Q) \leftrightarrow \sim P \wedge \sim Q$ |
| T67 | $\sim P \wedge \sim Q \rightarrow \sim(P \vee Q)$ |

| | |
|---|---|
| T68 | $P \leftrightarrow (P \wedge Q) \vee (P \wedge \sim Q)$ |
| T69 | $P \leftrightarrow (P \vee Q) \wedge (P \vee \sim Q)$ |
| T70 | $Q \rightarrow (P \wedge Q \leftrightarrow P)$ |
| T71 | $\sim Q \rightarrow (P \vee Q \leftrightarrow P)$ |
| T72 | $(P \rightarrow Q) \leftrightarrow (P \wedge Q \leftrightarrow P)$ |
| T73 | $(P \rightarrow Q) \leftrightarrow (P \vee Q \leftrightarrow Q)$ |
| T74 | $(P \leftrightarrow Q) \wedge P \rightarrow Q$ |
| T75 | $(P \leftrightarrow Q) \wedge Q \rightarrow P$ |
| T76 | $(P \leftrightarrow Q) \wedge \sim P \rightarrow \sim Q$ |
| T77 | $(P \leftrightarrow Q) \wedge \sim Q \rightarrow \sim P$ |
| T78 | $(P \rightarrow [Q \leftrightarrow R]) \leftrightarrow ([P \rightarrow Q] \leftrightarrow [P \rightarrow R])$ |
| T79 | $(P \rightarrow [Q \leftrightarrow R]) \leftrightarrow (P \wedge Q \leftrightarrow P \wedge R)$ |
| T80 | $(P \leftrightarrow Q) \vee (P \leftrightarrow \sim Q)$ |
| T81 | $(P \leftrightarrow Q) \leftrightarrow (P \rightarrow Q) \wedge (Q \rightarrow P)$ |
| T82 | $(P \leftrightarrow Q) \leftrightarrow \sim([P \rightarrow Q] \rightarrow \sim[Q \rightarrow P])$ |
| T83 | $(P \leftrightarrow Q) \leftrightarrow (P \wedge Q) \vee (\sim P \wedge \sim Q)$ |
| T84 | $P \wedge Q \rightarrow (P \leftrightarrow Q)$ |
| T85 | $\sim P \wedge \sim Q \rightarrow (P \leftrightarrow Q)$ |
| T86 | $([P \leftrightarrow Q] \rightarrow R) \leftrightarrow (P \wedge Q \rightarrow R) \wedge (\sim P \wedge \sim Q \rightarrow R)$ |
| T87 | $\sim(P \leftrightarrow Q) \leftrightarrow (P \wedge \sim Q) \vee (\sim P \wedge Q)$ |
| T88 | $P \wedge \sim Q \rightarrow \sim(P \leftrightarrow Q)$ |
| T89 | $\sim P \wedge Q \rightarrow \sim(P \leftrightarrow Q)$ |
| T90 | $\sim(P \leftrightarrow Q) \leftrightarrow (P \leftrightarrow \sim Q)$ |
| T91 | $P \leftrightarrow P$ |
| T92 | $(P \leftrightarrow Q) \leftrightarrow (Q \leftrightarrow P)$ |
| T93 | $(P \leftrightarrow Q) \wedge (Q \leftrightarrow R) \rightarrow (P \leftrightarrow R)$ |
| T94 | $(P \leftrightarrow [Q \leftrightarrow R]) \leftrightarrow ([P \leftrightarrow Q] \leftrightarrow R)$ |
| T95 | $(P \leftrightarrow Q) \leftrightarrow ([P \leftrightarrow R] \leftrightarrow [Q \leftrightarrow R])$ |
| T96 | $(P \leftrightarrow Q) \leftrightarrow (\sim P \leftrightarrow \sim Q)$ |

| | |
|---|---|
| T97 | $(P \leftrightarrow R) \wedge (Q \leftrightarrow S) \rightarrow ([P \rightarrow Q] \leftrightarrow [R \rightarrow S])$ |
| T98 | $(P \leftrightarrow R) \wedge (Q \leftrightarrow S) \rightarrow (P \wedge Q \leftrightarrow R \wedge S)$ |
| T99 | $(P \leftrightarrow R) \wedge (Q \leftrightarrow S) \rightarrow (P \vee Q \leftrightarrow R \vee S)$ |
| T100 | $(P \leftrightarrow R) \wedge (Q \leftrightarrow S) \rightarrow ([P \leftrightarrow Q] \leftrightarrow [R \leftrightarrow S])$ |
| T101 | $(Q \leftrightarrow S) \rightarrow ([P \rightarrow Q] \leftrightarrow [P \rightarrow S]) \wedge ([Q \rightarrow P] \leftrightarrow [S \rightarrow P])$ |
| T102 | $(Q \leftrightarrow S) \rightarrow (P \wedge Q \leftrightarrow P \wedge S)$ |
| T103 | $(Q \leftrightarrow S) \rightarrow (P \vee Q \leftrightarrow P \vee S)$ |
| T104 | $(Q \leftrightarrow S) \rightarrow ([P \leftrightarrow Q] \leftrightarrow [P \leftrightarrow S])$ |
| T105 | $P \wedge (Q \leftrightarrow R) \rightarrow (P \wedge Q \leftrightarrow R)$ |
| T106 | $(P \rightarrow [Q \rightarrow R]) \leftrightarrow ([P \rightarrow Q] \rightarrow [P \rightarrow R])$ |
| T107 | $(P \rightarrow [Q \rightarrow R]) \leftrightarrow (Q \rightarrow [P \rightarrow R])$ |
| T108 | $(P \rightarrow [P \rightarrow Q]) \leftrightarrow (P \rightarrow Q)$ |
| T109 | $([P \rightarrow Q] \rightarrow Q) \leftrightarrow ([Q \rightarrow P] \rightarrow P)$ |
| T110 | $P \leftrightarrow \sim\sim P$ |
| T111 | $(P \rightarrow Q) \leftrightarrow (\sim Q \rightarrow \sim P)$ |
| T112 | $(P \rightarrow \sim Q) \leftrightarrow (Q \rightarrow \sim P)$ |
| T113 | $(\sim P \rightarrow Q) \leftrightarrow (\sim Q \rightarrow P)$ |
| T114 | $(\sim P \rightarrow P) \leftrightarrow P$ |
| T115 | $(P \rightarrow \sim P) \leftrightarrow \sim P$ |
| T116 | $P \leftrightarrow P \wedge (Q \vee \sim P)$ |
| T117 | $P \leftrightarrow P \vee (Q \wedge \sim Q)$ |

# Chapter III

# 'ALL' and 'SOME'

**1. Variables, quantifiers, formulas.** Not all intuitively valid arguments can be reached by the procedures of chapters I and II. For example, the argument

(1) All Communists are Marxists. Some Communists are American.
∴ Some Marxists are American,

unlike the arguments in chapters I and II, depends for its validity on the meaning of the words 'all' and 'some'.

Before embarking on the logic of such words, we must engage in grammatical considerations. Of *sentences* we have already spoken. A *sentence of English* is an expression of English that is either true or false. An *English name* is a word or group of words that designates (at least within a given context) a single object. For example, 'Socrates', 'David Hume', '7', '7 + 5', and 'the author of *Waverley*' are names respectively designating Socrates, David Hume, 7, 12, and Sir Walter Scott. A *variable* is a lower-case Latin letter, with or without a numerical subscript. For example, '$x$', '$y$', '$z$', '$x_1$', '$y_3$', '$z_2$' are variables.

Using variables we can construct expressions that closely resemble English sentences but fail to be sentences. For instance, the expression

(2) $x$ is bald,

unlike

(3) Socrates is bald

and

(4) Samson is bald,

is neither true nor false, and hence is not a sentence.

Although we cannot ascribe truth or falsehood to (2), we can make other assertions about it. For instance, we can assert that a particular object, say, Socrates or Samson, *satisfies* (2); this amounts to asserting (3) or (4). Or we can say that every object, or at least one object, satisfies (2); these assertions amount respectively to

(5) For each $x$, $x$ is bald

and

(6) There is an object $x$ such that $x$ is bald.

(5) and (6), although not completely idiomatic, can nevertheless, like their more idiomatic counterparts 'Everything is bald' and 'Something is bald', be construed as sentences of English; in fact, (5) is false and (6) is true. We have thus discovered three ways of converting (2) into a sentence—by replacing its variable '$x$' by a name or by prefixing one of the phrases 'for each $x$' or 'there is an object $x$ such that'.

We shall abbreviate 'for each' by '$\bigwedge$' and 'there is an object . . . such that' by '$\bigvee$'. (5) and (6) then become

$$\bigwedge x \quad x \text{ is bald}$$

and

$$\bigvee x \quad x \text{ is bald} \quad .$$

The symbols '$\bigwedge$' and '$\bigvee$' are known respectively as the *universal quantifier* and the *existential quantifier*. A quantifier may be written before any variable to form a *quantifier phrase*, which is *universal* or *existential* according to the quantifier used. The English counterparts of quantifier phrases, that is, expressions of the form

for each $\alpha$

or

there is an object $\alpha$ such that ,

where $\alpha$ is a variable, are called *phrases of quantity*.

By a *formula of English* we shall understand either a sentence of English or an expression containing occurrences of variables which becomes a sentence of English when some or all of these occurrences are replaced by English names. (2) above, as well as each of the following, is a formula of English:

David Hume is human,
(7) $x$ loves $y$,
(8) there is an object $y$ such that $x$ loves y,
(9) $x + y = z$,
if $x$ loves $y$, then $y$ loves $x$.

To see, for example, that (8) is a formula, we need only replace the variable '$x$' by 'Romeo', and to see that (9) is a formula, we may replace the variables '$x$', '$y$', '$z$' by the names '2', '3', '5'.

In symbolic abbreviations, we shall henceforth employ not only sentence letters but also *predicate letters*. These are to be capitals from 'F' through

'O', with or without numerical subscripts; 'P' through 'Z' continue in their role as sentence letters, and 'A' through 'E' are reserved for later uses. A predicate letter accompanied by a variable will serve as an abbreviation of a formula of English containing that variable. For example, the formula (2) may be abbreviated by

$$\mathrm{F}x \quad .$$

The *language* with which we are now concerned is obtained by adding to English the following symbols:

(1) the sentential connectives;
(2) parentheses;
(3) the quantifiers, that is, '⋀' and '⋁';
(4) variables, that is, lower-case letters with or without subscripts;
(5) sentence letters, that is, capital letters 'P' through 'Z' with or without subscripts;
(6) predicate letters, that is, capital letters 'F' through 'O', with or without subscripts.

As *formulas* we count formulas of English, their symbolic counterparts, and mixed combinations. To be more explicit, the class of formulas can be exhaustively characterized as follows:

(*1*) *All formulas of English* (*that is, sentences of English or expressions like sentences of English except for the occurrence of variables*) *are formulas.*
(*2*) *Sentence letters are formulas.*
(*3*) *The result of writing a predicate letter followed by a variable is a formula.*
(*4*) *If* $\phi$ *and* $\psi$ *are formulas, then so are*

$$\sim \phi \quad ,$$
$$(\phi \rightarrow \psi) \quad ,$$
$$(\phi \wedge \psi) \quad ,$$
$$(\phi \vee \psi) \quad ,$$
$$(\phi \leftrightarrow \psi) \quad .$$

(*5*) *If* $\phi$ *is a formula and* $\alpha$ *a variable, then*

$$\bigwedge \alpha \phi \quad ,$$
$$\bigvee \alpha \phi$$

*are formulas.*

Clause (5) leads not only to

$$\bigwedge x \mathrm{F}x \quad ,$$

but also to seemingly meaningless combinations such as

$$\bigwedge x \mathrm{P} \quad .$$

It would be artificial to exclude these possibilities; their meaning will be explained in due course.

A *symbolic formula* is a formula constructed exclusively from variables, sentences letters, predicate letters, parentheses, sentential connectives, and quantifiers. More precisely, the class of symbolic formulas is exhaustively characterized by clauses (2) – (5) above, reading everywhere 'symbolic formula' for 'formula'.

A few more terms will be useful. The result of prefixing one or more universal quantifier phrases to a formula is called a *universal generalization* of that formula. Similarly, one forms *existential generalizations* of a formula. For example, the formulas

$$\bigwedge x \quad x \text{ loves } y \quad ,$$
$$\bigwedge x \bigwedge y \quad x \text{ loves } y$$

are universal generalizations of (7), and the formulas

$$\bigvee x \quad x \text{ loves } y \quad ,$$
$$\bigvee y \bigvee x \quad x \text{ loves } y$$

are existential generalizations of (7).

## EXERCISES

Which of the following are formulas? For each expression that is a formula, state whether it is a symbolic formula. Exercises 1 – 3 are solved for illustration.

1. (($\bigwedge x$($x$ is a man $\rightarrow$ $x$ is mortal) $\wedge$ Socrates is a man) $\rightarrow$ Socrates is mortal)

According to clause (1) of the characterization of the class of formulas, '$x$ is a man', '$x$ is mortal', 'Socrates is a man', and 'Socrates is mortal' are formulas; according to clause (4), then, '($x$ is a man $\rightarrow$ $x$ is mortal)' is a formula, and according to clause (5), '$\bigwedge x$($x$ is a man $\rightarrow$ $x$ is mortal)' is a formula. According to clause (4) again, '($\bigwedge x$($x$ is a man $\rightarrow$ $x$ is mortal) $\wedge$ Socrates is a man)' and hence No. 1 are formulas. Expression No. 1 is not a symbolic formula, however, for it contains words of English.

2. $\bigwedge z(\bigwedge x \bigvee y(\mathrm{F}x \vee \mathrm{G}y) \rightarrow \mathrm{H}z)$

According to clause (3), '$\mathrm{F}x$', '$\mathrm{G}y$', and '$\mathrm{H}z$' are formulas; thus according to clause (4), '$(\mathrm{F}x \vee \mathrm{G}y)$' is a formula, and according to clause (5), first '$\bigvee y(\mathrm{F}x \vee \mathrm{G}y)$' and then '$\bigwedge x \bigvee y(\mathrm{F}x \vee \mathrm{G}y)$' are formulas.

By clause (4) again, '$(\wedge x \vee y(Fx \vee Gy) \rightarrow Hz)$' is a formula, and according to clause (5) again, No. 2 is a formula. It is a symbolic formula, for it contains only variables, sentence letters, predicate letters, parentheses, sentential connectives, and quantifiers.

3. $\wedge x$ *x* is even $\vee$ $\wedge y$ *y* is odd

This could be a formula only of the kind introduced by clause (5), for it is not a formula of English (clause (1)), it is not a sentence letter (clause (2)), it is not a predicate letter followed by a variable (clause (3)), and its first symbol is neither a negation sign nor a left parenthesis (clause (4)). But clause (5) is applicable only if what follows the quantifier phrase '$\wedge x$' is a formula; that is, only if '*x* is even $\vee$ $\wedge y$ *y* is odd' is a formula, and, strictly speaking, this is not the case, as another review of clauses (1) – (5) would reveal. Thus No. 3 is not a formula.

4. $\sim(\sim \wedge x \sim Fx \vee Gy)$
5. $(Fxy \rightarrow Gyx)$
6. $\wedge a(Hx \leftrightarrow Gy)$
7. $\vee\wedge x(Hx \leftrightarrow Gx)$
8. $\wedge y \sim \vee \sim x(Hx \wedge Gy)$
9. $\wedge y(Hy \rightarrow \vee xHx)$
10. $\wedge x(\wedge x(Hx \rightarrow \sim \vee y \sim Gy) \vee$ *x* is between Berkeley and Los Angeles)

**2. Bondage and freedom.** A variable may occur more than once in a formula. For example, in each of the formulas

$$\vee x(Fx \wedge Gx) \quad , \qquad (1)$$
$$(\vee xFx \wedge Gx) \qquad (2)$$

there are three occurrences of the variable '*x*'. All three occurrences of '*x*' are *bound in* (1). Only the first and second occurrences of '*x*' are *bound in* (2); the third occurrence of '*x*' (that following 'G') is not *bound in* (2) but instead is *free in* (2).

In general, an *occurrence of a variable* $\alpha$ is *bound in* a symbolic formula $\phi$ just in case it stands within an occurrence in $\phi$ of a formula

$$\wedge\alpha\psi$$

or

$$\vee\alpha\psi \quad ,$$

where $\psi$ is a formula. An *occurrence of a variable* is *free in* a symbolic formula $\phi$ just in case it stands within $\phi$ but is not bound in $\phi$.

The claim that all three occurrences of '*x*' are bound in (1) is verified by noticing that each stands within an occurrence in (1) of '$\vee x(Fx \wedge Gx)$'. Likewise, the first and second occurrences of '*x*' in (2) stand within an occurrence in (2) of '$\vee xFx$'. However, the third occurrence of '*x*' in (2) does not stand within an occurrence of any formula beginning with a quantifier phrase; thus the third occurrence of '$x$' in (2) is free in (2). In the formula

(3) $$((Fx \wedge Gy) \vee \bigwedge x(Fx \wedge Gy))$$

the second and third occurrences of the variable '$x$' are bound, because they stand within an occurrence in (3) of

(4) $$\bigwedge x(Fx \wedge Gy) \quad .$$

The first occurrence of '$x$', however, as well as both occurrences of '$y$', is free in (3). Although the second occurrence of '$y$' in (3) stands within an occurrence of (4), it is still free in (3), because the quantifier phrase with which (4) begins does not contain '$y$'.

The previous considerations apply to *occurrences* of a variable. A *variable* itself is *bound in* a symbolic formula $\phi$ just in case some occurrence of it is bound in $\phi$. Similarly, a *variable* is *free in* $\phi$ just in case some occurrence of it is free in $\phi$. For example, the variable '$x$' is both bound and free in (3), while '$y$' is only free in (3).

By a *symbolic sentence* is understood a symbolic formula in which no variable is free. For example, (1) but not (2) is a symbolic sentence; this fact is important in connection with exercises 15 and 16 of section 4.

Notice that the definitions above apply only to symbolic formulas; it would be both unnecessary and extremely difficult to extend the notions of bondage and freedom to formulas containing English components.

## EXERCISES

Consider the formula

$$((\bigvee xFx \vee \bigwedge z((Gz \wedge Hx) \rightarrow (\bigvee zFz \vee Hz))) \leftrightarrow \bigvee z(Fy \vee Fz)) \quad .$$

11. In this formula identify each occurrence of a variable as bound or free.
12. Which variables are bound in the formula?
13. Which variables are free in the formula?

**3. Informal notational conventions.** We shall continue to use the conventions of chapter II (pp. 40, 51, 65) for omitting parentheses and replacing them by brackets. For practical purposes, then, we shall deal with a larger class of formulas than that given in section 1. It should be emphasized, however, that in theoretical discussions—for instance, in the criteria of bondage and freedom given in the last section and in the rules of derivation that will appear in section 5—the word 'formula' is always to be understood in the official sense characterized in section 1.

**4. Translation and symbolization.** The English sentence

For each $x$, $x$ is bald

may be paraphrased in a variety of ways, for instance, as

Everything is bald,
Each thing is bald,
All things are bald,
For all $x$, $x$ is bald;

and the English sentence

There is an object $x$ such that $x$ is bald

may be paraphrased as

For some $x$, $x$ is bald,
Something is bald,
At least one thing is bald,
There is a bald thing.

Using phrases of quantity together with the phrases of connection of chapters I and II, we can develop stylistic variants of a good number of English expressions in addition to those just mentioned. For instance,

Nothing is bald

may be paraphrased as either

For each $x$, it is not the case that $x$ is bald

or

It is not the case that there is an object $x$ such that $x$ is bald .

Other examples are afforded by sentences of the familiar Aristotelian forms.

(1) All men are mortal,
(2) Some cats are dogs,
(3) No men are mortal,
(4) Some cats are not dogs

are stylistic variants of

For each $x$ (if $x$ is a man, then $x$ is mortal) ,

There is an object $x$ such that ($x$ is a cat and $x$ is a dog) ,

For each $x$ (if $x$ is a man, then it is not the case that $x$ is mortal) ,

There is an object $x$ such that ($x$ is a cat and it is not the case that $x$ is a dog)

respectively. The supposition that

(5) For each $x$ ($x$ is a man and $x$ is mortal)

is a stylistic variant of the true sentence (1) can be rejected by observing that (5) is false, for it logically implies the falsehood 'for each $x$, $x$ is a man'. (See exercise 54.) Similarly, the supposition that

(6) There is an object $x$ such that (if $x$ is a cat, then $x$ is a dog)

is a stylistic variant of the false sentence (2) can be rejected by observing that (6) is true, for it is implied by the truth 'there is an object $x$ such that $x$ is a dog'. (See Exercise 55.)

Somewhat less obvious examples of stylistic variance are provided by the sentences

Only citizens are voters

and

None but citizens are voters,

both of which may be paraphrased as

For each $x$ (if $x$ is a voter, then $x$ is a citizen) .

It is impractical to list all the combinations that are expressible in terms of phrases of quantity. For example, context must be consulted to determine whether 'any' should pass into a universal or an existential phrase of quantity. Thus here, as in the sentential calculus, the notion of stylistic variance is employed in an intuitive rather than an exact manner.

The process of translation is somewhat more involved here than in the sentential calculus. We must first introduce an auxiliary notion: if $\phi$ is a formula of English, then $\alpha$ is said to be an *apparent variable* of $\phi$ if either $\phi$ is a sentence of English containing $\alpha$ or there is an occurrence of $\alpha$ in $\phi$ such that, upon replacement of some other occurrences of variables by English names, $\phi$ becomes a sentence of English. For example, '$a$' is an apparent variable of each of the formulas

'$a$' is the first letter of the alphabet
For each $a$, $a$ is bald
For some $a$, $a$ is father of $b$ .

The reason is that the first two formulas above are already sentences containing '$a$', and the third becomes a sentence upon replacement of '$b$' by any name.

We must also extend the notion of an abbreviation, for now formulas as well as sentences of English will require symbolic representation. Accordingly, we presently understand by an *abbreviation* either one in the earlier sense (that is, an ordered pair consisting of a sentence letter and a sentence of English) or an ordered pair of which the first member is a predicate letter and the second member a formula of English whose only variable is '*a*' and which contains no apparent variables. Again, a *scheme of abbreviation* is to be a collection of abbreviations such that no two abbreviations in the collection have the same first member.

For example, the following is a scheme of abbreviation:

(7) P : Moby Dick is a fish
F : *a* is a whale
G : *a* is a mammal .

The process of *literal translation into English on the basis of a given scheme of abbreviation* now begins with a symbolic formula and if successful ends with a formula of English. The process consists of the following steps:

(*i*) *Restore any parentheses that may have disappeared as a result of informal conventions.*

(*ii*) *Replace sentence letters by English sentences in accordance with the scheme of abbreviation.*

(*iii*) *Replace each predicate letter by the formula of English with which it is paired in the scheme of abbreviation, flanking the latter with a pair of braces.* (*The result of this step will not in general be a formula, for it will contain meaningless parts of the form*

$$\{\phi\}\,\alpha \quad ,$$

*with* $\phi$ *a formula and* $\alpha$ *a variable.*)

(*iv*) *Replace all parts of the form*

$$\{\phi\}\,\alpha \quad ,$$

*where* $\phi$ *is a formula of English and* $\alpha$ *a variable, by the result of replacing in* $\phi$ *all occurrences of the variable* '*a*' *by* $\alpha$.

(*v*) *Eliminate sentential connectives and quantifier phrases in favor of the corresponding phrases of connection and quantity, preserving all parentheses.*

As before, we say that an English formula is a *free translation* (or simply a *translation*) of a symbolic formula $\phi$ on the basis of a given scheme of abbreviation if it is a stylistic variant of the literal English translation of $\phi$ based on that scheme.

Consider, for example, the sentence

(8) $$\Lambda x(\mathrm{F}x \rightarrow \mathrm{G}x) \rightarrow \sim \mathrm{P} \quad .$$

Let us translate it into English on the basis of the scheme (7). In step (i) of the process of literal translation into English, (8) becomes

$$(\Lambda x(Fx \rightarrow Gx) \rightarrow \sim P) \quad ,$$

in step (ii)

$$(\Lambda x(Fx \rightarrow Gx) \rightarrow \sim \text{Moby Dick is a fish}) \quad ,$$

in step (iii)

$$(\Lambda x(\{a \text{ is a whale}\}\, x \rightarrow \{a \text{ is a mammal}\}\, x) \rightarrow \sim \text{Moby Dick is a fish}) \quad ,$$

in step (iv)

$$(\Lambda x(x \text{ is a whale} \rightarrow x \text{ is a mammal}) \rightarrow \sim \text{Moby Dick is a fish}) \quad ,$$

and in step (v)

(9) (If for each $x$ (if $x$ is a whale, then $x$ is a mammal), then it is not the case that Moby Dick is a fish) .

Thus (9) is the literal translation of (8) into English on the basis of the scheme (7), and the more idiomatic sentence,

If all whales are mammals, then Moby Dick is not a fish,

being a stylistic variant of (9), qualifies as a free translation of (8) on the basis of the same scheme.

We say that $\phi$ is a *symbolization* of a formula $\psi$ of English on the basis of a given scheme of abbreviation just in case $\phi$ is a symbolic formula which has $\psi$ as a translation on the basis of that scheme.

To find a symbolization of a given formula of English on the basis of a given scheme of abbreviation the reader will find it useful to proceed roughly as follows:

(1) *Introduce phrases of quantity and connection, the latter accompanied by parentheses and occurring canonically* (*that is, standing only before or between formulas*), *in place of their stylistic variants.*

(2) *Reverse the steps leading from a symbolic formula to a literal English translation.*

For example, consider the sentence

(10) If a professor is a Communist and all Communists are subversive, then he is subversive,

together with the scheme of abbreviation

F : $a$ is a professor
G : $a$ is a Communist
H : $a$ is subversive .

Some points should be noted before a symbolization is attempted. First, the indefinite article is often used as a stylistic variant of a phrase of quantity, and this is true of the first occurrence of 'a' in (10). Secondly, despite the fact that (10) begins with 'if', its symbolization should clearly be not a conditional but a universal generalization of a conditional. Thirdly, English pronouns often play a role like that of variables in our symbolic language; this is the case with 'he' in (10). These points suggest that (10) should become in step (1)

> For each $x$ (if (($x$ is a professor and $x$ is a Communist) and for each $y$ (if $y$ is a Communist, then $y$ is subversive)), then $x$ is subversive) .

Let us now perform step (2) of the process of symbolization. Reversing step (v) of the process of literal translation into English, we obtain

> $\bigwedge x$ ((($x$ is a professor $\wedge$ $x$ is a Communist) $\wedge$ $\bigwedge y$($y$ is a Communist $\rightarrow$ $y$ is subversive)) $\rightarrow$ $x$ is subversive) .

Reversing step (iv), this becomes

> $\bigwedge x$ ((({$a$ is a professor} $x$ $\wedge$ {$a$ is a Communist} $x$) $\wedge$ $\bigwedge y$({$a$ is a Communist} $y$ $\rightarrow$ {$a$ is subversive} $y$)) $\rightarrow$ {$a$ is subversive} $x$) .

Reversing step (iii), we obtain

$$\bigwedge x\,(((Fx \wedge Gx) \wedge \bigwedge y(Gy \rightarrow Hy)) \rightarrow Hx) \quad .$$

Finally, reversing step (i) (for step (ii) is irrelevant), we obtain

(11) $$\bigwedge x(Fx \wedge Gx \wedge \bigwedge y(Gy \rightarrow Hy) \rightarrow Hx) \quad .$$

(The use of two variables, '$x$' and '$y$', is not necessary for a symbolization of (10); it could equally well become '$\bigwedge x(Fx \wedge Gx \wedge \bigwedge x(Gx \rightarrow Hx) \rightarrow Hx)$', which is, however, somewhat less perspicuous than (11).)

By an *argument* we now understand a sequence of *formulas*, called its *premises*, together with another *formula*, called its *conclusion*. A *symbolic argument* or *English argument* is one whose premises and conclusion are respectively symbolic formulas or formulas of English. A *symbolization of an English argument on the basis of a given scheme of abbreviation* is a symbolic argument whose premises and conclusion are respective symbolizations, on the basis of that scheme, of the premises and conclusion of the English argument. A symbolic argument is called simply a *symbolization* of an English argument if there is some scheme of abbreviation on the basis of which it is a symbolization of that argument.

## EXERCISES

On the basis of the scheme of abbreviation,

> F : *a* is an even number
> G : *a* is a prime number

H : $a$ is honest
J : $a$ is a person
P : 2 is a prime number
Q : 4 is a prime number
R : the son of Lysimachus is honest ,

translate the following symbolic formulas into idiomatic English. Exercises 14 – 16 are solved for illustration.

14. $\bigwedge x(Jx \wedge Hx \rightarrow R)$

In steps (i) and (ii) of the process of literal translation into English, No. 14 becomes

$\bigwedge x((Jx \wedge Hx) \rightarrow$ the son of Lysimachus is honest) ,

in steps (iii) and (iv)

$\bigwedge x$(($x$ is a person $\wedge$ $x$ is honest) $\rightarrow$ the son of Lysimachus is honest) ,

and in step (v)

(12) For each $x$ (if ($x$ is a person and $x$ is honest), then the son of Lysimachus is honest) .

Thus (12) is the literal English translation of No. 14, and the following stylistic variant of (12) is a translation of that symbolic formula into idiomatic English:

If anyone is honest, then the son of Lysimachus is honest.

15. $P \rightarrow \bigvee x(Fx \wedge Gx)$
16. $P \rightarrow \bigvee xFx \wedge Gx$

On the basis of the given scheme, the literal translations into English of Nos. 15 and 16 are, respectively,

(If 2 is a prime number, then there is an object $x$ such that ($x$ is an even number and $x$ is a prime number))

and

(If 2 is a prime number, then (there is an object $x$ such that $x$ is an even number and $x$ is a prime number)) ;

and the following stylistic variants are respective translations into idiomatic English of the symbolic formulas:

If 2 is a prime number, then there is an even prime number,

If 2 is a prime number, then there is an even number and $x$ is a prime number.

To account for the fact that No. 15 has a different translation from No. 16, it is sufficient to note that the latter, in contrast with the former, is not a symbolic *sentence*.

17. $Q \leftrightarrow \bigwedge x(Fx \rightarrow Gx)$
18. $\bigvee x(Fx \wedge Gx) \rightarrow \bigvee xFx \wedge \bigvee xGx$
19. $Q \wedge \bigvee xFx \rightarrow \bigvee x(Fx \wedge Gx)$
20. $\bigwedge x(Fx \wedge Gx \rightarrow P)$
21. $\bigvee x(Fx \wedge Gx) \rightarrow P$

Symbolize each of the following sentences on the basis of the scheme of abbreviation that accompanies it. Exercises 22 – 24 are solved for illustration.

22. Among snakes, only copperheads and rattlers are poisonous. (F : *a* is a snake; G : *a* is poisonous; H : *a* is a copperhead; J : *a* is a rattler)

To symbolize No. 22 we must supplement our preceding comment on 'only' (p. 92) with an intuitive comprehension of English; this intuition suggests that the sentence pass in step (1) of the process of symbolization into either

> For each *x* (if *x* is a snake, then (if *x* is poisonous, then (*x* is a copperhead or *x* is a rattler)))

or

> For each *x* (if (*x* is a snake and *x* is poisonous), then (*x* is a copperhead or *x* is a rattler)) .

Let us now perform step (2) of the process of symbolization, using the first alternative above. Reversing step (v) of the process of literal translation into English, we obtain

> $\bigwedge x$(*x* is a snake $\rightarrow$ (*x* is poisonous $\rightarrow$ (*x* is a copperhead $\vee$ *x* is a rattler))) ;

reversing steps (iv) and (iii), we obtain

$$\bigwedge x(Fx \rightarrow (Gx \rightarrow (Hx \vee Jx))) \quad ;$$

and finally, reversing step (i) (for step (ii) is irrelevant), we obtain

$$\bigwedge x(Fx \rightarrow [Gx \rightarrow Hx \vee Jx]) \quad .$$

23. Nothing is a dog unless it is an animal. (F : *a* is a dog; G : *a* is an animal)

Intuition, aided by the preceding comments on 'nothing' (p. 91) and 'unless' (p. 43), suggests that the sentence should become in step (1) of the process of symbolization

For each $x$ (it is not the case that $x$ is a dog or $x$ is an animal) ,

and in step (2)

$$\bigwedge x(\sim Fx \vee Gx) \quad .$$

24. No one but a criminal is in a penitentiary. (F : *a* is in a penitentiary; G : *a* is a criminal)

It will sometimes be our practice (admittedly questionable) to treat 'no one', 'someone', 'everyone' as stylistic variants of 'nothing', 'something', 'everything'. Thus No. 24 may become in step (1) of the process of symbolization

For each $x$ (if $x$ is in a penitentiary, then $x$ is a criminal) ,

rather than

For each $x$ (if ($x$ is a person and $x$ is in a penitentiary), then $x$ is a criminal) ;

in step (2) it becomes

$$\bigwedge x(Fx \rightarrow Gx) \quad .$$

25. Drunkards are not admitted. (F : *a* is a drunkard; G : *a* is admitted)
26. None but the brave deserve the fair. (F : *a* is brave; G : *a* deserves the fair)
27. There is a round square. (F : *a* is round; G : *a* is square)
28. Something is round and something is square. (F : *a* is round; G : *a* is square)
29. Single women are decorous only if they are chaperoned. (F : *a* is a single woman; G : *a* is decorous; H : *a* is chaperoned)
30. Single women are decorous if they are chaperoned. (F : *a* is a single woman; G : *a* is decorous; H : *a* is chaperoned)
31. Some soldiers love war, but not all who love war are soldiers. (F : *a* is a soldier; G : *a* loves war)
32. If all men are mortal, then Christ is not a man. (F : *a* is a man; G : *a* is mortal; P : Christ is a man)
33. Men and women who are over twenty-one are permitted to vote. (F : *a* is a man; G : *a* is a woman; H : *a* is over twenty-one; I : *a* is permitted to vote)
34. Women without husbands are unhappy unless they have paramours. (F : *a* is a woman; G : *a* is without a husband; H : *a* is unhappy; I : *a* has a paramour)
35. If only Republicans support the incumbent and no Democrat supports the candidate, then if anyone is a Democrat, someone supports neither the incumbent nor the candidate. (F: *a* is a Republican; G: *a* supports the incumbent; H : *a* is a Democrat; I : *a* supports the candidate)

36. If those who believe in God have immortal souls, then, given that God exists, they will have eternal bliss. (F : *a* believes in God; G : *a* has an immortal soul; H : *a* will have eternal bliss; P : God exists)

**5. Inference rules and forms of derivation; theorems with unabbreviated proofs.** We say that a symbolic formula $\psi$ *comes from* a symbolic formula $\phi$ by *proper substitution of a variable* $\beta$ *for a variable* $\alpha$ if $\psi$ is like $\phi$ except for having free occurrences of $\beta$ wherever $\phi$ has free occurrences of $\alpha$. Consider, for example, the formula

$$\text{(1)} \qquad Fx \wedge Gy \rightarrow \bigvee x Hx \quad .$$

The formula

$$Fz \wedge Gy \rightarrow \bigvee x Hx$$

comes from (1) by proper substitution of '$z$' for '$x$', and

$$Fy \wedge Gy \rightarrow \bigvee x Hx$$

comes from (1) by proper substitution of '$y$' for '$x$'.

For the logic of quantifiers we must add three inference rules and a form of derivation to our original stock.

The rules of *universal instantiation* and *existential generalization* lead respectively from

$$\bigwedge \alpha \phi$$

to

$$\phi' \quad ,$$

and from

$$\phi'$$

to

$$\bigvee \alpha \phi \quad ,$$

where $\alpha$ is a variable, $\phi$ is a symbolic formula, and $\phi'$ comes from $\phi$ by proper substitution of some variable for $\alpha$. In both cases $\alpha$ is called the *variable of generalization* and the variable that replaces $\alpha$ the *variable of instantiation*. Universal instantiation corresponds to the intuitive principle that what is true of everything is true of any given thing, and existential generalization to the principle that what is true of a given thing is true of something.

For example, in the arguments

$$\bigwedge x Fx \quad \therefore \; Fy \quad ,$$
$$(Fx \wedge Gx) \quad \therefore \; \bigvee y\,(Fy \wedge Gy) \quad ,$$

the conclusion follows from the premise by the respective rules of universal instantiation and existential generalization. In the first case the variable of generalization is '*x*', and the variable of instantiation '*y*'; in the second case these roles are reversed.

Now consider the following informal derivation:

(1) Show that if $m$ is even, then $m \cdot n$ is even.
(2) Assume that $m$ is even.
(3) For some $k$, $m = 2 \cdot k$.
(4) Let $k_0$ be such a $k$; thus $m = 2 \cdot k_0$.
(5) Hence $m \cdot n = (2 \cdot k_0) \cdot n = 2 \cdot (k_0 \cdot n)$.
(6) Therefore $m \cdot n$ is even.

The third new inference rule, called the rule of *existential instantiation*, accounts for the intuitive transition from step (3) to step (4), and corresponds to the following principle; what is true of something may be asserted to hold for some particular object. In general, existential instantiation leads from

$$\bigvee \alpha \phi$$

to

$$\phi' \quad ,$$

where again $\alpha$ is a variable, $\phi$ is a symbolic formula, and $\phi'$ comes from $\phi$ by proper substitution of some variable for $\alpha$. To avoid fallacies, a restriction on the use of existential instantiation is incorporated into clause (5) of the directions given below for constructing a derivation: the *variable of instantiation* (that is, the variable which replaces $\alpha$) must be new. (Without such a restriction we should risk an unjustifiable identification of the object of which $\phi$ is asserted to hold with objects already mentioned in the derivation.)

Diagrammatically the three new rules appear as follows.

Universal instantiation (UI): $$\frac{\bigwedge \alpha \phi}{\phi'}$$

Existential generalization (EG): $$\frac{\phi'}{\bigvee \alpha \phi}$$

Existential instantiation (EI): $$\frac{\bigvee \alpha \phi}{\phi'}$$

In all three cases $\alpha$ is to be a variable, $\phi$ is to be a symbolic formula, and $\phi'$ is to come from $\phi$ by proper substitution of some variable for $\alpha$.

The new form of derivation is known as *universal derivation* and in its simplest form appears as follows:

*Show* $\wedge\alpha\phi$
$\chi_1$
.
.
.
$\chi_m$ ,

where $\phi$ occurs unboxed among $\chi_1$ through $\chi_m$. In a universal derivation one shows that everything has a certain property by showing that an arbitrary thing has that property. To ensure arbitrariness a restriction will be imposed: the variable $\alpha$, called the *variable of generalization*, must not be free in any antecedent line.

This form of derivation is familiar to every student of plane geometry, wherein one shows, for instance, that every triangle has a certain property by considering an arbitrary triangle and showing that it has the property in question.

It is natural to permit a more inclusive form of universal derivation, in which several variables of generalization occur. Thus *universal derivation* appears in general as follows:

*Show* $\wedge\alpha_1 \ldots \wedge\alpha_k\,\phi$
$\chi_1$
.
.
.
$\chi_m$ ,

where $\phi$ occurs unboxed among $\chi_1$ through $\chi_m$. Here one shows that all objects stand in a certain relation to one another by showing that arbitrary objects do so. To insure arbitrariness, the following restriction is imposed in clause (6) below: the *variables of generalization*, $\alpha_1$ through $\alpha_k$, may not be free in antecedent lines.

We shall continue to employ the inference rules of the sentential calculus, but they must be reconstrued in such a way as to admit as premises and conclusions any symbolic *formulas* of appropriate sentential structure (not merely symbolic *sentences*). For example, in the argument

$$(Fx \rightarrow Gx) \quad . \quad Fx \quad \therefore Gx \quad ,$$

the conclusion will now be considered to follow from the premises by MP.

We may call that branch of logic which essentially involves quantifiers as well as sentential connectives the *quantifier calculus*. That part of the quantifier calculus which concerns the rather restricted symbolic language of the present chapter is called the *monadic quantifier calculus*. The directions for constructing a derivation within this discipline (which in the next chapter will be extended only slightly in order to arrive at the full quantifier calculus) transcend in two ways those given for the sentential calculus. On the one hand, the earlier procedures are extended so as to apply not only to symbolic sentences but to arbitrary symbolic formulas; on the other hand, additional provisions are made for the accommodation of quantifiers. The directions for constructing a *derivation* from given symbolic premises become, then, the following.

(1) *If $\phi$ is any symbolic formula, then*

$$Show\ \phi$$

*may occur as a line.* (*Annotation:* '*Assertion*'.)

(2) *Any one of the premises may occur as a line.* (*Annotation:* '*Premise*'.)

(3) *If $\phi$, $\psi$ are symbolic formulas such that*

$$Show\ (\phi \rightarrow \psi)$$

*occurs as a line, then $\phi$ may occur as the next line.* (*Annotation:* '*Assumption*'.)

(4) *If $\phi$ is a symbolic formula such that*

$$Show\ \phi$$

*occurs as a line, then*

$$\sim \phi$$

*may occur as the next line; if $\phi$ is a symbolic formula such that*

$$Show \sim \phi$$

*occurs as a line, then $\phi$ may occur as the next line.* (*Annotation:* '*Assumption*'.)

(5a) *A symbolic formula may occur as a line if it follows from antecedent lines by an inference rule of the sentential calculus* (*that is, MP, MT, DN, R, S, Adj, Add, MTP, BC, or CB*), *by UI, or by EG.*

(5b) *A symbolic formula may occur as a line if it follows from an antecedent line by EI, provided that the variable of instantiation does not occur in any preceding line.* (*The annotation for* (5a) *and* (5b) *should refer to the inference rule employed and the numbers of the antecedent lines involved.*)

(6) *When the following arrangement of lines has appeared:*

$Show\ \phi$

$\chi_1$

.

.

.

$\chi_m$ ,

*where none of* $\chi_1$ *through* $\chi_m$ *contains uncancelled 'Show' and either*

(*i*) $\phi$ *occurs unboxed among* $\chi_1$ *through* $\chi_m$,

(*ii*) $\phi$ *is of the form*

$$(\psi_1 \rightarrow \psi_2)$$

*and* $\psi_2$ *occurs unboxed among* $\chi_1$ *through* $\chi_m$,

(*iii*) *for some formula* $\chi$, *both* $\chi$ *and its negation occur unboxed among* $\chi_1$ *through* $\chi_m$, *or*

(*iv*) $\phi$ *is of the form*

$$\wedge\alpha_1 \ldots \wedge\alpha_k \psi \quad ,$$

$\psi$ *occurs unboxed among the lines* $\chi_1$ *through* $\chi_m$, *and the variables* $\alpha_1$ *through* $\alpha_k$ *are not free in lines antecedent to the displayed occurrence of*

*Show* $\phi$ ,

*then one may simultaneously cancel the displayed occurrence of 'Show' and box all subsequent lines.*

As before, a derivation is *complete* just in case every line either is boxed or contains cancelled '*Show*'. A symbolic formula $\phi$ is *derivable* (in the quantifier calculus) from given symbolic premises just in case, by using only clauses (1) through (6), a complete derivation from those premises can be constructed in which

~~*Show*~~ $\phi$

occurs as an unboxed line. A *proof* is a derivation from an empty class of formulas (that is, a derivation employing no premises), and a *theorem* is a symbolic formula derivable from an empty class of premises.

We turn next to some theorems and their proofs. For clerical convenience, we now assign numbers beginning with 201.

T201 is established by a conditional derivation with subsidiary conditional and universal derivations.

T201

| | | |
|---|---|---|
| 1. | ~~*Show*~~ $\wedge x(\mathrm{F}x \rightarrow \mathrm{G}x) \rightarrow (\wedge x\mathrm{F}x \rightarrow \wedge x\mathrm{G}x)$ | |
| 2. | $\wedge x(\mathrm{F}x \rightarrow \mathrm{G}x)$ | |
| 3. | ~~*Show*~~ $\wedge x\mathrm{F}x \rightarrow \wedge x\mathrm{G}x$ | |
| 4. | $\wedge x\mathrm{F}x$ | |
| 5. | ~~*Show*~~ $\wedge x\mathrm{G}x$ | |
| 6. | $\mathrm{F}x$ | 4, UI |
| 7. | $\mathrm{F}x \rightarrow \mathrm{G}x$ | 2, UI |
| 8. | $\mathrm{G}x$ | 6, 7, MP |

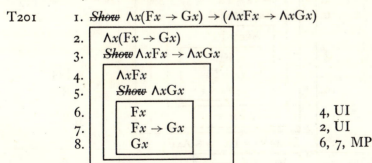

If, in this derivation, line 6 had preceded line 5, we should not have been able to complete the universal derivation of line 5; for then the variable of generalization, '$x$', would have been free in an antecedent line. As a general practice, if a universal derivation is to be employed, it is advisable to begin it before using UI.

In the proof of T201, only one of our new inference rules, UI, was employed; T202 illustrates the application of EI and EG.

T202

| | | |
|---|---|---|
| 1. | ~~*Show*~~ $\bigwedge x(Fx \rightarrow Gx) \rightarrow (\bigvee xFx \rightarrow \bigvee xGx)$ | |
| 2. | $\bigwedge x(Fx \rightarrow Gx)$ | |
| 3. | ~~*Show*~~ $\bigvee xFx \rightarrow \bigvee xGx$ | |
| 4. | $\bigvee xFx$ | |
| 5. | $Fy$ | 4, EI |
| 6. | $Fy \rightarrow Gy$ | 2, UI |
| 7. | $Gy$ | 5, 6, MP |
| 8. | $\bigvee xGx$ | 7, EG |

In this derivation the order of lines is again important. If line 6 had preceded line 5, we should have been unable to choose '$y$' as the variable of instantiation in line 5. As a general practice, it is advisable to use EI before UI.

Two symbolic formulas $\phi$ and $\psi$ are said to be *equivalent* if the formula

$$\phi \leftrightarrow \psi$$

is a theorem. T203–T206 provide equivalent expressions for formulas beginning with various combinations of quantifier phrases and negation signs.

T203

| | | |
|---|---|---|
| 1. | ~~*Show*~~ $\sim \bigwedge xFx \leftrightarrow \bigvee x \sim Fx$ | |
| 2. | ~~*Show*~~ $\sim \bigwedge xFx \rightarrow \bigvee x \sim Fx$ | |
| 3. | $\sim \bigwedge xFx$ | |
| 4. | ~~*Show*~~ $\bigvee x \sim Fx$ | |
| 5. | $\sim \bigvee x \sim Fx$ | |
| 6. | ~~*Show*~~ $\bigwedge xFx$ | |
| 7. | ~~*Show*~~ $Fx$ | |
| 8. | $\sim Fx$ | |
| 9. | $\bigvee x \sim Fx$ | 8, EG |
| 10. | $\sim \bigvee x \sim Fx$ | 5, R |
| 11. | $\sim \bigwedge xFx$ | 3, R |
| 12. | ~~*Show*~~ $\bigvee x \sim Fx \rightarrow \sim \bigwedge xFx$ | |
| 13. | $\bigvee x \sim Fx$ | |
| 14. | ~~*Show*~~ $\sim \bigwedge xFx$ | |

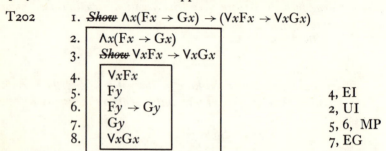

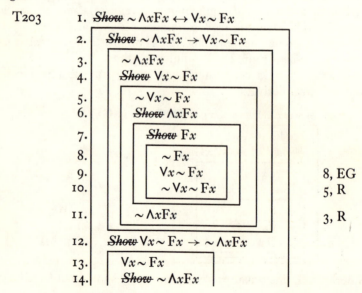

15. $\bigwedge x Fx$
16. $\sim Fy$ 13, EI
17. $Fy$ 15, UI
18. $\sim \bigwedge x Fx \leftrightarrow \bigvee x \sim Fx$ 2, 12, CB

T204

1. ~~*Show*~~ $\sim \bigvee x Fx \leftrightarrow \bigwedge x \sim Fx$
2. ~~*Show*~~ $\sim \bigvee x Fx \rightarrow \bigwedge x \sim Fx$
3. $\sim \bigvee x Fx$
4. ~~*Show*~~ $\bigwedge x \sim Fx$
5. ~~*Show*~~ $\sim Fx$
6. $Fx$
7. $\bigvee x Fx$ 6, EG
8. $\sim \bigvee x Fx$ 3, R
9. ~~*Show*~~ $\bigwedge x \sim Fx \rightarrow \sim \bigvee x Fx$
10. $\bigwedge x \sim Fx$
11. ~~*Show*~~ $\sim \bigvee x Fx$
12. $\bigvee x Fx$
13. $Fy$ 12, EI
14. $\sim Fy$ 10, UI
15. $\sim \bigvee x Fx \leftrightarrow \bigwedge x \sim Fx$ 2, 9, CB

T205 $\bigwedge x Fx \leftrightarrow \sim \bigvee x \sim Fx$

T206 $\bigvee x Fx \leftrightarrow \sim \bigwedge x \sim Fx$

## EXERCISES

37. Annotate each line of the following derivation.

1. ~~*Show*~~ $\bigwedge x Fx \wedge \bigvee x Gx \rightarrow \bigvee x (Fx \wedge Gx)$ ASSERTION
2. $\bigwedge x Fx \wedge \bigvee x Gx$ ASSUMP C.D
3. $\bigvee x Gx$ 2 S
4. $Gy$ 3 EI.
5. $\bigwedge x Fx$ 2 S
6. $Fy$ 5 UI.
7. $Fy \wedge Gy$ 4,6 ADJ
8. $\bigvee x (Fx \wedge Gx)$ 7 EG

38. Prove T205 and T206.

The reader will find careful study of the proofs of T203 and T204 useful in solving exercise 38.

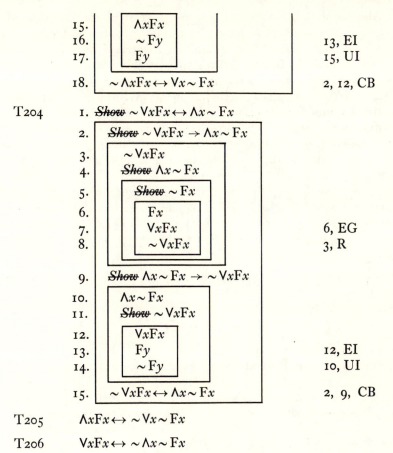

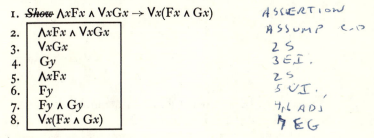

**6. Abbreviated derivations; theorems with abbreviated proofs.** It is useful at this point to introduce counterparts to the methods for abbreviating derivations that appeared in chapter II and, in addition, one new method. As before, the methods of abbreviation are obtained by supplementing the directions for constructing derivations. The new clauses, (7) – (9), will satisfy the requirements given on page 58; that is, they will be theoretically dispensable, and there will be an automatic procedure for checking the correctness of an abbreviated derivation (that is, a derivation constructed on the basis of clauses (1) – (9)), at least when membership in the class of premises is automatically decidable and annotations are present.

We must modify slightly the notion of an instance of a sentential theorem, now that formulas as well as sentences are at hand. By an *instance* of a *sentential theorem* $\phi$ we now understand any symbolic *formula* obtained from $\phi$ by replacing sentence letters uniformly by symbolic formulas. Thus

$$Q \rightarrow (P \rightarrow Q)$$

has as an instance

$$Fx \rightarrow (Gy \rightarrow Fx) \quad ,$$

even though the latter is not a sentence.

We introduce now the counterpart to clause (7) of chapter II.

(7) *If* $\phi$ *is an instance of a theorem of the* sentential *calculus that has already been proved, then* $\phi$ *may occur as a line.* (*Annotation as in Chapter II.*)

We do not admit at this point instances of theorems of the *quantifier* calculus. The relevant notion of an instance is rather complex and will be reserved for chapter IV.

It is convenient, however, to add to our resources the following derived rule involving quantifiers, whose various forms correspond to T203 and T204.

Quantifier negation (QN), in four forms:

$$\frac{\sim \bigwedge \alpha \phi}{\bigvee \alpha \sim \phi} \qquad \frac{\bigvee \alpha \sim \phi}{\sim \bigwedge \alpha \phi} \qquad \frac{\sim \bigvee \alpha \phi}{\bigwedge \alpha \sim \phi} \qquad \frac{\bigwedge \alpha \sim \phi}{\sim \bigvee \alpha \phi}$$

Here $\alpha$ is to be a variable and $\phi$ a symbolic formula. QN corresponds to the following intuitive principles: to deny that every object satisfies a given condition is to assert that some object satisfies its negation, and to deny that there is an object satisfying a given condition is to assert that every object satisfies its negation.

For example, in each of the arguments

$$\sim \bigwedge x(Fx \rightarrow Gx) \quad \therefore \bigvee x \sim (Fx \rightarrow Gx) \quad ,$$
$$\bigvee x \sim (Fx \rightarrow Gx) \quad \therefore \sim \bigwedge x(Fx \rightarrow Gx) \quad ,$$
$$\sim \bigvee x(Fx \wedge Gx) \quad \therefore \bigwedge x \sim (Fx \wedge Gx) \quad ,$$
$$\bigwedge x \sim (Fx \wedge Gx) \quad \therefore \sim \bigvee x(Fx \wedge Gx) \quad ,$$

the conclusion follows from the premise by QN.

We incorporate QN into our system by adding the following clause to the directions for constructing a derivation.

(8) *A symbolic formula may occur as a line if it follows from an antecedent line by QN. (Annotation: 'QN', together with the number of the antecedent line involved.)*

In the quantifier calculus, as in the sentential calculus, it is again convenient, when constructing a derivation, to compress several steps into one, omitting some lines which an unabbreviated derivation would require. As in chapter II, such compression will be allowed only in certain cases, specifically, only when the omitted steps can be justified on the basis of clause (2) (premises), (5a) (inference rules other than EI), (7) (instances of previously proved theorems of the sentential calculus), or (8) (quantifier negation). We legitimize such compressions by clause (9)—the counterpart to clause (8) of chapter II.

(9) *A symbolic formula may occur as a line if it follows from antecedent lines by a succession of steps, and each intermediate step can be justified by one of clauses (2), (5a), (7), or (8). (The annotation should determine unambiguously the succession of steps leading to the line in question. This can be done by indicating, in order of application, the antecedent lines, the premises, the inference rules, and the previously proved theorems employed. Also, in connection with the rules Add, UI, and EG, the added disjunct, the variable of instantiation, and the variable of generalization should respectively be indicated whenever there is a chance of ambiguity; and when an instance of a previously proved theorem is involved, the relevant replacement, if not obvious, should be indicated. The special annotations 'SC' and 'CD', which were introduced on pages 64 and 65 to indicate a certain kind of compression, may again be used.)*

For example, clause (9) provides the justification for line 4 in the following abbreviated derivation:

1. ~~*Show*~~ $\sim \bigvee x Fx \rightarrow \bigwedge x(Fx \rightarrow Gx)$
2. $\sim \bigvee x Fx$
3. ~~*Show*~~ $\bigwedge x(Fx \rightarrow Gx)$
4. $Fx \rightarrow Gx$     2, QN, UI, T18, MP

The omitted lines are, in order:

| | | |
|---|---|---|
| (i) | $\bigwedge x \sim Fx$ | 2, QN |
| (ii) | $\sim Fx$ | (i), UI |
| (iii) | $\sim Fx \rightarrow (Fx \rightarrow Gx)$ | T18 |

Line 4 follows from (ii) and (iii) by MP.

Another illustration of the application of clause (9) is provided by the derivation

| | | |
|---|---|---|
| 1. | ~~*Show*~~ $\sim \bigwedge x Fx \rightarrow \bigvee x(Fx \rightarrow Gx)$ | |
| 2. | $\sim \bigwedge x Fx$ | |
| 3. | $\sim Fy$ | 2, QN, EI |
| 4. | $\bigvee x(Fx \rightarrow Gx)$ | 3, T18, MP, EG |

(The reader should note that the omission of line 3 would not be justified by clause (9); clause (5b), pertaining to EI, is not listed as one of the clauses by which an *omitted* step may be justified.)

The reader should recall that abbreviated derivations, unlike unabbreviated derivations, require as an essential part their annotations—at least those given in connection with clause (9). Without an indication of omitted steps, it would be impossible to give an automatic procedure for checking the correctness of an abbreviated derivation. (See the remarks on page 80.)

T201 and T202 are called *distribution laws* for quantifiers; we state now some additional distribution laws.

T207

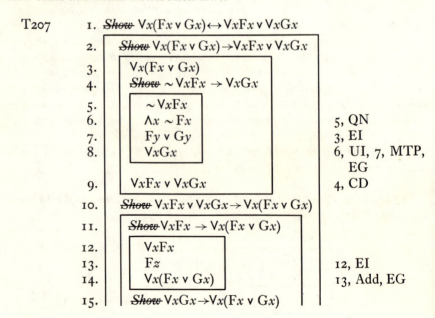

| | | |
|---|---|---|
| 1. | ~~*Show*~~ $\bigvee x(Fx \vee Gx) \leftrightarrow \bigvee x Fx \vee \bigvee x Gx$ | |
| 2. | ~~*Show*~~ $\bigvee x(Fx \vee Gx) \rightarrow \bigvee x Fx \vee \bigvee x Gx$ | |
| 3. | $\bigvee x(Fx \vee Gx)$ | |
| 4. | ~~*Show*~~ $\sim \bigvee x Fx \rightarrow \bigvee x Gx$ | |
| 5. | $\sim \bigvee x Fx$ | |
| 6. | $\bigwedge x \sim Fx$ | 5, QN |
| 7. | $Fy \vee Gy$ | 3, EI |
| 8. | $\bigvee x Gx$ | 6, UI, 7, MTP, EG |
| 9. | $\bigvee x Fx \vee \bigvee x Gx$ | 4, CD |
| 10. | ~~*Show*~~ $\bigvee x Fx \vee \bigvee x Gx \rightarrow \bigvee x(Fx \vee Gx)$ | |
| 11. | ~~*Show*~~ $\bigvee x Fx \rightarrow \bigvee x(Fx \vee Gx)$ | |
| 12. | $\bigvee x Fx$ | |
| 13. | $Fz$ | 12, EI |
| 14. | $\bigvee x(Fx \vee Gx)$ | 13, Add, EG |
| 15. | ~~*Show*~~ $\bigvee x Gx \rightarrow \bigvee x(Fx \vee Gx)$ | |

| | | |
|---|---|---|
| 16. | $\bigvee x Gx$ | |
| 17. | $Gw$ | 16, EI |
| 18. | $\bigvee x(Fx \lor Gx)$ | 17, Add, EG |
| 19. | $\bigvee x Fx \lor \bigvee x Gx \rightarrow \bigvee x(Fx \lor Gx)$ | 11, 15, SC |
| 20. | $\bigvee x(Fx \lor Gx) \leftrightarrow \bigvee x Fx \lor \bigvee x Gx$ | 2, 10, CB |

It is often convenient to move in an initial negation sign by means of QN, as in the transition from line 5 to line 6 in the proof of T207.

T208 $\bigwedge x(Fx \land Gx) \leftrightarrow \bigwedge x Fx \land \bigwedge x Gx$

T209 $\bigvee x(Fx \land Gx) \rightarrow \bigvee x Fx \land \bigvee x Gx$

T210 $\bigwedge x Fx \lor \bigwedge x Gx \rightarrow \bigwedge x(Fx \lor Gx)$

T211 $(\bigvee x Fx \rightarrow \bigvee x Gx) \rightarrow \bigvee x(Fx \rightarrow Gx)$

T212 $(\bigwedge x Fx \rightarrow \bigwedge x Gx) \rightarrow \bigvee x(Fx \rightarrow Gx)$

T213 $\bigwedge x(Fx \leftrightarrow Gx) \rightarrow (\bigwedge x Fx \leftrightarrow \bigwedge x Gx)$

T214 $\bigwedge x(Fx \leftrightarrow Gx) \rightarrow (\bigvee x Fx \leftrightarrow \bigvee x Gx)$

(None of the biconditionals corresponding to T201, T202, T209 – T214 can be proved; see section 9.)

The intuitive plausibility of logical laws can often be seen by translation into English. For example, on the basis of the scheme,

F : *a* is a man
G : *a* is a mortal ,

the distribution laws T201 and T202 can be translated, respectively, into

> If all men are mortal, then if everything is a man, then everything is mortal

and

> If all men are mortal, then if something is a man, then something is mortal.

T215 – T222 are called *confinement laws.* According to these theorems, a generalization of a conjunction, disjunction, or conditional in which only one component contains the relevant variable is equivalent to a formula in which the quantifier phrase is confined to that component. Note that when the relevant component is the antecedent of a conditional (T221 and T222), the quantifier phrase must be changed from universal to existential, or conversely.

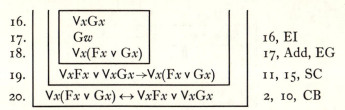

| | | | |
|---|---|---|---|
| T215 | 1. | ~~*Show*~~ $\bigwedge x(P \wedge Fx) \leftrightarrow P \wedge \bigwedge xFx$ | |
| | 2. | ~~*Show*~~ $\bigwedge x(P \wedge Fx) \rightarrow P \wedge \bigwedge xFx$ | |
| | 3. | $\bigwedge x(P \wedge Fx)$ | |
| | 4. | ~~*Show*~~ $\bigwedge xFx$ | |
| | 5. | $Fx$ | 3, UI, S |
| | 6. | $P \wedge \bigwedge xFx$ | 3, UI, S, 4, Adj |
| | 7. | ~~*Show*~~ $P \wedge \bigwedge xFx \rightarrow \bigwedge x(P \wedge Fx)$ | |
| | 8. | $P \wedge \bigwedge xFx$ | |
| | 9. | ~~*Show*~~ $\bigwedge x(P \wedge Fx)$ | |
| | 10. | $P \wedge Fx$ | 8, S, S, UI, Adj |
| | 11. | $\bigwedge x(P \wedge Fx) \leftrightarrow P \wedge \bigwedge xFx$ | 2, 7, CB |
| T216 | | $\bigvee x(P \wedge Fx) \leftrightarrow P \wedge \bigvee xFx$ | |
| T217 | | $\bigwedge x(P \vee Fx) \leftrightarrow P \vee \bigwedge xFx$ | |
| T218 | | $\bigvee x(P \vee Fx) \leftrightarrow P \vee \bigvee xFx$ | |
| T219 | | $\bigwedge x(P \rightarrow Fx) \leftrightarrow (P \rightarrow \bigwedge xFx)$ | |
| T220 | | $\bigvee x(P \rightarrow Fx) \leftrightarrow (P \rightarrow \bigvee xFx)$ | |
| T221 | | $\bigwedge x(Fx \rightarrow P) \leftrightarrow (\bigvee xFx \rightarrow P)$ | |
| T222 | 1. | ~~*Show*~~ $\bigvee x(Fx \rightarrow P) \leftrightarrow (\bigwedge xFx \rightarrow P)$ | |
| | 2. | ~~*Show*~~ $\bigvee x(Fx \rightarrow P) \rightarrow (\bigwedge xFx \rightarrow P)$ | |
| | 3. | $\bigvee x(Fx \rightarrow P)$ | |
| | 4. | ~~*Show*~~ $\bigwedge xFx \rightarrow P$ | |
| | 5. | $\bigwedge xFx$ | |
| | 6. | $Fy \rightarrow P$ | 3, EI |
| | 7. | $P$ | 5, UI, 6, MP |
| | 8. | ~~*Show*~~ $\sim \bigwedge xFx \rightarrow \bigvee x(Fx \rightarrow P)$ | |
| | 9. | $\sim \bigwedge xFx$ | |
| | 10. | $\sim Fz$ | 9, QN, EI |
| | 11. | $\bigvee x(Fx \rightarrow P)$ | 10, T18, MP, EG |
| | 12. | ~~*Show*~~ $P \rightarrow \bigvee x(Fx \rightarrow P)$ | |
| | 13. | $P$ | |
| | 14. | $\bigvee x(Fx \rightarrow P)$ | 13, T2, MP, EG |
| | 15. | $(\bigwedge xFx \rightarrow P) \rightarrow \bigvee x(Fx \rightarrow P)$ | 8, 12, Adj, T35, BC, MP |
| | 16. | $\bigvee x(Fx \rightarrow P) \leftrightarrow (\bigwedge xFx \rightarrow P)$ | 2, 15, CB |

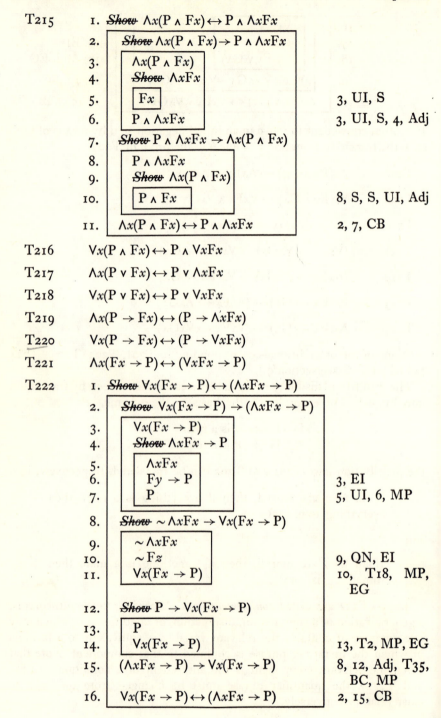

In idiomatic English, counterparts to unconfined quantifier phrases seldom occur. On the basis, however, of a suitable scheme of abbreviation, the two constituents of T221 can be translated into

If anyone is honest, then the son of Lysimachus is honest

and

If someone is honest, then the son of Lysimachus is honest.

There are no simple confinement laws for biconditionals. We can say no more than is expressed in the following four theorems. In no case is the converse also a theorem; this is substantiated in section 9.

T223 $\quad \bigwedge x(Fx \leftrightarrow P) \rightarrow (\bigwedge xFx \leftrightarrow P)$

T224 $\quad \bigwedge x(Fx \leftrightarrow P) \rightarrow (\bigvee xFx \leftrightarrow P)$

T225 $\quad (\bigvee xFx \leftrightarrow P) \rightarrow \bigvee x(Fx \leftrightarrow P)$

T226

| | | |
|---|---|---|
| 1. | ~~Show~~ $(\bigwedge xFx \leftrightarrow P) \rightarrow \bigvee x(Fx \leftrightarrow P)$ | |
| 2. | ~~Show~~ $\bigwedge xFx \wedge P \rightarrow \bigvee x(Fx \leftrightarrow P)$ | |
| 3. | $\bigwedge xFx \wedge P$ | |
| 4. | $Fx$ | 3, S, UI |
| 5. | $Fx \wedge P$ | 3, S, 4, Adj |
| 6. | $Fx \leftrightarrow P$ | 5, T84, MP |
| 7. | $\bigvee x(Fx \leftrightarrow P)$ | 6, EG |
| 8. | ~~Show~~ $\sim\bigwedge xFx \wedge \sim P \rightarrow \bigvee x(Fx \leftrightarrow P)$ | |
| 9. | $\sim\bigwedge xFx \wedge \sim P$ | |
| 10. | $\sim Fy$ | 9, S, QN, EI |
| 11. | $\sim Fy \wedge \sim P$ | 9, S, 10, Adj |
| 12. | $Fy \leftrightarrow P$ | 11, T85, MP |
| 13. | $\bigvee x(Fx \leftrightarrow P)$ | 12, EG |
| 14. | $(\bigwedge xFx \leftrightarrow P) \rightarrow \bigvee x(Fx \leftrightarrow P)$ | 2, 8, Adj, T86, BC, MP |

T227 and T228 are the laws of *vacuous quantification;* by T227 the seemingly meaningless formula '$\bigwedge x$P' is equivalent to 'P'.

T227

| | | |
|---|---|---|
| 1. | ~~Show~~ $\bigwedge xP \leftrightarrow P$ | |
| 2. | ~~Show~~ $\bigwedge xP \rightarrow P$ | |
| 3. | $\bigwedge xP$ | |
| 4. | $P$ | 3, UI |
| 5. | ~~Show~~ $P \rightarrow \bigwedge xP$ | |
| 6. | $P$ | |

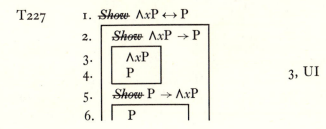

7. ~~*Show*~~ $\wedge x$P
8. P — 6, R
9. $\wedge x\text{P} \leftrightarrow \text{P}$ — 2, 5, CB

T228

1. ~~*Show*~~ $\vee x\text{P} \leftrightarrow \text{P}$
2. ~~*Show*~~ $\vee x\text{P} \rightarrow \text{P}$
3. $\vee x$P
4. P — 3, EI
5. ~~*Show*~~ $\text{P} \rightarrow \vee x\text{P}$
6. P
7. $\vee x$P — 6, EG
8. $\vee x\text{P} \leftrightarrow \text{P}$ — 2, 5, CB

The application of UI, EI, and EG in the proofs of T227 and T228 may seem suspect. For instance, may '$\vee x$P' and 'P' be taken as

$$\vee \alpha \phi$$

and $\phi'$ in the formulation on page 99? Yes; for examination of the quantifier rules will reveal that $\alpha$ need not occur in $\phi$, and that if it does not, $\phi'$ is $\phi$.

T229 and T230 have a slightly paradoxical character. Their validity depends on the fact that '$x$' is not free in '$\vee x\text{F}x$' or '$\wedge x\text{F}x$', just as the validity of the confinement laws depends on the fact that '$x$' is not free in 'P'.

T229

1. ~~*Show*~~ $\vee x(\vee x\text{F}x \rightarrow \text{F}x)$
2. $\sim \vee x(\vee x\text{F}x \rightarrow \text{F}x)$
3. $\sim(\vee x\text{F}x \rightarrow \text{F}x)$ — 2, QN, UI
4. F$y$ — 3, T21, MP, EI
5. $\sim(\vee x\text{F}x \rightarrow \text{F}y)$ — 2, QN, UI
6. $\sim \text{F}y$ — 5, T22, MP

T230

1. ~~*Show*~~ $\vee x(\text{F}x \rightarrow \wedge x\text{F}x)$
2. $\sim \vee x(\text{F}x \rightarrow \wedge x\text{F}x)$
3. $\sim(\text{F}x \rightarrow \wedge x\text{F}x)$ — 2, QN, UI
4. $\sim \text{F}y$ — 3, T22, MP, QN, EI
5. $\sim(\text{F}y \rightarrow \wedge x\text{F}x)$ — 2, QN, UI
6. F$y$ — 5, T21, MP

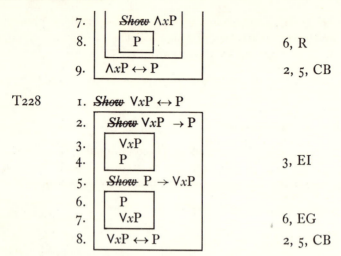

T231 and T232 are called *laws of alphabetic variance* (for bound variables). They reflect the fact that a generalization such as 'Everything is material' may pass, on the basis of a suitable scheme of abbreviation, into either '$\wedge x Fx$' or '$\wedge y Fy$'; the choice of variable is unimportant. Detailed consideration will be given to these laws in the next chapter.

T231

| | | |
|---|---|---|
| 1. | ~~Show~~ $\wedge x Fx \leftrightarrow \wedge y Fy$ | |
| 2. | ~~Show~~ $\wedge x Fx \rightarrow \wedge y Fy$ | |
| 3. | $\wedge x Fx$ | |
| 4. | ~~Show~~ $\wedge y Fy$ | |
| 5. | $Fy$ | 3, UI |
| 6. | ~~Show~~ $\wedge y Fy \rightarrow \wedge x Fx$ | |
| 7. | $\wedge y Fy$ | |
| 8. | ~~Show~~ $\wedge x Fx$ | |
| 9. | $Fx$ | 7, UI |
| 10. | $\wedge x Fx \leftrightarrow \wedge y Fy$ | 2, 6, CB |

T232

| | | |
|---|---|---|
| 1. | ~~Show~~ $\vee x Fx \leftrightarrow \vee y Fy$ | |
| 2. | ~~Show~~ $\vee x Fx \rightarrow \vee y Fy$ | |
| 3. | $\vee x Fx$ | |
| 4. | $Fz$ | 3, EI |
| 5. | $\vee y Fy$ | 4, EG |
| 6. | ~~Show~~ $\vee y Fy \rightarrow \vee x Fx$ | |
| 7. | $\vee y Fy$ | |
| 8. | $Fw$ | 7, EI |
| 9. | $\vee x Fx$ | 8, EG |
| 10. | $\vee x Fx \leftrightarrow \vee y Fy$ | 2, 6, CB |

Any instance, as well as any universal generalization of an instance, of a theorem of the sentential calculus will clearly be a theorem of the quantifier calculus. Further, certain ways of distributing the universal quantifier through such theorems will again lead to theorems. T233 – T237 are examples. (T235 corresponds to the Aristotelian syllogism in *Barbara*.) Further examples are left to the imagination of the reader.

T233 $(Fx \rightarrow Gx) \wedge (Gx \rightarrow Hx) \rightarrow (Fx \rightarrow Hx)$

T234 $\wedge x[(Fx \rightarrow Gx) \wedge (Gx \rightarrow Hx) \rightarrow (Fx \rightarrow Hx)]$

T235 $\wedge x(Fx \rightarrow Gx) \wedge \wedge x(Gx \rightarrow Hx) \rightarrow \wedge x(Fx \rightarrow Hx)$

T236 $\wedge x(Fx \leftrightarrow Gx) \wedge \wedge x(Gx \leftrightarrow Hx) \rightarrow \wedge x(Fx \leftrightarrow Hx)$

T237 1. ~~Show~~ $\wedge x(Fx \rightarrow Gx) \wedge \wedge x(Fx \rightarrow Hx) \rightarrow \wedge x(Fx \rightarrow Gx \wedge Hx)$

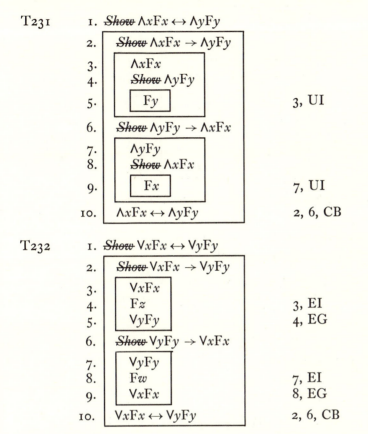

| | | |
|---|---|---|
| 2. | $\Lambda x(Fx \to Gx) \wedge \Lambda x(Fx \to Hx)$ | |
| 3. | ~~*Show*~~ $\Lambda x(Fx \to Gx \wedge Hx)$ | |
| 4. | ~~*Show*~~ $Fx \to Gx \wedge Hx$ | |
| 5. | $Fx$ | |
| 6. | $\Lambda x(Fx \to Gx)$ | 2, S |
| 7. | $Fx \to Gx$ | 6, UI |
| 8. | $Gx$ | 5, 7, MP |
| 9. | $\Lambda x(Fx \to Hx)$ | 2, S |
| 10. | $Fx \to Hx$ | 9, UI |
| 11. | $Hx$ | 5, 10, MP |
| 12. | $Gx \wedge Hx$ | 8, 11, Adj |

We list eleven more theorems involving quantifiers.

T238 $\Lambda xFx \to \mathrm{V}xFx$

T239 $\Lambda xFx \wedge \mathrm{V}xGx \to \mathrm{V}x(Fx \wedge Gx)$

T240 $\Lambda x(Fx \to Gx) \wedge \mathrm{V}x\,(Fx \wedge Hx) \to \mathrm{V}x(Gx \wedge Hx)$

T241 $\Lambda x(Fx \to Gx \vee Hx) \to \Lambda x(Fx \to Gx) \vee \mathrm{V}x(Fx \wedge Hx)$

T242 $\sim\Lambda x(Fx \to Gx) \leftrightarrow \mathrm{V}x(Fx \wedge \sim Gx)$

T243 $\sim\mathrm{V}x(Fx \wedge Gx) \leftrightarrow \Lambda x(Fx \to \sim Gx)$

T244 $\sim\mathrm{V}xFx \to \Lambda x(Fx \to Gx)$

T245 $\sim\mathrm{V}xFx \leftrightarrow \Lambda x(Fx \to Gx) \wedge \Lambda x(Fx \to \sim Gx)$

T246 $\Lambda xFx \leftrightarrow \Lambda x\Lambda y(Fx \wedge Fy)$

T247 $\mathrm{V}xFx \wedge \mathrm{V}xGx \to (\Lambda x[Fx \to Hx] \wedge \Lambda x\,[Gx \to Jx] \leftrightarrow$
$\Lambda x\,\Lambda y[Fx \wedge Gy \to Hx \wedge Jy])$

T248 $(\mathrm{V}xFx \leftrightarrow \mathrm{V}xGx) \wedge \Lambda x\Lambda y(Fx \wedge Gy \to [Hx \leftrightarrow Jy]) \to$
$(\Lambda x[Fx \to Hx] \leftrightarrow \Lambda x[Gx \to Jx])$

## EXERCISES

39. Fill in the lines that have been omitted from the following abbreviated derivation.

1. ~~*Show*~~ $\Lambda x(Fx \to Gx \vee Hx) \wedge \sim\Lambda x(Fx \to Gx) \to$
$\mathrm{V}x(Fx \wedge Hx)$

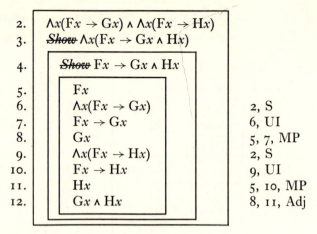

| | | |
|---|---|---|
| 2. | $\bigwedge x(Fx \to Gx \lor Hx) \land \sim \bigwedge x(Fx \to Gx)$ | |
| 3. | $\sim(Fy \to Gy)$ | 2, S, QN, EI |
| 4. | $\bigvee x(Fx \land Hx)$ | 3, T21, MP, 3, T22, MP, 2, S, UI, MP, MTP, Adj, EG |

(The reader should note that clause (9) would not justify the omission of line 3. This derivation constitutes, essentially, a proof of T241.)

40. Annotate the following abbreviated derivation.

| | |
|---|---|
| 1. | ~~*Show*~~ $\sim \bigvee x(Fx \land Gx) \leftrightarrow \bigwedge x(Fx \to \sim Gx)$ |
| 2. | ~~*Show*~~ $\sim \bigvee x(Fx \land Gx) \to \bigwedge x(Fx \to \sim Gx)$ |
| 3. | $\sim \bigvee x(Fx \land Gx)$ |
| 4. | ~~*Show*~~ $\bigwedge x(Fx \to \sim Gx)$ |
| 5. | $Fx \to \sim Gx$ |
| 6. | ~~*Show*~~ $\bigwedge x(Fx \to \sim Gx) \to \sim \bigvee x(Fx \land Gx)$ |
| 7. | $\bigwedge x(Fx \to \sim Gx)$ |
| 8. | ~~*Show*~~ $\sim \bigvee x(Fx \land Gx)$ |
| 9. | $\bigvee x(Fx \land Gx)$ |
| 10. | $Fy \land Gy$ |
| 11. | $\sim(Fy \land Gy)$ |
| 12. | $\sim \bigvee x(Fx \land Gx) \leftrightarrow \bigwedge x(Fx \to \sim Gx)$ |

The reader should note the order of steps in this derivation, specifically, that a universal derivation is started before UI is employed, and that application of EI precedes that of UI. (This derivation is a proof of T243.)

41. Prove T208 – T214.
42. Prove T218, T219, T221.
43. Prove T223 – T225.
44. Prove T240, T242, T245.
45. Prove T246 – T248.

In solving exercises 41 – 45 (as well as later exercises) the reader will find the following informal suggestions helpful (though again not infallible).

(*1*) *To derive a formula*

$$\phi \lor \psi \to \chi$$

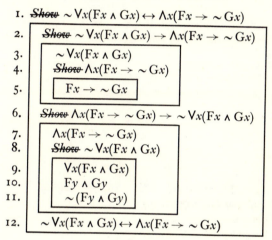

*derive first*

$$\phi \rightarrow \chi$$

*and*

$$\psi \rightarrow \chi \quad ,$$

*and then use SC.*

(2) *To derive*

$$(\phi \rightarrow \psi) \rightarrow \chi$$

*derive first*

$$\sim\phi \rightarrow \chi$$

*and*

$$\psi \rightarrow \chi \quad ,$$

*and then use Adj, T35, BC, MP. (See lines 8 – 15 of the proof of T222, p. 110, for an application of this suggestion.)*

(3) *To derive*

$$(\phi \leftrightarrow \psi) \rightarrow \chi$$

*derive first*

$$\phi \wedge \psi \rightarrow \chi$$

*and*

$$\sim\phi \wedge \sim \psi \rightarrow \chi \quad ,$$

*and then use Adj, T86, BC, MP. (See the proof of T226, p. 111, for an application of this suggestion.)*

(4) *To derive*

$$\phi \rightarrow \psi \quad ,$$

*where* $\phi$ *is neither a disjunction, a conditional, nor a biconditional, use conditional derivation.*

(5) *To derive a conjunction, derive first both conjuncts, and then use Adj.*

(6) *To derive*

$$\phi \vee \psi$$

*derive first*

$$\sim\phi \rightarrow \psi \quad ,$$

*and then use CD.*

(7) *To derive a biconditional, derive first the two corresponding conditionals, and then use CB.*

(8) *To derive a formula*

$$\bigwedge\alpha\phi \quad ,$$

*use universal derivation.*

(9) *To derive*

$$\bigvee \alpha \phi \quad ,$$

*either derive first* $\phi'$, *where* $\phi'$ *comes from* $\phi$ *by proper substitution of some variable for* $\alpha$, *and then use EG, or else use indirect derivation. (See lines 8 – 13 of the proof of T226, p. 111, for an application of the first alternative and the proof of T229, p. 112, for an application of the second.)*

(10) *To derive anything else, either use indirect derivation or separate cases.*

The reader should also review the remarks accompanying T201 (p. 103), T202 (p. 104), and T207 (p. 108).

**7. Arguments.** Let us pass now to derivations involving premises. As before, we call a symbolic argument *valid* if its conclusion is derivable from its premises, and an English argument *valid* (in the monadic quantifier calculus) if it has a valid symbolization.

Consider, for example, the argument (1) of page 85:

> All Communists are Marxists. Some Communists are American. ∴ Some Marxists are American.

Its validity is established by the following symbolization and derivation

$$\bigwedge x(Fx \rightarrow Gx) \quad . \quad \bigvee x(Fx \wedge Hx) \quad \therefore \bigvee x(Gx \wedge Hx)$$

1. ~~Show~~ $\bigvee x(Gx \wedge Hx)$

| | | |
|---|---|---|
| 2. | $Fy \wedge Hy$ | 2nd premise, EI |
| 3. | $Fy \rightarrow Gy$ | 1st premise, UI |
| 4. | $Gy$ | 2, S, 3, MP |
| 5. | $\bigvee x(Gx \wedge Hx)$ | 2, S, 4, Adj, EG |

## EXERCISES

For each of the following arguments, derive its conclusion from its premises. (The reader will again find useful the suggestions made on pages 115 – 17.)

46. $\bigwedge y(Fx \wedge Gy) \quad \therefore Fx \wedge Gx$

47. $Fx \wedge Gx \quad \therefore \bigwedge y(Fx \wedge Gy)$

48. $\bigwedge x(Fx \rightarrow \bigwedge xGx) \quad . \quad \bigwedge x(Gx \vee Hx) \rightarrow \bigwedge xJx \quad \therefore \bigwedge x(Fx \rightarrow Jx)$

49. $\bigvee x(Fx \wedge {\sim}Gx) \quad . \quad \bigwedge x(Fx \rightarrow Hx) \quad .$
$\bigwedge x(Jx \wedge Ix \rightarrow Fx) \quad . \quad \bigvee x(Hx \wedge {\sim}Gx) \rightarrow \bigwedge x(Ix \rightarrow {\sim}Hx)$
$\therefore \bigwedge x(Jx \rightarrow {\sim}Ix)$

50. $\bigwedge x(Fx \rightarrow Gx \vee Hx) \quad . \quad \bigwedge x(Gx \vee Hx \rightarrow Ix) \quad .$
${\sim}\bigvee x(Ix \wedge Gx) \quad . \quad {\sim}\bigvee xFx \rightarrow \bigvee xGx \quad \therefore \bigvee x(Fx \wedge Hx)$

51. $\bigwedge x(Fx \rightarrow Gx) \vee \bigvee x(Fx \wedge Hx)$ . $\bigwedge x(Ix \rightarrow \sim Jx \vee \sim Hx)$ . $\bigwedge x(Fx \rightarrow Ix \wedge Jx)$ . $\bigvee xFx$ ∴ $\bigvee x(Gx \wedge Fx)$

52. $\bigwedge x(Fx \rightarrow \bigwedge xGx)$ . $\bigwedge x\,(Gx \vee Hx) \rightarrow \bigvee x(Gx \wedge Ix)$ . $\bigvee xIx \rightarrow \bigwedge x(Jx \rightarrow Kx)$ ∴ $\bigwedge x(Fx \wedge Jx \rightarrow Kx)$

53. $\bigvee x(Fx \rightarrow P)$ . $\bigvee x(P \rightarrow Fx)$ ∴ $\bigvee x(Fx \leftrightarrow P)$

Show the following arguments valid by constructing symbolizations and deriving the conclusions of the symbolizations from their premises. Indicate in each case the scheme of abbreviation used. (Before symbolizing, the reader should review the remarks made in connection with the solution of exercise 19 of chapter I, p. 29. Exercises 54 and 55 substantiate claims made on page 92.)

54. For each *x* (*x* is a man and *x* is mortal). ∴ For each *x*, *x* is a man.

55. There is an object *x* such that *x* is a dog. ∴ There is an object *x* such that (if *x* is a cat, then *x* is a dog).

56. A gentleman does not prefer blondes only if he is blond. ∴ Every gentleman either is blond or prefers blondes.

57. There is not a single Communist who either likes logic or is able to construct derivations correctly. Some impartial seekers of truth are Communists. Anyone who is not able to construct derivations correctly eschews Philosophy. ∴ Some impartial seekers of truth eschew Philosophy.

58. Everyone who signed the loyalty oath is an honest citizen. If someone signed the loyalty oath and was convicted of perjury, or signed the loyalty oath and is a Communist, then not all who signed the loyalty oath are honest citizens. ∴ No one who signed the loyalty oath is a Communist.

59. All men who have either a sense of humor or the spirit of adventure seek the company of women. Anyone who seeks the company of women and has the spirit of adventure finds life exciting. Whoever gives in to temptation has the spirit of adventure. ∴ Every man who gives in to temptation finds life exciting.

60. No egghead is a good security risk. Every professor lives in an ivory tower. If there is someone who lives in an ivory tower and is not a good security risk, then no one who is either a professor or an egghead should be trusted with confidential information. ∴ If some professor is an egghead, then no professor should be trusted with confidential information.

61. No one who is either a skeptic or an atheist hates God. (For who can hate that which he doubts or believes nonexistent?) Everyone is such that if he is an atheist only if he will not go to heaven, then he is a skeptic and hates God. ∴ Everyone will go to heaven.

**8. *Fallacies.** In stating rules of inference and directions for constructing derivations, we have again imposed a number of restrictions whose significance is perhaps not immediately obvious. The restrictions are all introduced in order to prevent the validation of *false English arguments* (arguments whose premises are true sentences of English and whose conclusions are false sentences of English), that is to say, in order to prevent *fallacies*.

Of the restrictions not pertaining to variables only one needs to be mentioned here. (For the others, see section 4 of chapter I.) Accor[illegible] to clauses (3) and (4) of the directions for constructing a deri[illegible] assumption may be introduced only in the line immediatel[illegible]he inception of the derivation to which it is relevant. T[illegible]ng false argument, symbolization, and derivation esta[illegible] need for this restriction.

Argument:

Something is wise. ∴ Everything is wise.

Symbolization:

⋁*x*F*x* ∴ ⋀*x*F*x*

Derivation:

| | | |
|---|---|---|
| 1. | ~~*Show*~~ ⋀*x*F*x* | |
| 2. | ~~*Show*~~ F*x* | |
| 3. | ~~*Show*~~ ⋀*x* ~ F*x* | |
| 4. | ~ F*x* | Fallacious assumption for indirect derivation |
| 5. | ⋁*x*F*x* | Premise |
| 6. | F*y* | 5, EI |
| 7. | ~ F*y* | 3, UI |

In this derivation line 4 is incorrect; if line 2 is to begin an indirect derivation, its assumption must occur as the next line. The reader can easily construct for himself a similar example involving a fallacious assumption for conditional derivation.

The restriction imposed on the variable of instantiation of EI, that it be entirely new to the derivation, is more stringent than necessary. It would be sufficient, but tediously complex, to impose the following three prohibitions: (i) the variable of instantiation may not occur *free* in an antecedent line; (ii) it may not occur *bound* in preceding lines containing uncancelled '*Show*'; and (iii) it may not occur *free* in preceding lines containing uncancelled '*Show*'. The following argument, symbolization, and derivation show the need for (i), and exercises 62 and 63 the need for (ii) and (iii).

Argument:

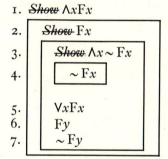

There is an even number. There is an odd number. ∴ Something is both an even number and an odd number.

Symbolization:

$$\bigvee x Fx \quad . \quad \bigvee x Gx \quad \therefore \bigvee x(Fx \wedge Gx)$$

Derivation:

| | | |
|---|---|---|
| 1. | ~~*Show*~~ $\bigvee x(Fx \wedge Gx)$ | |
| 2. | $\bigvee x Fx$ | Premise |
| 3. | $Fy$ | 2, EI |
| 4. | $\bigvee x Gx$ | Premise |
| 5. | $Gy$ | 4, EI |
| 6. | $\bigvee x(Fx \wedge Gx)$ | 3, 5, Adj, EG |

In this derivation line 5 is fallaciously inferred from the second premise. The variable of instantiation, '*y*', is free in line 3, which is antecedent to line 5.

The restriction imposed on the variable of generalization in a universal derivation prohibits a free occurrence of that variable in lines antecedent to that derivation. The need for this restriction is shown by the following argument, symbolization, and derivation.

Argument:

There is an odd number. ∴ Everything is an odd number.

Symbolization:

$$\bigvee x Fx \quad \therefore \bigwedge x Fx$$

Derivation:

| | | |
|---|---|---|
| 1. | ~~*Show*~~ $\bigwedge x Fx$ | |
| 2. | $\bigvee x Fx$ | Premise |
| 3. | $Fy$ | 2, EI |
| 4. | ~~*Show*~~ $\bigwedge y Fy$ | |
| 5. | $Fy$ | 3, R |
| 6. | $Fx$ | 4, UI |

In this derivation the cancellation of '*Show*' in line 4, along with the boxing of line 5, is incorrect. The variable of generalization in the subsidiary universal derivation, '*y*', is free in line 3, which is antecedent to the subsidiary derivation.

The notion of ***proper substitution*** is involved in the formulation of each of the rules UI, EG, and EI. In order for $\psi$ to come from a formula $\phi$ by proper substitution of a variable $\beta$ for a variable $\alpha$, two conditions must be fulfilled: (i) *every* free occurrence of $\alpha$ must be replaced by an occurrence of $\beta$, and (ii) each of the occurrences of $\beta$ that replaces an occurrence of $\alpha$ must be *free*. The false argument (1) below shows the need for (i) with respect to UI; exercises 64 and 65, the need for (i) with respect to EG and EI; the false argument (2), the need for (ii) with respect to EG; and exercise 66, the need for (ii) with respect to UI. The requirement for EI that the variable of instantiation be new already makes impossible a legitimate application of that rule in which a free occurrence of $\alpha$ is replaced by a bound occurrence of $\beta$.

Argument:

(1) All dogs are animals. There is a dog. ∴ Everything is an animal.

Symbolization:

$$\bigwedge x(Fx \to Gx) \quad . \quad \bigvee xFx \quad \therefore \bigwedge xGx$$

Derivation:

| | | |
|---|---|---|
| 1. | ~~*Show*~~ $\bigwedge xGx$ | |
| 2. | $\bigvee xFx$ | Premise |
| 3. | $Fy$ | 2, EI |
| 4. | $\bigwedge x(Fx \to Gx)$ | Premise |
| 5. | $(Fy \to Gx)$ | 4, incorrect application of UI |
| 6. | $Gx$ | 3, 5, MP |

In this derivation line 5 comes from line 4 incorrectly; not *all* free occurrences of '$x$' in '$(Fx \to Gx)$' have been replaced by '$y$'.

Argument:

(2) There is a human. ∴ Everything is human.

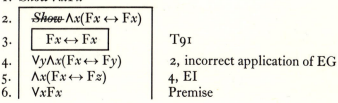

Symbolization:

$$\bigvee xFx \quad \therefore \bigwedge xFx$$

Derivation:

| | | |
|---|---|---|
| 1. | ~~*Show*~~ $\bigwedge xFx$ | |
| 2. | ~~*Show*~~ $\bigwedge x(Fx \leftrightarrow Fx)$ | |
| 3. | $Fx \leftrightarrow Fx$ | T91 |
| 4. | $\bigvee y\bigwedge x(Fx \leftrightarrow Fy)$ | 2, incorrect application of EG |
| 5. | $\bigwedge x(Fx \leftrightarrow Fz)$ | 4, EI |
| 6. | $\bigvee xFx$ | Premise |

| | | |
|---|---|---|
| 7. | F$w$ | 6, EI |
| 8. | F$z$ | 5, UI, BC, 7, MP |
| 9. | F$x$ | 5, UI, BC, 8, MP |

In this derivation line 4 comes from line 2 incorrectly; the occurrence of '$x$' that replaces the free occurrence of '$y$' in '$\bigwedge x(Fx \rightarrow Fy)$' is not *free.*

It might appear that fallacies can occur even if all the restrictions imposed in this chapter on the process of derivation are observed, for consider the argument

> '$a$' is the first letter of the alphabet. ∴ '$x$' is the first letter of the alphabet .

This argument is certainly false, and can be considered (making allowance for vacuous quantification) a stylistic variant of

> '$a$' is the first letter of the alphabet. ∴ There is an object $x$ such that '$x$' is the first letter of the alphabet .

Further, on the basis of the scheme

(3) F : '$a$' is the first letter of the alphabet ,

the latter argument seems to be a translation of the valid symbolic argument

$$Fa \quad \therefore \bigvee xFx \quad .$$

The fault in this line of reasoning is that (3) is not a scheme of abbreviation: the formula

> '$a$' is the first letter of the alphabet

contains an apparent variable, as was observed on page 92.

### EXERCISES

For each of the following false arguments annotate the derivation accompanying its symbolization, indicating the fallacious step and the error that led to the fallacy.

62. Something is wise. ∴ Everything is wise.

Symbolization:

$$\bigvee yFy \quad \therefore \bigwedge xFx$$

Derivation:

1. ~~*Show*~~ $\bigwedge xFx$
2. $\bigvee yFy$
3. F$x$

63. There is a human. ∴ Everything is human.

Symbolization:

$$\bigvee x Fx \quad \therefore \bigwedge x Fx$$

Derivation:

1. ~~*Show*~~ $\bigwedge x Fx$
2. ~~*Show*~~ $\bigwedge x(Fx \leftrightarrow Fy)$
3. $(Fx \leftrightarrow Fx)$
4. $\bigvee z(Fx \leftrightarrow Fz)$
5. $Fx \leftrightarrow Fy$
6. $\bigvee x Fx$
7. $Fw$
8. $Fy$
9. $Fx$

64. There is an even number. There is an odd number. ∴ Something is both an even number and an odd number.

Symbolization:

$$\bigvee x Fx \quad . \quad \bigvee x Gx \quad \therefore \bigvee x(Fx \wedge Gx)$$

Derivation:

1. ~~*Show*~~ $\bigvee x(Fx \wedge Gx)$
2. $\bigvee x Fx$
3. $Fy$
4. $\bigvee x Gx$
5. $Fz$
6. $Fy \wedge Fz$
7. $\bigvee y(Fy \wedge Gy)$
8. $Fw \wedge Gw$
9. $\bigvee x(Fx \wedge Gx)$

65. There is a prime number. ∴ Everything is a number.

Symbolization:

$$\bigvee x(Fx \wedge Gx) \quad \therefore \bigwedge x Gx$$

Derivation:

1. ~~*Show*~~ $\bigwedge x Gx$
2. $\bigvee x(Fx \wedge Gx)$
3. $Fy \wedge Gx$
4. $Gx$

66. Something is human. Something is not human. ∴ Something is such that it is human just in case it is not human.

Symbolization:

$$\bigvee x Fx \quad . \quad \bigvee x \sim Fx \quad \therefore \bigvee x(Fx \leftrightarrow \sim Fx)$$

Derivation:

1. ~~Show~~ $\bigvee x(Fx \leftrightarrow \sim Fx)$
2. ~~Show~~ $\bigwedge y \bigvee x(Fx \leftrightarrow \sim Fy)$
3. $\sim \bigwedge y \bigvee x(Fx \leftrightarrow \sim Fy)$
4. $\bigvee x Fx$
5. $Fz$
6. $\bigvee x \sim Fx$
7. $\sim Fw$
8. $\sim \bigvee x(Fx \leftrightarrow \sim Fu)$
9. $\bigwedge x \sim (Fx \leftrightarrow \sim Fu)$
10. $\sim \sim Fu$
11. $Fw$
12. $\bigvee x(Fx \leftrightarrow \sim Fx)$

67. It is not the case that everything is male. ∴ Every male is female.

Symbolization:

$$\sim \bigwedge x Fx \quad \therefore \bigwedge x(Fx \rightarrow Gx)$$

Derivation:

1. ~~Show~~ $\bigwedge x(Fx \rightarrow Gx)$
2. ~~Show~~ $Fx \rightarrow Gx$
3. $Fx$
4. $\sim \bigwedge x Fx$
5. $\bigwedge x Fx \rightarrow Gx$
6. $Fx \rightarrow Gx$
7. $Gx$

(In connection with this exercise the reader should review the comments on parentheses in section 3, p. 90.)

**9.** * **Invalidity.** In chapter II truth tables were used to show arguments invalid. The technique employed there can be extended to a method for showing invalidity within the present symbolic language. We shall call the new method that of *truth-functional expansions.*

Suppose we are presented with a symbolic argument whose premises and conclusion are sentences. To obtain a *truth-functional expansion* of such an argument, proceed as follows:

1. Choose a sequence of distinct variables, say $\alpha_1, \ldots, \alpha_n$, which do not occur in the given argument.

2. Throughout the argument replace each occurrence of a formula

$$\bigwedge \alpha \phi$$

or

$$\bigvee \alpha \phi \quad ,$$

where $\alpha$ is a variable and $\phi$ a formula, by

$$\phi_1 \wedge \ldots \wedge \phi_n$$

or

$$\phi_1 \vee \ldots \vee \phi_n$$

respectively, where $\phi_1$ comes from $\phi$ by proper substitution of $\alpha_1$ for $\alpha$, $\phi_2$ comes from $\phi$ by proper substitution of $\alpha_2$ for $\alpha$, and so on.

3. Throughout the resulting argument replace formulas of the form

$$\pi\, \alpha \quad ,$$

where $\pi$ is a predicate letter and $\alpha$ is a variable, by sentence letters not occurring in the argument. The replacement must be *biunique;* that is, distinct formulas are to pass into distinct sentence letters.

For example, consider the argument

(1) $\bigvee x\mathrm{F}x \to \bigvee x\mathrm{G}x \quad \therefore \bigwedge x(\mathrm{F}x \to \mathrm{G}x) \quad .$

In the first step toward obtaining a truth-functional expansion, we may choose the variables '$a$', '$b$'. In the second step, the argument (1) becomes

$$\mathrm{F}a \vee \mathrm{F}b \to \mathrm{G}a \vee \mathrm{G}b \quad \therefore (\mathrm{F}a \to \mathrm{G}a) \wedge (\mathrm{F}b \to \mathrm{G}b) \quad ;$$

and in the third step,

(2) $\mathrm{P} \vee \mathrm{Q} \to \mathrm{R} \vee \mathrm{S} \quad \therefore (\mathrm{P} \to \mathrm{R}) \wedge (\mathrm{Q} \to \mathrm{S}) \quad .$

The argument (1) has, of course, truth-functional expansions other than (2). The significant differences from (2) would result from a choice of more or fewer variables in step 1.

It happens that a symbolic argument whose premises and conclusion are sentences is valid if and only if, in each of its truth-functional expansions, the premises tautologically imply the conclusion. Thus to show such an argument invalid it is sufficient to exhibit a truth-functional expansion of it, together with a truth table showing that the premises of that expansion do not tautologically imply its conclusion.

For instance, the argument (1) is seen to be invalid in view of the truth-functional expansion (2) and the line of a truth table for (2) that follows:

| P | Q | R | S | P ∨ Q | R ∨ S | P → R | Q → S | P ∨ Q → R ∨ S | (P → R) ∧ (Q → S) |
|---|---|---|---|---|---|---|---|---|---|
| *T* | *T* | *F* | *T* | *T* | *T* | *F* | *T* | *T* | *F* |

With different truth-functional expansions of a given argument, the test for tautological implication will, of course, in general lead to different results. It happens, however, that if we obtain tautological implication for a truth-functional expansion reached by choosing $2^n$ variables in step 1, where $n$ is the number of different predicates and sentence letters appearing in the given argument, then that argument is valid. Thus the method of truth-functional expansions provides not only a means of showing invalidity but also an automatic test of validity for arguments of the kind under consideration.

The method given above is directly applicable only to those symbolic arguments whose premises and conclusions are sentences. We can, however, extend the method to arbitrary symbolic arguments.

We say that a formula $\psi$ is a *closure of a formula* $\phi$ just in case $\psi$ is a sentence and either $\psi$ is $\phi$ or $\psi$ is a universal generalization of $\phi$; and we understand by a *closure of an argument* any argument obtained from it by replacing each formula comprised in it by a closure of that formula.

It happens that a symbolic argument is valid just in case each of its closures is valid, and furthermore that one closure of it is valid if and only if any other is. Thus to test an arbitrary symbolic argument for validity it is sufficient to construct a closure of that argument and to test the closure for validity.

Another method of showing invalidity is suggested by the discussion of fallacies in the preceding section. It happens that if a symbolic argument has a false English translation it cannot be valid. The method of showing invalidity based on this principle, unlike the method of truth-functional expansions, does not directly give rise to an automatic test for validity and, in addition, suffers from several drawbacks that will be discussed in chapter IV.

### EXERCISES

Using the method of truth-functional expansions show the following arguments invalid. (From the invalidity of (1) on page 125 and exercises 68 – 78 it follows that the converses of T201, T202, T209 – T214, T223 – T226 are not theorems.)

68. $\therefore (\bigwedge x Fx \to \bigwedge x Gx) \to \bigwedge x(Fx \to Gx)$

69. $\therefore \bigvee x Fx \wedge \bigvee x Gx \to \bigvee x(Fx \wedge Gx)$

70. $\therefore \bigwedge x(Fx \vee Gx) \to (\bigwedge x Fx \vee \bigwedge x Gx)$

71. $\therefore \bigvee x(Fx \to Gx) \to (\bigvee xFx \to \bigvee xGx)$

72. $\therefore \bigvee x(Fx \to Gx) \to (\bigwedge xFx \to \bigwedge xGx)$

73. $\therefore (\bigwedge xFx \leftrightarrow \bigwedge xGx) \to \bigwedge x(Fx \leftrightarrow Gx)$

74. $\therefore (\bigvee xFx \leftrightarrow \bigvee xGx) \to \bigwedge x(Fx \leftrightarrow Gx)$

75. $\therefore (\bigwedge xFx \leftrightarrow P) \to \bigwedge x(Fx \leftrightarrow P)$

76. $\therefore (\bigvee xFx \leftrightarrow P) \to \bigwedge x(Fx \leftrightarrow P)$

77. $\therefore \bigvee x(Fx \leftrightarrow P) \to (\bigvee xFx \leftrightarrow P)$

78. $\therefore \bigvee x(Fx \leftrightarrow P) \to (\bigwedge xFx \leftrightarrow P)$

For each of the following arguments either derive the conclusion from the premises or show it invalid by the method of truth-functional expansions.

79. $\bigwedge x(Fx \to P) \quad \therefore \bigwedge xFx \to P$

80. $\bigwedge xFx \to P \quad \therefore \bigwedge x(Fx \to P)$

81. $\bigvee x(Fx \to P) \quad \therefore \bigvee xFx \to P$

82. $\bigvee xFx \to P \quad \therefore \bigvee x(Fx \to P)$

83. $\bigwedge xFx \quad \therefore Fx$

84. $Fx \quad \therefore \bigwedge xFx$

85. $\therefore Fx \leftrightarrow \bigwedge xFx$

86. $\bigwedge x(Fx \to Lx \vee Kx)$ . $\bigvee x(Gx \wedge \sim Hx \wedge \sim Jx)$ . $\bigwedge x(Lx \to \sim Ix)$ . $\bigvee x(Kx \wedge Jx)$ . $\bigwedge x(Kx \to Ix)$ . $\bigwedge x(\sim Fx \vee \sim Gx \to \sim Lx \wedge \sim Kx) \quad \therefore \sim \bigvee x(Lx \wedge Hx)$

Give an interesting symbolization of each of the following arguments, and for each symbolization either derive the conclusion from the premises or show it invalid by the method of truth-functional expansions. (In this connection the reader should review the remarks on page 77, preceding exercise 53 of chapter II.)

87. All men are mortal. ∴ Some men are mortal.

88. All dogs are animals. All mammals are animals. ∴ All dogs are mammals.

89. Dogs are expensive if and only if they are pedigreed. ∴ If there is a dog, then something is expensive if and only if something is pedigreed.

90. If none but the gracious deserve the respect of their countrymen, then Coriolanus deserved his fate. Only the magnanimous deserve the respect of their countrymen. All who are magnanimous are gracious. ∴ Coriolanus deserved his fate.

91. If none but the gracious deserve the respect of their countrymen, then Coriolanus deserved his fate. All who are magnanimous deserve the respect of their countrymen. Only the gracious are magnanimous. ∴ Coriolanus deserved his fate.

**10. *Historical remarks.** What we call the quantifier calculus has also been called the functional calculus, the predicate calculus, the first-order functional calculus, the lower predicate calculus, and the restricted predicate calculus.

The essential ideas of the quantifier calculus occur first in Frege [1]. The quantifier calculus was extensively developed, though in a rather informal way, in Whitehead and Russell [1] and received its first completely explicit formulation, containing one error corrected in a later edition and another corrected in Pager [1], in Hilbert and Ackermann [1].

The symbols used here are due to Tarski. The quantificational symbols most frequent in the literature are those of Whitehead and Russell [1]:

'$(x)$' for '$\bigwedge x$'
'$(\exists x)$' for '$\bigvee x$' .

The first automatic test of validity for the monadic quantifier calculus was given in Behmann [1]. The present test is based on a result in Bernays and Schönfinkel [1].

Our formulation of the quantifier calculus admits of one simplification in addition to those indicated in connection with the sentential calculus. The general form of universal derivation may be replaced by the simple form (see p. 101).

**11. Appendix: list of theorems of chapter III.**

LAWS OF DISTRIBUTION:

T201 $\bigwedge x(Fx \rightarrow Gx) \rightarrow (\bigwedge xFx \rightarrow \bigwedge xGx)$

T202 $\bigwedge x(Fx \rightarrow Gx) \rightarrow (\bigvee xFx \rightarrow \bigvee xGx)$

T207 $\bigvee x(Fx \vee Gx) \leftrightarrow \bigvee xFx \vee \bigvee xGx$

T208 $\bigwedge x(Fx \wedge Gx) \leftrightarrow \bigwedge xFx \wedge \bigwedge xGx$

T209 $\bigvee x(Fx \wedge Gx) \rightarrow \bigvee xFx \wedge \bigvee xGx$

T210 $\bigwedge xFx \vee \bigwedge xGx \rightarrow \bigwedge x(Fx \vee Gx)$

T211 $(\bigvee xFx \rightarrow \bigvee xGx) \rightarrow \bigvee x(Fx \rightarrow Gx)$

T212 $(\bigwedge xFx \rightarrow \bigwedge xGx) \rightarrow \bigvee x(Fx \rightarrow Gx)$

T213 $\bigwedge x(Fx \leftrightarrow Gx) \rightarrow (\bigwedge xFx \leftrightarrow \bigwedge xGx)$

T214 $\bigwedge x(Fx \leftrightarrow Gx) \rightarrow (\bigvee xFx \leftrightarrow \bigvee xGx)$

LAWS OF QUANTIFIER NEGATION:

T203 $\sim \bigwedge x Fx \leftrightarrow \bigvee x \sim Fx$

T204 $\sim \bigvee x Fx \leftrightarrow \bigwedge x \sim Fx$

T205 $\bigwedge x Fx \leftrightarrow \sim \bigvee x \sim Fx$

T206 $\bigvee x Fx \leftrightarrow \sim \bigwedge x \sim Fx$

LAWS OF CONFINEMENT:

T215 $\bigwedge x(P \wedge Fx) \leftrightarrow P \wedge \bigwedge x Fx$

T216 $\bigvee x(P \wedge Fx) \leftrightarrow P \wedge \bigvee x Fx$

T217 $\bigwedge x(P \vee Fx) \leftrightarrow P \vee \bigwedge x Fx$

T218 $\bigvee x(P \vee Fx) \leftrightarrow P \vee \bigvee x Fx$

T219 $\bigwedge x(P \to Fx) \leftrightarrow (P \to \bigwedge x Fx)$

T220 $\bigvee x(P \to Fx) \leftrightarrow (P \to \bigvee x Fx)$

T221 $\bigwedge x(Fx \to P) \leftrightarrow (\bigvee x Fx \to P)$

T222 $\bigvee x(Fx \to P) \leftrightarrow (\bigwedge x Fx \to P)$

T223 $\bigwedge x(Fx \leftrightarrow P) \to (\bigwedge x Fx \leftrightarrow P)$

T224 $\bigwedge x(Fx \leftrightarrow P) \to (\bigvee x Fx \leftrightarrow P)$

T225 $(\bigvee x Fx \leftrightarrow P) \to \bigvee x(Fx \leftrightarrow P)$

T226 $(\bigwedge x Fx \leftrightarrow P) \to \bigvee x(Fx \leftrightarrow P)$

LAWS OF VACUOUS QUANTIFICATION:

T227 $\bigwedge x P \leftrightarrow P$

T228 $\bigvee x P \leftrightarrow P$

T229 $\bigvee x(\bigvee x Fx \to Fx)$

T230 $\bigvee x(Fx \to \bigwedge x Fx)$

LAWS OF ALPHABETIC VARIANCE:

T231 $\bigwedge x Fx \leftrightarrow \bigwedge y Fy$

T232 $\bigvee x Fx \leftrightarrow \bigvee y Fy$

T233 $(Fx \to Gx) \wedge (Gx \to Hx) \to (Fx \to Hx)$

| | |
|---|---|
| T234 | $\bigwedge x[(Fx \rightarrow Gx) \wedge (Gx \rightarrow Hx) \rightarrow (Fx \rightarrow Hx)]$ |
| T235 | $\bigwedge x(Fx \rightarrow Gx) \wedge \bigwedge x(Gx \rightarrow Hx) \rightarrow \bigwedge x(Fx \rightarrow Hx)$ |
| T236 | $\bigwedge x(Fx \leftrightarrow Gx) \wedge \bigwedge x(Gx \leftrightarrow Hx) \rightarrow \bigwedge x(Fx \leftrightarrow Hx)$ |
| T237 | $\bigwedge x(Fx \rightarrow Gx) \wedge \bigwedge x(Fx \rightarrow Hx) \rightarrow \bigwedge x(Fx \rightarrow Gx \wedge Hx)$ |
| T238 | $\bigwedge xFx \rightarrow \bigvee xFx$ |
| T239 | $\bigwedge xFx \wedge \bigvee xGx \rightarrow \bigvee x(Fx \wedge Gx)$ |
| T240 | $\bigwedge x(Fx \rightarrow Gx) \wedge \bigvee x(Fx \wedge Hx) \rightarrow \bigvee x(Gx \wedge Hx)$ |
| T241 | $\bigwedge x(Fx \rightarrow Gx \vee Hx) \rightarrow \bigwedge x(Fx \rightarrow Gx) \vee \bigvee x(Fx \wedge Hx)$ |
| T242 | $\sim \bigwedge x(Fx \rightarrow Gx) \leftrightarrow \bigvee x(Fx \wedge \sim Gx)$ |
| T243 | $\sim \bigvee x(Fx \wedge Gx) \leftrightarrow \bigwedge x(Fx \rightarrow \sim Gx)$ |
| T244 | $\sim \bigvee xFx \rightarrow \bigwedge x(Fx \rightarrow Gx)$ |
| T245 | $\sim \bigvee xFx \leftrightarrow \bigwedge x(Fx \rightarrow Gx) \wedge \bigwedge x(Fx \rightarrow \sim Gx)$ |
| T246 | $\bigwedge xFx \leftrightarrow \bigwedge x\bigwedge y(Fx \wedge Fy)$ |
| T247 | $\bigvee xFx \wedge \bigvee xGx \rightarrow (\bigwedge x[Fx \rightarrow Hx] \wedge \bigwedge x[Gx \rightarrow Jx] \leftrightarrow \bigwedge x\bigwedge y[Fx \wedge Gy \rightarrow Hx \wedge Jy])$ |
| T248 | $(\bigvee xFx \leftrightarrow \bigvee xGx) \wedge \bigwedge x\bigwedge y(Fx \wedge Gy \rightarrow [Hx \leftrightarrow Jy]) \rightarrow (\bigwedge x[Fx \rightarrow Hx] \leftrightarrow \bigwedge x[Gx \rightarrow Jx])$ |

# Chapter IV

# 'ALL' and 'SOME', *continued*

**1. Terms and formulas.** There are arguments whose validity depends essentially on phrases of quantity and their stylistic variants, but which cannot be reached by the theory of chapter III. For example, the arguments

(1) All men are mortal. Socrates is a man. Therefore Socrates is mortal,

(2) All horses are animals. Therefore every head of a horse is a head of an animal,

though intuitively valid, cannot be validated by the methods of the previous chapter. Before extending the apparatus of chapter III to cover such cases, we must supplement the grammatical considerations given there.

Of *sentences of English*, *formulas of English*, and *names of English* we have already spoken (pp. 85 and 86). There are expressions that resemble names in the manner in which formulas resemble sentences. For example,

(3) $7 + x$

is related to the names '$7 + 5$' and '$7 + 6$' but is not itself a name of English; it designates neither 12, 13, nor any other object. Expressions such as (3) are called *terms*. As a limiting case, a name itself will also be regarded as a term. Accordingly, a *term of English* is either a name of English or an expression containing occurrences of variables that becomes a name of English when some or all of these occurrences are replaced by names of English. Each of the following, then, is a term of English:

(4) Sir Walter Scott,
(5) the native state of $z$,
(6) the square of $x$,
(7) the product of $x$ and $y$;

for (4) is a name, and by replacing '$z$' by 'Lincoln', '$x$' by '2', and '$y$' by '3', (5) – (7) become, respectively, the names

the native state of Lincoln,
the square of 2,
the product of 2 and 3,

which designate Kentucky, 4, and 6, respectively.

In addition to terms of English, we admit abbreviations of such terms, constructed with the help of *operation letters*. The symbols

$$A^0 ,\quad B^0 ,\quad C^0 ,\quad D^0 ,\quad E^0$$

are *0-place operation letters;*

$$A^1 ,\ \ldots,\ E^1$$

are *1-place operation letters;* and

$$A^2 ,\ \ldots,\ E^2$$

are *2-place operation letters*. Additional *0-place*, *1-place*, or *2-place operation letters* may be obtained by adding numerical subscripts to the symbols above. For any nonnegative integer *k*, we may characterize *k-place operation letters* in a similar way. Thus an *operation letter* consists of one of the letters 'A' through 'E' accompanied by either a numerical superscript and subscript or only a superscript. The superscript indicates the number of places of the operation letter.

Names of English will be abbreviated by 0-place operation letters. Other terms of English will be symbolized by operation letters together with variables. For instance, the expressions

(8) $A^0$ ,
(9) $B^1z$ ,
(10) $C^1x$ ,
(11) $D^2xy$

may be construed as symbolizing the respective terms (4) – (7).

As in chapter II, we shall admit abbreviations of formulas of English, but here we shall provide means of abbreviating formulas with arbitrarily many variables. To this end, we introduce *predicate letters* of arbitrarily many places in analogy with the operation letters introduced above. The symbols

$$\begin{array}{c} F^0 \ , \\ \cdot \\ \cdot \\ \cdot \\ Z^0 \ , \end{array}$$

with or without subscripts, are *0-place predicate letters;*

$$\begin{array}{c} F^1 \ , \\ \cdot \\ \cdot \\ \cdot \\ Z^1 \ , \end{array}$$

again with or without subscripts, are *1-place predicate letters;* and so on. Thus a *predicate letter* consists of one of the letters 'F' through 'Z' accompanied by either a superscript and subscript or only a superscript; the number of places of the letter in question is denoted by its superscript.

0-place predicate letters will assume the role played by sentence letters in earlier chapters, and 1-place predicate letters that of the predicate letters of chapter III. Formulas of English containing $k$ variables may be symbolized by $k$-place predicate letters accompanied by those variables. For instance, the formulas

$$x \text{ loves } y$$

and

$$x \text{ lies between } y \text{ and } z$$

may be symbolized by

$$F^2xy$$

and

$$G^3xyz$$

respectively.

The extended *language* with which we are now concerned is obtained by adding to English the following symbols:

(1) the sentential connectives;
(2) parentheses;
(3) the quantifiers, that is, '$\wedge$' and '$\vee$';

(4) variables, that is, lower-case letters with or without subscripts;
(5) predicate letters; and
(6) operation letters.

As *terms* we reckon terms of English, their symbolic counterparts, and mixed combinations. To be more explicit, the class of terms can be exhaustively characterized as follows:

(*i*) *Terms of English (that is, names of English or expressions that are like names of English except for the occurrence of variables) are terms. (In particular, variables are terms.)*

(*ii*) *The result of writing a* k*-place operation letter followed by* k *terms is a term. (In particular, each* 0*-place operation letter is itself a term.)*

By *formulas* we understand formulas of English, their symbolic counterparts, and mixed combinations. Thus the class of formulas can be exhaustively characterized as follows:

(*1*) *All formulas of English (that is, sentences of English or expressions like sentences of English except for the occurrence of variables) are formulas.*

(*2*) *The result of writing a* k*-place predicate letter followed by* k *terms is a formula. (In particular, a* 0*-place predicate letter is itself a formula.)*

(*3*) *If* $\phi$ *and* $\psi$ *are formulas, so are*

$$\begin{array}{l} \sim\phi \ , \\ (\phi \rightarrow \psi) \ , \\ (\phi \wedge \psi) \ , \\ (\phi \vee \psi) \ , \\ (\phi \leftrightarrow \psi) \ . \end{array}$$

(*4*) *If* $\phi$ *is a formula and* $\alpha$ *a variable, then*

$$\begin{array}{l} \bigwedge\alpha\phi \ , \\ \bigvee\alpha\phi \end{array}$$

*are formulas.*

The *symbolic terms and formulas* of our present language are those terms and formulas that can be constructed exclusively from variables, predicate letters, operation letters, parentheses, sentential connectives, and quantifiers. More precisely, the class of *symbolic terms* can be exhaustively characterized as follows:

(*a*) *All variables are symbolic terms.*

(*b*) *The result of writing a* k*-place operation letter followed by* k *symbolic terms is a symbolic term. (In particular, a* 0*-place operation letter is itself a symbolic term.)*

The class of *symbolic formulas* can be exhaustively characterized as follows:

(*A*) *The result of writing a* k*-place predicate letter followed by* k *symbolic terms is a symbolic formula.* (*In particular, a* 0*-place predicate letter is itself a symbolic formula.*)

(*B*) *If* $\phi$ *and* $\psi$ *are symbolic formulas, so are*

$$\sim\phi \quad ,$$
$$(\phi \rightarrow \psi) \quad ,$$
$$(\phi \wedge \psi) \quad ,$$
$$(\phi \vee \psi) \quad ,$$
$$(\phi \leftrightarrow \psi) \quad .$$

(*C*) *If* $\phi$ *is a symbolic formula and* $\alpha$ *a variable, then*

$$\bigwedge\alpha\phi \quad ,$$
$$\bigvee\alpha\phi$$

*are symbolic formulas.*

The only symbolic terms occurring in the language of chapter III were variables. We now countenance symbolic terms such as (8) – (11) above, as well as more complex terms like

(12) $$C^1 D^2 xy$$

and

(13) $$D^2 C^1 x y \quad ,$$

which may respectively be regarded as symbolizing

the square of the product of *x* and *y*

and

the product of $x^2$ and *y*.

To see that (12) is a term, we observe first that '*x*' and '*y*' are terms (clause (a)), and hence (clause (b)) that '$D^2xy$' is a term. Applying clause (b) again we see that '$C^1 D^2xy$' is a term. The analysis of (13) is slightly different; this time we consider the following sequence of terms: '*x*', '$C^1x$', '*y*', '$D^2 C^1x y$'.

The class of symbolic formulas also expands. Clause (A) leads not only to formulas like

$$P^0$$

and

$$F^1x \quad ,$$

but also to formulas like

(14) $G^2xy$ ,
(15) $F^1A^0$ ,
(16) $G^2A^0x$ ,
(17) $G^2\,x\,B^1y$ ;

and clauses (B) and (C) provide us with compounds of these.

## EXERCISES

For each of the following expressions, state whether it is (a) a term, (b) a formula, (c) a symbolic term, (d) a symbolic formula, or (e) no one of (a) – (d). Exercise 1 is solved for illustration.

1. $(\Lambda xF^2xA^0 \rightarrow G^1A^1B^1x)$

This expression is a formula and also a symbolic formula. We refer to the clauses characterizing terms and formulas (p. 134). '$x$' is a term (clause (i)) and '$A^0$' is a term (clause (ii)); hence '$F^2xA^0$' is a formula (clause (2)) and '$B^1x$' is a term (clause (ii)). Since '$B^1x$' is a term, '$A^1B^1x$' is a term (clause (ii)), and hence '$G^1A^1B^1x$' is a formula (clause (2)). Since '$F^2xA^0$' is a formula, so is '$\Lambda xF^2xA^0$' (clause (4)). Finally, since '$\Lambda xF^2xA^0$' and '$G^1A^1B^1x$' are formulas, No. 1 is a formula (clause (3)). In addition, since it contains only predicate letters, operation letters, parentheses, sentential connectives, and quantifiers, it is a symbolic formula.

2. The individual identical with the author of *Waverley*
3. $A^1$ the father of $x$
4. the son of $x$ loves the daughter of $y$
5. $\Lambda A^0(F^1A^0 \rightarrow G^1A^0)$
6. $A^1x_1B^2yz$
7. the square of the product of $x$
8. $G^3C^1yD^1xA^1A^0$
9. $E^1F^1x$
10. $\sim(\mathrm{V}x(\Lambda y(F^2yA^0 \rightarrow G^2xB^0) \rightarrow H^1A^1B^3A^0xB^0) \leftrightarrow F^1z)$

**2. Bondage and freedom.** With the addition of terms other than variables to our language, we must reconsider the notions of bondage and freedom; they are now to apply not only to variables but to arbitrary symbolic terms.

An *occurrence* of a symbolic term $\zeta$ is said to be *bound in* a symbolic formula $\phi$ if it stands within an occurrence in $\phi$ of a formula

$$\Lambda\alpha\psi$$

or

$$\mathrm{V}\alpha\psi \quad ,$$

where $\psi$ is a formula and $\alpha$ is a variable occurring in $\zeta$; an *occurrence* of a symbolic term is *free in* $\phi$ if and only if it stands within $\phi$ but is not bound in $\phi$. A symbolic *term* is itself *bound* or *free in* $\phi$ according as there is a bound or free occurrence of it in $\phi$.

For example, consider the formula

$$(\bigvee x\ F^1A^1x \lor G^1A^1x) \quad .$$

In this formula the first occurrence of the term '$A^1x$' is bound, the second occurrence is free, and the term itself is both bound and free.

The reader should note that bondage and freedom of occurrences of variables and of variables, as characterized in the preceding chapter, are simply special cases of the present notions, which apply to arbitrary symbolic terms and their occurrences.

As in chapter III, a *symbolic sentence* may be characterized as a symbolic formula in which no variable is free. We can also characterize a *symbolic name* as a symbolic term in which no variable occurs.

The reader should observe that the notions of bondage and freedom apply only to *symbolic* terms and formulas; as before, applications of these notions to nonsymbolic expressions will not be required for the formulation of our deductive system.

## EXERCISES

Consider the following formula:

$$(\bigwedge y\ (\bigvee x\ G^2\ A^0\ A^1x \lor G^2\ B^2yz\ A^1x) \rightarrow \bigwedge z\ (G^2\ B^1y\ z \lor G^1\ B^2yz))$$

11. In this formula identify each occurrence of a term as bound or free.

12. Which terms are bound in the formula? Which terms are free in the formula?

**3. Informal notational conventions.** We shall again employ the conventions of chapter II (pp. 40, 51, and 65) for omitting parentheses and replacing them by brackets.

In addition, it will be our usual practice to omit superscripts from operation letters and predicate letters, inserting parentheses and brackets to avoid ambiguity. In the case of 0-place operation letters, and 0-place and 1-place predicate letters, no parentheses or brackets will be employed. In the case of operation letters of one or more places, and predicate letters of two or more places, we shall enclose the terms accompanying the letter in question in a pair of parentheses or brackets. For instance, the terms (12) and (13) of section 1 (p. 135) may become

$$C[D(xy)] \quad ,$$
$$D[C(x)y]$$

respectively, and the formulas (14) – (17) of page 136 may become

$$\begin{array}{l} G(xy) \quad , \\ F\,A \quad , \\ G(A\,x) \quad , \\ G[x\,B(y)] \end{array}$$

respectively. Restoration of official notation is completely automatic; the superscript of a letter is determined by the sequence of terms following it.

Again, we must emphasize that in theoretical discussions—for instance, the definitions of bondage and freedom and the ensuing characterization of derivability—the words 'term' and 'formula' are always to be understood in the official sense.

### EXERCISES

13. For each of the following formulas in unofficial notation, delete the parentheses and brackets inserted, and restore the superscripts and parentheses omitted, in accordance with the informal notational conventions introduced in this section.

$$\begin{array}{l} F\,A(x) \\ G(B\,x) \\ H[C(B\,B)\,A(B)\,B] \\ \bigwedge x\,Fx \lor P \rightarrow G(xy) \land H(xyz) \end{array}$$

**4. Translation and symbolization.** Many symbolic sentences of our present language cannot be translated into English on the basis of the schemes of abbreviation used in chapter III; an example is

(1) $$\sim \bigvee x F(xA) \quad .$$

For present purposes we must admit, in addition to the abbreviations of the last chapter, abbreviations involving operation letters as well as predicate letters of more than one place.

It is convenient to establish a standard order of variables, in fact, the order represented by the following list:

$$a, \ldots, z, a_0, \ldots, z_0, \ldots, a_1, \ldots, z_1, \ldots \quad .$$

Thus the *first variable* will be '*a*', the *second variable* '*b*', the *twenty-sixth variable* '*z*', and so on.

If $\phi$ is a *formula* of English, then $\alpha$ is said to be an *apparent variable* of $\phi$ if either $\phi$ is a sentence of English containing $\alpha$ or there is an occurrence of $\alpha$ in $\phi$ such that, upon replacement of some other occurrences of variables by English names, $\phi$ becomes a sentence of English. Similarly, if $\zeta$ is a *term* of English, then $\alpha$ is an *apparent variable* of $\zeta$ if either $\zeta$ is an English name containing $\alpha$ or there is an occurrence of $\alpha$ in $\zeta$ such that, upon replacement of some occurrences of variables by English names, $\zeta$ becomes an English name. For example, '*a*' is an apparent variable of the name

the letter following '*a*'

and of the term

the object *a* such that *a* wrote *b* .

We shall admit abbreviations only of those formulas and terms of English whose variables are the first *k* variables, for some number *k*, and which contain no apparent variables. (The reason for this limitation is discussed in section 11 below.) We shall abbreviate such a formula or term by a *k*-place predicate letter or operation letter.

Thus an *abbreviation* will now be either (1) an ordered pair whose first member is a *k*-place predicate letter, for some $k \geqslant 0$, and whose second member is a formula of English containing exactly the first *k* variables and having no apparent variables, or (2) an ordered pair whose first member is a *k*-place operation letter, for some number $k \geqslant 0$, and whose second member is a term of English containing exactly the first *k* variables and having no apparent variables.

As in earlier chapters, a *scheme of abbreviation* is to be a collection of abbreviations such that no two abbreviations in the collection have the same first member. The following is an example:

(2) $A^0$ : Adam
$F^2$ : *a* is father of *b*

The process of *literal translation into English on the basis of a given scheme of abbreviation* begins as before with a symbolic formula and if successful ends with a formula of English. The process consists of the following steps:

(*i*) *Restore official notation by reversing the conventions of the preceding section.*

(*ii*) *Replace 0-place operation letters and 0-place predicate letters by English names and sentences in accordance with the scheme of abbreviation.*

(*iii*) *Replace each operation letter and predicate letter of one or more places by the term or formula of English with which it is paired in the scheme of abbreviation, flanking the latter with a pair of braces.* (*As before, the result of this step will not in general be a formula.*)

(*iv*) *Successively replace all parts of the form*

$$\{\phi\}\zeta_1 \ldots \zeta_k \quad ,$$

*where $\phi$ is a term or formula of English containing exactly the first* k *variables and $\zeta_1, \ldots, \zeta_k$ are terms, by the term or formula obtained from*

*$\phi$ by replacing all occurrences of 'a' by $\zeta_1$, 'b' by $\zeta_2$, etc., up to the* kth *variable, which is to be replaced by $\zeta_k$.*

(*v*) *Eliminate sentential connectives and quantifier phrases in favor of the corresponding phrases of connection and quantity, preserving all parentheses.*

As before, we say that an English formula is a *free translation* (or simply a *translation*) of a symbolic formula $\phi$ on the basis of a given scheme of abbreviation if it is a stylistic variant of the literal English translation of $\phi$ based on that scheme.

Let us, for example, translate the sentence (1) into English on the basis of the scheme (2). In step (i) of the process of literal translation into English, (1) becomes

$$\sim \bigvee x\ F^2\ x\ A^0 \quad ,$$

in step (ii)

$$\sim \bigvee x\ F^2\ x\ \text{Adam} \quad ,$$

in step (iii)

$$\sim \bigvee x\ \{a \text{ is father of } b\}\ x\ \text{Adam} \quad ,$$

in step (iv)

$$\sim \bigvee x\ x \text{ is father of Adam} \quad ,$$

and in step (v)

(3) It is not the case that there is an object $x$ such that $x$ is father of Adam.

Thus (3) is the literal translation of (1) into English on the basis of the scheme (2); and the more idiomatic stylistic variant of (3),

Adam has no father,

qualifies as a free translation of (1) on the basis of the same scheme.

As a slightly more involved illustration, consider the symbolic sentence

(4) $$\bigwedge x \bigwedge y(Fx \wedge Fy \wedge Gy \rightarrow G\ A(xy))$$

and the scheme of abbreviation

(5)
$F^1$ : $a$ is a number
$G^1$ : $a$ is even
$A^2$ : the product of $a$ and $b$ .

Let us find the literal English translation of (4) based on the scheme (5). Applying step (i), we obtain

$$\bigwedge x \bigwedge y(((F^1x \wedge F^1y) \wedge G^1y) \rightarrow G^1\ A^2xy) \quad .$$

Step (ii) is irrelevant. In step (iii) we obtain

$$\Lambda x \Lambda y(((\{a \text{ is a number}\}\, x \wedge \{a \text{ is a number}\}\, y) \wedge \{a \text{ is even}\}\, y) \rightarrow \{a \text{ is even}\}\{\text{the product of } a \text{ and } b\}\, x\, y) \quad .$$

Applying step (iv), we obtain first

$$\Lambda x \Lambda y(((\{a \text{ is a number}\}\, x \wedge \{a \text{ is a number}\}\, y) \wedge \{a \text{ is even}\}\, y) \rightarrow \{a \text{ is even}\} \text{ the product of } x \text{ and } y) \quad ,$$

and then

$$\Lambda x \Lambda y(((x \text{ is a number} \wedge y \text{ is a number}) \wedge y \text{ is even}) \rightarrow \text{the product of } x \text{ and } y \text{ is even}) \quad .$$

Finally, by step (v), we obtain as the literal translation of (4) the English sentence

(6) For each $x$, for each $y$ (if (($x$ is a number and $y$ is a number) and $y$ is even), then the product of $x$ and $y$ is even) .

Moreover,

The product of any number and any even number is even,

being a stylistic variant of (6), is a free English translation of (4) on the basis of (5).

As before, $\phi$ is said to be a *symbolization* of an English formula $\psi$ on the basis of a given scheme of abbreviation just in case $\phi$ is a symbolic formula that has $\psi$ as a translation on the basis of that scheme.

To find a symbolization of a given formula of English on the basis of a given scheme of abbreviation the reader will again find it useful to proceed roughly as follows:

(*1*) *Introduce phrases of quantity and connection, the latter accompanied by parentheses and occurring canonically, in place of their stylistic variants.*

(*2*) *Reverse the steps leading from a symbolic formula to a literal English translation.*

For an example of the process of symbolization, consider the sentence

(7) There is no composer whose contrapuntal ingenuity surpasses that of Bach,

together with the scheme of abbreviation

$F^1$ : $a$ is a composer
$G^2$ : $a$ surpasses $b$
$A^1$ : the contrapuntal ingenuity of $a$
$B^0$ : Bach .

In step (1), with intuition as guide, we transform (7) into

> It is not the case that there is an object *x* such that (*x* is a composer and the contrapuntal ingenuity of *x* surpasses the contrapuntal ingenuity of Bach) .

Let us now perform step (2) of the process of symbolization. Reversing step (v) of the process of literal translation into English, we obtain

> $\sim \bigvee x$(*x* is a composer $\wedge$ the contrapuntal ingenuity of *x* surpasses the contrapuntal ingenuity of Bach) .

Reversing step (iv), this becomes first

> $\sim \bigvee x$({*a* is a composer} *x* $\wedge$ {*a* surpasses *b*} the contrapuntal ingenuity of *x* the contrapuntal ingenuity of Bach) ,

and then

> $\sim \bigvee x$({*a* is a composer} *x* $\wedge$ {*a* surpasses *b*} {the contrapuntal ingenuity of *a*} *x* {the contrapuntal ingenuity of *a*} Bach) .

Reversing step (iii), we obtain

$$\sim \bigvee x(F^1 x \wedge G^2\ A^1 x\ A^1 \text{Bach}) \quad ;$$

reversing step (ii),

$$\sim \bigvee x(F^1 x \wedge G^2\ A^1 x\ A^1 B^0) \quad ;$$

and reversing step (i),

$$\sim \bigvee x(Fx \wedge G[A(x)A(B)]) \quad .$$

An example of stylistic variance of rather frequent incidence is the mutual conversion of active and passive voice. Consider, for example, the sentence

> If someone is loved by one whom he does not love and by nobody else, then he does not love his lover,

together with the following scheme of abbreviation:

| | | |
|---|---|---|
| $L^2$ | : | *a* loves *b* |
| $T^2$ | : | *a* differs from *b* |
| $A^1$ | : | the lover of *a* . |

In step (1), construing '*x* is loved by *y*' as a stylistic variant of '*y* loves *x*', the sentence above becomes

> For each *x* (if there is an object *y* such that ((*y* loves *x* and it is not the case that *x* loves *y*) and it is not the case that there is an object *z* such that (*z* differs from *y* and *z* loves *x*)), then it is not the case that *x* loves the lover of *x*) ;

and carrying through the successive parts of step (2), we obtain the symbolization

$$\bigwedge x[\bigvee y(L(yx) \wedge \sim L(xy) \wedge \sim \bigvee z[T(zy) \wedge L(zx)]) \rightarrow \sim L(x\ A(x))] \quad .$$

The order of quantifiers in a symbolization is now of particular importance. For example, if we adopt the scheme of abbreviation

$T^2$ : *a* differs from *b* ,

then

$$\bigwedge x \bigvee y\ T(xy)$$

is a symbolization of the true sentence

Each thing differs from something,

whereas

$$\bigvee y \bigwedge x\ T(xy)$$

is a symbolization of the false sentence

Something is such that everything differs from it.

## EXERCISES

14. On the basis of the scheme of abbreviation

$F^1$ : *a* is a person
$G^2$ : *a* loves *b* ,

find for each sentence of group A a sentence of group B that is a symbolization of it.

### GROUP A

(1) Everyone loves someone.
(2) Someone loves someone.
(3) If anything is a person, then someone loves himself.
(4) Someone loves everyone.
(5) Everyone loves everyone.

### GROUP B

(a) $\bigvee x(Fx \wedge \bigwedge y[Fy \rightarrow G(xy)])$
(b) $\bigwedge x(Fx \wedge \bigwedge y[Fy \rightarrow G(xy)])$
(c) $\bigvee y(Fy \wedge \bigvee x[Fx \wedge G(yx)])$
(d) $\bigwedge x(Fx \rightarrow \bigvee x[Fx \wedge G(xx)])$
(e) $\bigwedge y(Fy \rightarrow \bigwedge x[Fx \rightarrow G(yx)])$
(f) $\bigvee x(Fx \rightarrow \bigvee y[Fy \wedge G(xy)])$
(g) $\bigwedge x(Fx \rightarrow \bigvee y[Fy \wedge G(xy)])$

15. On the basis of the scheme of abbreviation

$$
\begin{array}{ll}
F^1 : & a \text{ is a student} \\
G^1 : & a \text{ is a teacher} \\
H^1 : & a \text{ is a subject} \\
S^3 : & a \text{ studies } b \text{ with } c \quad ,
\end{array}
$$

find for each sentence of group B a sentence of group A that is a translation of it.

GROUP A

(1) Every student studies every subject with every teacher.
(2) Every student studies some subject with some teacher.
(3) No student studies every subject with a teacher.
(4) Only students study every subject with a teacher.
(5) No subject is such that every student studies it with every teacher.
(6) Every teacher has some subject that some student studies with him.
(7) Each subject has some student who studies it with some teacher.
(8) None but teachers are such that all students study all subjects with them.
(9) Anyone who studies any subject with any teacher is a student.
(10) There is a teacher such that every student studies some subject with him.
(11) There is a teacher such that some student studies every subject with him.
(12) Teachers who study some subject with a teacher are students.
(13) No student who does not study every subject with a teacher is a teacher.
(14) Any student who studies any subject with himself is a teacher.
(15) Some teacher who studies every subject with himself is a student.
(16) There is no teacher with whom any student studies all subjects.

GROUP B

(a) $\bigwedge x(\bigwedge y \bigwedge z[Fy \land Hz \to S(yzx)] \to Gx)$
(b) $\sim \bigvee x(Gx \land \bigvee y[Fy \land \bigwedge z(Hz \to S(yzx))])$
(c) $\bigwedge x(Fx \to \bigwedge y \bigwedge z[Hy \land Gz \to S(xyz)])$
(d) $\bigwedge x(Hx \to \bigvee y \bigvee z[Fy \land Gz \land S(yxz)])$
(e) $\bigwedge x(Fx \to \bigvee y \bigvee z[Hy \land Gz \land S(xyz)])$
(f) $\bigwedge x(Fx \land \bigvee y[Hy \land S(xyx)] \to Gx)$
(g) $\bigvee x(Gx \land \bigwedge y[Fy \to \bigvee z(Hz \land S(yzx))])$
(h) $\bigwedge x \bigwedge y \bigwedge z(Fx \land Hy \land Gz \to S(xyz))$
(i) $\sim \bigvee x(Hx \land \bigwedge y \bigwedge z[Fy \land Gz \to S(yxz)])$
(j) $\bigwedge x(Fx \to \sim \bigwedge y[Hy \to \bigvee z(Gz \land S(xyz))])$
(k) $\bigwedge x(Fx \land \sim \bigwedge y[Hy \to \bigvee z(Gz \land S(xyz))] \to \sim Gx)$
(l) $\bigwedge x(Fx \to \bigwedge y[Hy \to \bigwedge z(Gz \to S(xyz))])$

$\mathcal{9}$ (m)   $\Lambda x (\vee y \vee z [Hy \wedge Gz \wedge S(xyz)] \rightarrow Fx)$

$\mathcal{15}$ (n)   $\vee x (Gx \wedge \Lambda y [Hy \rightarrow S(xyx)] \wedge Fx)$

$\mathcal{2}$ (o)   $\Lambda x (Fx \rightarrow \vee y (Hy \wedge \vee z (Gz \wedge S(xyz))])$

$\mathcal{4}$ (p)   $\Lambda x (\Lambda y [Hy \rightarrow \vee z (Gz \wedge S(xyz))] \rightarrow Fx)$

$\mathcal{13}$ (q)   $\sim \vee x (Fx \wedge \vee y [Hy \wedge \sim \vee z (Gz \wedge S(xyz))] \wedge Gx)$

$\mathcal{11}$ (r)   $\vee x \vee y (Gx \wedge Fy \wedge \Lambda z [Hz \rightarrow S(yzx)])$

$\mathcal{6}$ (s)   $\Lambda x (Gx \rightarrow \vee y \vee z [Fy \wedge Hz \wedge S(yzx)])$

$\mathcal{3}$ (t)   $\sim \vee x (Fx \wedge \Lambda y [Hy \rightarrow \vee z (Gz \wedge S(xyz))])$

$\mathcal{12}$ (u)   $\Lambda x (Gx \wedge \vee y \vee z [Hy \wedge Gz \wedge S(xyz)] \rightarrow Fx)$

Translate each of the following symbolic formulas into idiomatic English on the basis of the scheme of abbreviation that accompanies it. Exercise 16 is solved for illustration.

16.   $\Lambda x \Lambda y \Lambda z \Lambda w [Fx \wedge G(yx) \wedge G(zx) \wedge G(wx) \wedge H(yz) \wedge H(yw) \wedge H(zw) \rightarrow I(A(B[y] \; B[z]) \; B[w])]$

($F^1$ : *a* is a triangle;   $G^2$ : *a* is a side of *b*;   $H^2$ : *a* is different from *b*;   $I^2$ : *a* is greater than *b*;   $A^2$ : the sum of *a* and *b*;   $B^1$ : the length of *a*)

In step (i) of the process of literal translation into English, No. 16 becomes

$\Lambda x \Lambda y \Lambda z \Lambda w ((((((F^1 x \wedge G^2 yx) \wedge G^2 zx) \wedge G^2 wx) \wedge H^2 yz) \wedge H^2 yw) \wedge H^2 zw) \rightarrow I^2 \; A^2 B^1 y B^1 z \; B^1 w)$   ;

in step (iii), for no change occurs in step (ii),

$\Lambda x \Lambda y \Lambda z \Lambda w ((((((\{a$ is a triangle$\}$ *x* $\wedge$ $\{a$ is a side of $b\}$ *y* *x*) $\wedge$ $\{a$ is a side of $b\}$ *z* *x*) $\wedge$ $\{a$ is a side of $b\}$ *w* *x*) $\wedge$ $\{a$ is different from $b\}$ *y* *z*) $\wedge$ $\{a$ is different from $b\}$ *y* *w*) $\wedge$ $\{a$ is different from $b\}$ *z* *w*) $\rightarrow$ $\{a$ is greater than $b\}$ $\{$the sum of *a* and *b*$\}$ $\{$the length of *a*$\}$ *y* $\{$the length of *a*$\}$ *z* $\{$the length of *a*$\}$ *w*)   ;

in step (iv), first

$\Lambda x \Lambda y \Lambda z \Lambda w ((((((x$ is a triangle $\wedge$ *y* is a side of *x*) $\wedge$ *z* is a side of *x*) $\wedge$ *w* is a side of *x*) $\wedge$ *y* is different from *z*) $\wedge$ *y* is different from *w*) $\wedge$ *z* is different from *w*) $\rightarrow$ $\{a$ is greater than $b\}$ $\{$the sum of *a* and *b*$\}$ the length of *y* the length of *z* the length of *w*) ,

next

$\Lambda x \Lambda y \Lambda z \Lambda w ((((((x$ is a triangle $\wedge$ *y* is a side of *x*) $\wedge$ *z* is a side of *x*) $\wedge$ *w* is a side of *x*) $\wedge$ *y* is different from *z*) $\wedge$ *y* is different from *w*) $\wedge$ *z* is different from *w*) $\rightarrow$ $\{a$ is greater than $b\}$ the sum of the length of *y* and the length of *z* the length of *w*) , and finally

> $\bigwedge x \bigwedge y \bigwedge z \bigwedge w$((((((($x$ is a triangle $\wedge$ $y$ is a side of $x$) $\wedge$ $z$ is a side of $x$) $\wedge$ $w$ is a side of $x$) $\wedge$ $y$ is different from $z$) $\wedge$ $y$ is different from $w$) $\wedge$ $z$ is different from $w$) $\rightarrow$ the sum of the length of $y$ and the length of $z$ is greater than the length of $w$) ;

and in step (v)

> For each $x$, for each $y$, for each $z$, for each $w$ (if ($x$ is a triangle and $y$ is a side of $x$) and $z$ is a side of $x$) and $w$ is a side of $x$) and $y$ is different from $z$) and $y$ is different from $w$) and $z$ is different from $w$), then the sum of the length of $y$ and the length of $z$ is greater than the length of $w$) ,

which is the literal translation into English of No. 16; and the following stylistic variant is a free translation:

> The sum of the lengths of any two sides of a triangle is greater than the length of the third side.

17. $\bigvee x(Fx \wedge \bigwedge y[Fy \rightarrow GA(yx)]) \rightarrow \bigvee x(Fx \wedge Gx)$

($F^1$ : *a* is a number; $G^1$ : *a* is even; $A^2$ : the product of *a* and *b*)

18. $\bigwedge x(Fx \rightarrow \bigvee yG(yx) \wedge \bigvee yH(yx)) \wedge \bigvee x(Fx \wedge \sim[\bigvee yG(xy) \vee \bigvee yH(xy)])$

($F^1$ : *a* is a person; $G^2$ : *a* is father of *b*; $H^2$ : *a* is mother of *b*)

19. $\bigwedge x(F(x\ E) \rightarrow \bigvee y[G(y\ E) \wedge H(xy) \wedge I(y\ B(C(E)))])$

($F^2$ : *a* is a student of *b*; $E^0$ : the course; $G^2$ : *a* is a quiz section of *b*; $H^2$ : *a* attends *b*; $I^2$ : *a* is taught by *b*; $B^1$ : the teaching assistant of *a*; $C^1$ : the instructor of *a*)

Symbolize each of the following formulas on the basis of the scheme of abbreviation that accompanies it. Exercise 20 is solved for illustration.

20. A person pays a dollar for the banquet of the club only if he belongs to at least one of the club's committees and attends every meeting of that committee. ($F^1$ : *a* is a person; $G^2$ : *a* pays a dollar for *b*; $H^2$ : *a* is a committee of *b*; $I^2$ : *a* belongs to *b*; $J^2$ : *a* is a meeting of b; $K^2$ : *a* attends *b*; $A^1$ : the banquet of *a*; $B^0$ : the club)

Linguistic insight would suggest that No. 20 become in step (1) of the process of symbolization

> For each $x$ (if ($x$ is a person and $x$ pays a dollar for the banquet of the club), then there is an object $y$ such that (($y$ is a committee of the club and $x$ belongs to $y$) and for each $z$ (if $z$ is a meeting of $y$, then $x$ attends $z$))) .

By performing the successive stages of step (2), we obtain first

$\bigwedge x((x$ is a person $\wedge$ $x$ pays a dollar for the banquet of the club) $\rightarrow \bigvee y((y$ is a committee of the club $\wedge$ $x$ belongs to $y)$ $\wedge \bigwedge z(z$ is a meeting of $y \rightarrow x$ attends $z)))$ ,

next

$\bigwedge x((\{a$ is a person$\}\ x \wedge \{a$ pays a dollar for $b\}\ x\ \{$the banquet of $a\}$ the club$) \rightarrow \bigvee y((\{a$ is a committee of $b\}\ y$ the club $\wedge \{a$ belongs to $b\}\ x\, y) \wedge \bigwedge z(\{a$ is a meeting of $b\}\ z\, y \wedge \{a$ attends $b\}\ x\, z)))$ ,

next

$\bigwedge x((F^1 x \wedge G^2\, x\, A^1$ the club$) \rightarrow \bigvee y((H^2\, y$ the club $\wedge I^2 xy) \wedge \bigwedge z(J^2 zy \rightarrow K^2 xz)))$ ,

next

$\bigwedge x((F^1 x \wedge G^2\, x\, A^1 B^0) \rightarrow \bigvee y((H^2 y\, B^0 \wedge I^2 xy) \wedge \bigwedge z(J^2 zy \rightarrow K^2 xz)))$ ,

and finally

$\bigwedge x[Fx \wedge G(x\ A(B)) \rightarrow \bigvee y(H(y\ B) \wedge I(xy) \wedge \bigwedge z[J(zy) \rightarrow K(xz)])]$ .

21. Every husband and wife has a spouse. ($F^1$ : *a* is a husband; $G^1$ : *a* is a wife; $H^2$ : *a* is a spouse of *b*)

22. No one lacks a father, but not everyone is a father. ($F^1$ : *a* is a person; $G^2$ : *a* is father of *b*)

23. Every number has some number as its successor. ($F^1$ : *a* is a number; $G^2$ : *a* is a successor of *b*)

24. Some number is a successor of every number. ($F^1$ : *a* is a number; $G^2$ : *a* is a successor of *b*)

25. A teacher has no scruples if he assigns a problem that has no solution. ($F^1$ : *a* is a teacher; $G^1$ : *a* has scruples; $I^2$ : *a* assigns *b*; $J^1$ : *a* is a problem; $K^2$ : *a* is a solution of *b*)

26. The net force acting on a particle is equal to the product of its mass and its acceleration. ($F^1$ : *a* is a particle; $A^1$ : the net force acting on *a*; $G^2$ : *a* is equal to *b*; $B^2$ : the product of *a* and *b*; $C^1$ : the mass of *a*; $D^1$: the acceleration of *a*)

27. No principle is innate unless everyone who hears it gives his assent. ($F^1$ : *a* is a principle; $G^1$ : *a* is innate; $H^1$ : *a* is a person; $I^2$: *a* hears *b*; $J^2$ : *a* gives his assent to *b*)

28. The cube of the square of 2 is even. ($F^1$ : *a* is even; $A^1$ : the cube of *a*; $B^1$ : the square of *a*; $C^0$ : 2)

29. No student likes every course he takes unless the only courses he takes are philosophical studies. ($F^1$ : *a* is a student; $G^1$ : *a* is a course; $H^1$ : *a* is a philosophical study; $I^2$ : *a* takes *b*; $J^2$ : *a* likes *b*)

30. A person who loves himself more than he loves anyone else is not loved by anyone other than himself. ($F^1$ : *a* is a person; $G^2$ : *a* loves *b*; $H^2$ : *a* is different from *b*; $L^4$ : *a* loves *b* more than *c* loves *d*)

31. If a father has only male children, then he does not have to provide a dowry for any one of them. ($F^1$ : *a* is a father; $G^1$ : *a* is male; $H^2$ : *a* is a child of *b*; $J^2$ : *a* has to provide a dowry for *b*)

32. The wife of anyone who marries the daughter of the brother of his father marries the son of the brother of the husband of her mother. ($F^2$ : *a* marries *b*; $A^1$ : the wife of *a*; $B^1$ : the daughter of *a*; $C^1$ : the brother of *a*; $D^1$ : the father of *a*; $E^1$ : the son of *a*; $A^1_1$ : the husband of *a*; $B^1_1$ : the mother of *a*)

33. If *x* is an integer greater than or equal to zero and every integer is divisible by *x*, then *x* is equal to 1. ($F^1$ : *a* is an integer; $G^2$ : *a* is greater than *b*; $H^2$ : *a* is equal to *b*; $I^2$ : *a* is divisible by *b*; $A^0$ : zero; $B^0$ : 1)

34. The square of the hypotenuse of a right triangle is equal to the sum of the squares of the other two sides. ($F^1$ : *a* is a right triangle; $G^2$ : *a* is a side of *b*; $H^2$ : *a* is equal to *b*; $A^1$ : the square of *a*; $B^1$ : the hypotenuse of *a*; $C^2$ : the sum of *a* and *b*)

35. The gravitational force exerted by one particle on another is directly proportional to the mass of each particle and inversely proportional to the square of the distance between the two particles. ($F^1$ : *a* is a particle; $G^2$ : *a* is directly proportional to *b*; $H^2$ : *a* is inversely proportional to *b*; $A^1$ : the mass of *a*; $B^1$ : the square of *a*; $C^2$ : the distance between *a* and *b*; $D^2$ : the gravitational force exerted by *a* on *b*)

36. Someone is hit by a car every day. ($F^1$ : *a* is a day; $H^1$ : *a* is a car; $I^3$ : *a* is hit by *b* on *c*)

**5. Revised inference rules.** We must extend the notion of proper substitution for a variable. In chapter III only variables could be substituted for variables; now we shall admit the substitution of arbitrary symbolic terms for variables. Thus we say that a symbolic formula $\psi$ *comes from* a symbolic formula $\phi$ *by proper substitution of a symbolic term* $\zeta$ *for a variable* $\alpha$ if $\psi$ is like $\phi$ except for having free occurrences of $\zeta$ wherever $\phi$ has free occurrence of $\alpha$.

In chapter III the only terms to which UI and EG could be applied were variables. We now reformulate these two rules so as to admit an arbitrary symbolic term in place of the variable of instantiation.

Universal instantiation (UI): $$\frac{\Lambda\alpha\phi}{\psi}$$

Existential generalization (EG): $$\frac{\psi}{\mathrm{V}\alpha\phi}$$

In both cases $\alpha$ is to be a variable, $\phi$ is to be a symbolic formula, and $\psi$ is to come from $\phi$ by proper substitution of *some symbolic term* for $\alpha$.

All previous applications of UI and EG are comprehended under the present formulation; there are additional applications of the following sort. From the sentence

$$\bigwedge x \mathrm{F} x$$

we can now infer such formulas as

$$\begin{array}{ll} (1) & \mathrm{F\,A} \quad , \\ (2) & \mathrm{F\,B(A)} \quad , \\ (3) & \mathrm{F\,B}(x) \quad , \\ (4) & \mathrm{F\,C(A}\,x) \quad . \end{array}$$

Further, from any of (1) – (4) the sentence

$$\bigvee x \mathrm{F} x$$

follows by EG. From (4) we can also infer by EG the formulas

$$\begin{array}{l} \bigvee y\ \mathrm{F\,C}(y\,x) \quad , \\ \bigvee x\ \mathrm{F\,C(A}\,x) \quad , \\ \bigvee y\ \mathrm{F\,C(A}\,y) \quad ; \end{array}$$

but we cannot infer

$$\bigvee x\ \mathrm{F\,C}(x\,x) \quad ,$$

because (4) does not come from 'F C($x$ $x$)' by proper substitution for '$x$'.

It should be emphasized that the present extensions concern only UI and EG. The formulation of EI remains exactly as before; that is, from one formula another follows by EI just in case the two formulas have the respective forms

$$\bigvee \alpha \phi$$

and

$$\phi' \quad ,$$

where $\alpha$ is a variable, $\phi$ is a symbolic formula, and $\phi'$ comes from $\phi$ by proper substitution of some *variable* for $\alpha$.

The directions for constructing an *unabbreviated* and (for the moment) an *abbreviated derivation* remain as in chapter III. For an unabbreviated derivation we use clauses (1) – (6) of pages 102 – 03; for an abbreviated derivation we supplement these with clauses (7) – (9) of pages 106 – 07. Throughout these clauses, however, we must now understand the terms 'formula', 'UI', and 'EG' in the broader sense of the present chapter. We thus arrive at a characterization of a *derivation* in the full (rather than the monadic) quantifier calculus.

The characterizations of a *complete* derivation, *derivability*, a *proof*, a *theorem*, an *argument*, a *symbolic argument*, a *valid symbolic argument*, an *English argument*, a *symbolization* of an English argument, and a *valid English argument* remain exactly as before. (See pp. 95, 103, and 117.)

**EXERCISES**

37. Which of the following formulas follow from

$$\bigwedge x \bigvee z F(A(xyx)z)$$

by UI?

(a) $\bigvee z F(zz)$
(b) $\bigvee z F(A(xyx)z)$
(c) $\bigvee z F(A(yyy)z)$
(d) $\bigvee z F(A(xxx)z)$
(e) $\bigvee z F(A(CyB)z)$
(f) $\bigvee z F(A(CyC)z)$
(g) $\bigvee z F(A(ByB)z)$
(h) $\bigvee z F(A(A(xyx)yA(xyx))z)$
(i) $\bigvee z F(A(B(y)yB(y))z)$
(j) $\bigvee z F(A(B(C(z))yB(C(z)))z)$
(k) $\bigvee z F(A(C(B(w))yC(B(w)))z)$

38. Which of the following formulas follow from

$$\bigwedge x FA(xB(y)) \vee G(Cy)$$

by EG?

(a) $\bigvee w[\bigwedge x FA(xB(y)) \vee G(wy)]$
(b) $\bigvee w[\bigwedge x Fw \vee G(Cy)]$
(c) $\bigvee w[\bigwedge x FA(xB(w)) \vee G(Cw)]$
(d) $\bigvee w[\bigwedge x FA(xw) \vee G(Cy)]$
(e) $\bigvee x[\bigwedge x FA(xB(y)) \vee G(xy)]$
(f) $\bigvee x[\bigwedge x FA(xB(x)) \vee G(Cx)]$
(g) $\bigvee y[\bigwedge x FA(xB(y)) \vee G(Cy)]$
(h) $\bigvee y[\bigwedge x FA(xB(y)) \vee G(yy)]$
(i) $\bigvee x[\bigwedge x FA(xB(y)) \vee G(Cy)]$
(j) $\bigvee w[\bigwedge x FA(xB(y)) \vee Gw]$

For each of the following arguments give an unabbreviated derivation of its conclusion from its premise.

39. $\bigwedge x\ F\ A(x\ B) \quad \therefore \bigvee x\ F\ A(xx)$
40. $\bigvee x\ F\ A(xx) \quad \therefore \bigvee x Fx$
41. $\bigwedge x \bigwedge y\ F\ A[x\ B(y)\ C] \quad \therefore \bigvee x\ F\ A[B(x)\ B(C)\ x]$
42. $\bigwedge x \bigwedge y\ F\ A[x\ B(y)\ C] \quad \therefore \bigvee x\ F\ A[B(x)\ x\ C]$

Show the following arguments valid by constructing symbolizations and giving derivations (either abbreviated or unabbreviated) of the conclusions of the symbolizations from their premises. Exercise 43 is discussed for illustration.

43. Argument (2) of page 131.

On the basis of the scheme of abbreviation

$F^1$ : *a* is a horse
$G^1$ : *a* is an animal
$H^2$ : *a* is a head of *b* ,

one obtains the symbolization

$$\bigwedge x(Fx \rightarrow Gx) \quad \therefore \bigwedge x(\bigvee y[H(xy) \wedge Fy] \rightarrow \bigvee y[H(xy) \wedge Gy]) \quad ,$$

whose validity the reader can easily establish.

44. Argument (1) of page 131.
45. The square of 2 is even. ∴ The square of something is even.
46. The square of something is even. ∴ Something is even.
47. The product of numbers is a number. There is a number such that the product of it and any number is even. ∴ There is an even number.

**6. Theorems.** Symbolic formulas corresponding to those of chapter III are known as *monadic formulas*. More exactly, a *monadic formula* is a symbolic formula that contains no predicate letters of more than one place and no operation letters at all.

For each of the theorems of chapter III a nonmonadic analogue may be obtained by multiplying variables. For example,

T249 $\bigwedge x \bigwedge y\ F(xy) \rightarrow \bigvee x \bigvee y\ F(xy)$

is a two-variable analogue of T238, and

T250 $\bigwedge x \bigwedge y \bigwedge z\ F(xyz) \leftrightarrow \sim \bigvee x \bigvee y \bigvee z \sim F(xyz)$

is a three-variable analogue of T205. Proofs of these two theorems can be obtained without essential change from the proofs of their earlier counterparts, as the reader can easily confirm; and this applies to all multiple-variable analogues of theorems of chapter III.

There are nonmonadic theorems that cannot be obtained in this way, for example T251 – T253, the laws of *commutation of quantifiers*. According to T251 and T252, a pair of like quantifiers may be commuted. If, however, the quantifiers are unlike, only the conditional, T253, holds.

T251 1. ~~Show~~ $\bigwedge x \bigwedge y F(xy) \leftrightarrow \bigwedge y \bigwedge x F(xy)$
2. ~~Show~~ $\bigwedge x \bigwedge y F(xy) \rightarrow \bigwedge y \bigwedge x F(xy)$

| | | |
|---|---|---|
| 3. | $\bigwedge x \bigwedge y F(xy)$ | |
| 4. | ~~*Show*~~ $\bigwedge y \bigwedge x F(xy)$ | |
| 5. | $\bigwedge y F(xy)$ | 3, UI |
| 6. | $F(xy)$ | 5, UI |
| 7. | ~~*Show*~~ $\bigwedge y \bigwedge x F(xy) \rightarrow \bigwedge x \bigwedge y F(xy)$ | |
| 8. | $\bigwedge y \bigwedge x F(xy)$ | |
| 9. | ~~*Show*~~ $\bigwedge x \bigwedge y F(xy)$ | |
| 10. | $\bigwedge x F(xy)$ | 8, UI |
| 11. | $F(xy)$ | 10, UI |
| 12. | $\bigwedge x \bigwedge y F(xy) \leftrightarrow \bigwedge y \bigwedge x F(xy)$ | 2, 7, CB |

T252 $\bigvee x \bigvee y F(xy) \leftrightarrow \bigvee y \bigvee x F(xy)$

T253 1. ~~*Show*~~ $\bigvee x \bigwedge y F(xy) \rightarrow \bigwedge y \bigvee x F(xy)$

| | | |
|---|---|---|
| 2. | $\bigvee x \bigwedge y F(xy)$ | |
| 3. | $\bigwedge y F(wy)$ | 2, EI |
| 4. | ~~*Show*~~ $\bigwedge y \bigvee x F(xy)$ | |
| 5. | $F(wy)$ | 3, UI |
| 6. | $\bigvee x F(xy)$ | 5, EG |

The reader will find an effort to prove the converse of T253 (which is not a theorem; see exercise 73) an instructive review of the restrictions on variables that must be observed in constructing a derivation.

We list a few more nonmonadic theorems. The reader who is familiar with Russell's paradox or with axiomatic set theory will perhaps find diversion in proving T255 and T256.

T254 $\bigvee x \bigvee y F(xy) \leftrightarrow \bigvee x \bigvee y [F(xy) \vee F(yx)]$

T255 $\sim \bigvee y \bigwedge x [F(xy) \leftrightarrow \sim F(xx)]$

T256 $\bigwedge z \bigvee y \bigwedge x [F(xy) \leftrightarrow F(xz) \wedge \sim F(xx)] \rightarrow \sim \bigvee z \bigwedge x F(xz)$

T257 $\bigwedge x \bigwedge y F(xy) \rightarrow \bigwedge y \bigwedge x F(yx)$

T258 $F(xA(x)) \leftrightarrow \bigvee y [\bigwedge z [F(zy) \rightarrow F(zA(x))] \wedge F(xy)]$

We consider now some monadic theorems whose intuitive content is less clear than that of the theorems of chapter III. By a *generalization* we understand a formula that is either a universal generalization or an existential generalization, that is, which has the form

$$\bigwedge \alpha \phi$$

or

$$\bigvee \alpha \phi \quad ,$$

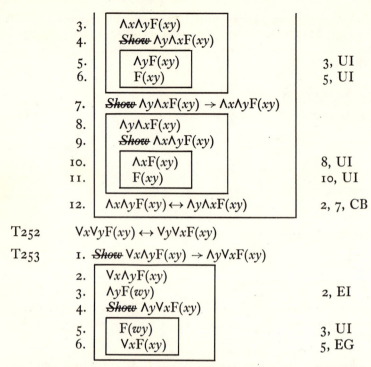

where $\alpha$ is a variable and $\phi$ a formula. A symbolic formula is said to be *without overlay* if it contains no generalization that in turn contains another generalization. T259 – T261 illustrate the following fact: every monadic formula is equivalent to a monadic formula without overlay.

T259

| | | |
|---|---|---|
| 1. | ~~Show~~ $\bigwedge y \bigvee x(Fx \leftrightarrow \sim Fy) \leftrightarrow \bigvee xFx \wedge \bigvee x \sim Fx$ | |
| 2. | ~~Show~~ $\bigvee xFx \wedge \bigvee x \sim Fx \rightarrow \bigwedge y \bigvee x(Fx \leftrightarrow \sim Fy)$ | |
| 3. | $\bigvee xFx \wedge \bigvee x \sim Fx$ | |
| 4. | $Fw$ | 3, S, EI |
| 5. | $\sim Fz$ | 3, S, EI |
| 6. | ~~Show~~ $\bigwedge y \bigvee x(Fx \leftrightarrow \sim Fy)$ | |
| 7. | $\sim \bigwedge y \bigvee x(Fx \leftrightarrow \sim Fy)$ | |
| 8. | $\sim \bigvee x(Fx \leftrightarrow \sim Fu)$ | 7, QN, EI |
| 9. | $Fw \leftrightarrow \sim \sim Fu$ | 8, QN, UI, T90, BC, MP |
| 10. | $\sim \sim Fu$ | 9, BC, 4, MP |
| 11. | $Fz \leftrightarrow \sim \sim Fu$ | 8, QN, UI, T90, BC, MP |
| 12. | $\sim \sim \sim Fu$ | 11, BC, 5, MT |
| 13. | ~~Show~~ $\bigwedge y \bigvee x(Fx \leftrightarrow \sim Fy) \rightarrow \bigvee xFx \wedge \bigvee x \sim Fx$ | |
| 14. | $\bigwedge y \bigvee x(Fx \leftrightarrow \sim Fy)$ | |
| 15. | $Fv \leftrightarrow \sim Fy$ | 14, UI, EI |
| 16. | ~~Show~~ $\bigvee xFx$ | |
| 17. | $\sim \bigvee xFx$ | |
| 18. | $\sim Fy$ | 17, QN, UI |
| 19. | $Fv$ | 15, BC, 18, MP |
| 20. | $\sim Fv$ | 17, QN, UI |
| 21. | ~~Show~~ $\bigvee x \sim Fx$ | |
| 22. | $\sim \bigvee x \sim Fx$ | |
| 23. | $Fv$ | 22, QN, UI, DN |
| 24. | $\sim Fy$ | 15, BC, 23, MP |
| 25. | $\sim \sim Fy$ | 22, QN, UI |
| 26. | $\bigvee xFx \wedge \bigvee x \sim Fx$ | 16, 21, Adj |
| 27. | $\bigwedge y \bigvee x(Fx \leftrightarrow \sim Fy) \leftrightarrow \bigvee xFx \wedge \bigvee x \sim Fx$ | 2, 13, CB |

T260 $\bigvee x \bigwedge y(Fx \leftrightarrow Fy) \leftrightarrow \sim \bigvee xFx \vee \bigwedge xFx$

T261 $\bigwedge x(Fx \rightarrow \bigvee y[Gy \wedge (Hy \vee Hx)]) \leftrightarrow \bigvee x(Gx \wedge Hx) \vee \sim \bigvee xFx \vee (\bigvee xGx \wedge \bigwedge x[Fx \rightarrow Hx])$

In T259 – T261, we were interested in 'disentangling' quantifiers. The following theorems, T262 – T266, illustrate the reverse process, that of 'consolidating' quantifiers. More precisely, let us say that a formula is in *prenex normal form* just in case it is a symbolic formula consisting of a string of quantifier phrases followed by a formula without quantifiers. T262 – T266 illustrate the fact that every symbolic formula is equivalent to one in prenex normal form. (In connection with T265 the reader should reconsider exercise 43.)

T262

| | | |
|---|---|---|
| 1. | ~~Show~~ $(\bigwedge x Fx \rightarrow \bigwedge x Gx) \leftrightarrow \bigvee x \bigwedge y(Fx \rightarrow Gy)$ | |
| 2. | ~~Show~~ $\sim \bigwedge x Fx \rightarrow \bigvee x \bigwedge y(Fx \rightarrow Gy)$ | |
| 3. | $\sim \bigwedge x Fx$ | |
| 4. | $\sim Fz$ | 3, QN, EI |
| 5. | ~~Show~~ $\bigwedge y(Fz \rightarrow Gy)$ | |
| 6. | $Fz \rightarrow Gy$ | 4, T18, MP |
| 7. | $\bigvee x \bigwedge y(Fx \rightarrow Gy)$ | 5, EG |
| 8. | ~~Show~~ $\bigwedge x Gx \rightarrow \bigvee x \bigwedge y(Fx \rightarrow Gy)$ | |
| 9. | $\bigwedge x Gx$ | |
| 10. | ~~Show~~ $\bigwedge y(Fx \rightarrow Gy)$ | |
| 11. | $Fx \rightarrow Gy$ | 9, UI, T2, MP |
| 12. | $\bigvee x \bigwedge y(Fx \rightarrow Gy)$ | 10, EG |
| 13. | $(\bigwedge x Fx \rightarrow \bigwedge x Gx) \rightarrow \bigvee x \bigwedge y(Fx \rightarrow Gy)$ | 2, 8, Adj, T35, BC, MP |
| 14. | ~~Show~~ $\bigvee x \bigwedge y(Fx \rightarrow Gy) \rightarrow (\bigwedge x Fx \rightarrow \bigwedge x Gx)$ | |
| 15. | $\bigvee x \bigwedge y(Fx \rightarrow Gy)$ | |
| 16. | $\bigwedge y\ (Fu \rightarrow Gy)$ | 15, EI |
| 17. | ~~Show~~ $\bigwedge x Fx \rightarrow \bigwedge x Gx$ | |
| 18. | $\bigwedge x Fx$ | |
| 19. | ~~Show~~ $\bigwedge x Gx$ | |
| 20. | $Gx$ | 16, UI, 18, UI, MP |
| 21. | $(\bigwedge x Fx \rightarrow \bigwedge x Gx) \leftrightarrow \bigvee x \bigwedge y(Fx \rightarrow Gy)$ | 13, 14, CB |

T263 $(\bigwedge x Fx \leftrightarrow \bigvee x Gx) \leftrightarrow \bigvee x \bigvee y \bigwedge z \bigwedge w([Fx \rightarrow Gy] \wedge [Gz \rightarrow Fw])$

T264 $(\bigvee x Fx \rightarrow [\bigvee x Gx \rightarrow \bigwedge x Hx]) \leftrightarrow \bigwedge x \bigwedge y \bigwedge z\ (Fx \wedge Gy \rightarrow Hz)$

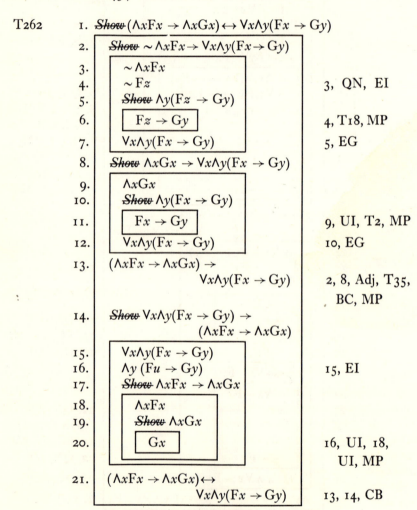

T265 $\bigwedge x(\bigvee y[H(xy) \wedge Fy] \rightarrow \bigvee y[H(xy) \wedge Gy]) \leftrightarrow$
$\bigwedge x \bigwedge y \bigvee z(H(xy) \wedge Fy \rightarrow H(xz) \wedge Gz)$

T266 $\bigwedge x(Fx \wedge \bigvee y G(xy) \rightarrow \bigvee y[H(xy) \wedge \bigwedge z J(xyz)]) \leftrightarrow$
$\bigwedge x \bigwedge y \bigvee w \bigwedge z[Fx \wedge G(xy) \rightarrow H(xw) \wedge J(xwz)]$

A universal and an existential quantifier can in some cases, though not of course in general, be commuted; T267 – T271 are illustrations.

T267

1. ~~*Show*~~ $\bigwedge x \bigvee y(Fx \rightarrow Gy) \rightarrow \bigvee y \bigwedge x(Fx \rightarrow Gy)$
2. $\bigwedge x \bigvee y\ (Fx \rightarrow Gy)$
3. ~~*Show*~~ $\bigvee y \bigwedge x(Fx \rightarrow Gy)$
4. $\sim \bigvee y \bigwedge x(Fx \rightarrow Gy)$
5. $\sim(Fz \rightarrow Gy)$ 4, QN, UI, QN, EI
6. $Fz \rightarrow Gw$ 2, UI, EI
7. $Gw$ 5, T21, MP, 6, MP
8. $\sim(Fv \rightarrow Gw)$ 4, QN, UI, QN, EI
9. $\sim Gw$ 8, T22, MP

T268 $\bigvee y \bigwedge x(Fx \rightarrow Gy) \rightarrow \bigwedge x \bigvee y(Fx \rightarrow Gy)$

T269 $\bigvee y \bigwedge x(Fx \vee Gy) \leftrightarrow \bigwedge x \bigvee y(Fx \vee Gy)$

T270 $\bigvee y \bigwedge x(Fx \wedge Gy) \leftrightarrow \bigwedge x \bigvee y(Fx \wedge Gy)$

T271 $\bigwedge x \bigwedge y \bigvee z(Fx \wedge Gy \rightarrow Hz) \leftrightarrow \bigwedge y \bigvee z \bigwedge x(Fx \wedge Gy \rightarrow Hz)$

## EXERCISES

In solving these exercises, the reader will still find useful the informal suggestions made on pages 115 – 17 of chapter III.

48. Prove T254 – T258.
49. Prove T260 and T261.
50. Prove T264.
51. Prove T269.

**7. Alphabetic variance.** In chapter III the equivalence of two generalizations differing only in their initial bound variables was observed. We may refer to symbolic formulas related in this way as *immediate alphabetic variants* of one another. Symbolic formulas differing only in immediate alphabetic variants are also equivalent and may be called *alphabetic variants.*

More precisely, we call two symbolic formulas *immediate alphabetic variants* if they have the forms

$$\bigwedge \alpha \phi$$

and

$$\bigwedge \alpha' \phi' \quad ,$$

or else

$$\bigvee \alpha \phi$$

and

$$\bigvee \alpha' \phi' \quad ,$$

where $\alpha$ and $\alpha'$ are variables, $\phi$ and $\phi'$ are symbolic formulas, $\phi'$ comes from $\phi$ by proper substitution of $\alpha'$ for $\alpha$, and $\phi$ comes from $\phi'$ by proper substitution of $\alpha$ for $\alpha'$. We say that a symbolic formula $\psi$ is an *alphabetic variant* of a symbolic formula $\psi'$ if $\psi$ is like $\psi'$ except for having one or more occurrences of a formula $\phi$ where $\psi'$ has some immediate alphabetic variant of $\phi$.

For example,

$$\bigvee x(Gx \wedge Hx)$$

is an immediate alphabetic variant of

$$\bigvee y(Gy \wedge Hy) \quad ,$$

and therefore

$$\bigwedge z[Fz \rightarrow \bigvee x(Gx \wedge Hx)]$$

is an alphabetic variant of

$$\bigwedge z[Fz \rightarrow \bigvee y\,(Gy \wedge Hy)] \quad .$$

Alphabetic variance will play an important role in section 9, in connection with abbreviatory clauses.

## EXERCISES

52. For each of the following pairs of formulas, state whether the second formula is an alphabetic variant of the first. (a) is answered for illustration.

(a) $\bigwedge x \bigvee y\ G(xy)$
$\bigwedge x \bigvee x\ G(xx)$

These formulas are not alphabetic variants; although '$G(xx)$' comes from '$G(xy)$' by proper substitution of '$x$' for '$y$', '$G(xy)$' does not come from '$G(xx)$' by proper substitution of '$y$' for '$x$'.

(b) $\bigwedge x(Fx \rightarrow \bigvee x Gx \wedge Hy)$
$\bigwedge x(Fx \rightarrow \bigvee y Gy \wedge Hy)$

(c) $\bigwedge x(Fx \rightarrow \bigvee x[Gx \wedge Hy])$
$\bigwedge x(Fx \rightarrow \bigvee y[Gy \wedge Hy])$

(d) $\bigvee z(\bigvee x\mathrm{H}x \wedge \mathrm{G}z)$
$\bigvee z(\bigvee z\mathrm{H}z \wedge \mathrm{G}z)$

**8. Substitution.** Present abbreviations make available, as lines of a derivation, instances of *sentential* theorems that have previously been proved. The reader may already have considered the possibility of extending this practice to *all* theorems previously proved, that is, the possibility of using instances of theorems of the quantifier as well as the sentential calculus in constructing derivations. Such an extension is indeed possible, but it has been deferred to the present point because the relevant notion of an *instance* exhibits complexities that did not arise in connection with sentential theorems.

Let us take as an example the theorem

(1) $$\bigwedge x\mathrm{F}x \to \mathrm{FA}(z) \quad .$$

(For perspicuity we employ here and in other illustrations the informal notation permitted by our conventions on superscripts and parentheses; in general formulations, however, the words 'term' and 'formula' are always to be understood in the official sense.) In forming instances of (1), we must make replacements that do not change its logical structure, that is, replacements of components other than sentential connectives, quantifiers, and parentheses. The replaceable components fall then into three categories: predicate letters, operation letters, and variables.

Let us first consider the replacement of predicate letters. We wish to say that the formula

(2) $$\bigwedge x \sim \mathrm{G}(xy) \to \sim \mathrm{G}(\mathrm{A}(z)\,y)$$

comes from (1) by *substitution* for the *predicate letter* 'F'. We may think of (2) as obtained from (1) in two steps.

We first consider the formula

(3) $$\sim \mathrm{G}(ay)$$

(which may be called the *substituend*, and in which, the reader will note, the variable '$a$' is free), and in (1) replace all occurrences of 'F' by (3) enclosed in braces. Thus we obtain

(4) $$\bigwedge x\,\{\sim \mathrm{G}(ay)\}\,x \to \{\sim \mathrm{G}(ay)\}\,A(z)$$

(which is not a formula).

Next we replace each part

$$\{\sim \mathrm{G}(ay)\}\zeta$$

of (4), where $\zeta$ is a term, by

$$\sim \mathrm{G}(\zeta y) \quad ,$$

that is, the result of replacing in (3) all free occurrence of '*a*' by $\zeta$. Thus we obtain the formula (2).

We consider now the general situation. Let $\delta$ be a $k$-place predicate letter occurring in a symbolic formula $\phi$. Let $\chi$ be a symbolic formula that is to be substituted for $\delta$. (In the example above, $\delta$ is 'F', $\phi$ is (1), and $\chi$ is (3).) The *substitution* takes place in two steps, as follows:

I. Throughout $\phi$, replace $\delta$ by $\chi$ enclosed in braces.

II. In the expression resulting from I (which will in general not be a formula), successively consider each part of the form

$$\{\chi\}\zeta_1 \ldots \zeta_k \quad ,$$

where $\zeta_1, \ldots, \zeta_k$ are terms. Replace each such part by a certain formula—in fact, by the formula obtained from $\chi$ by replacing all free occurrences of '*a*' by $\zeta_1$, '*b*' by $\zeta_2$, etc., up to the $k$th variable (in the standard ordering of variables), whose free occurrences are to be replaced by $\zeta_k$.

Certain restrictions are necessary in connection with substitution. Otherwise we could obtain nontheorems by substitution on theorems.

For example, let $\phi$, the formula in which the substitution is to take place, be

(5) $$\Lambda x Fx \rightarrow \Lambda x FB \quad .$$

((5) is obviously a theorem.) By substituting 'G(*xa*)' for 'F', we secure first

$$\Lambda x\{G(xa)\}\, x \rightarrow \Lambda x\{G(xa)\}\, B \quad ,$$

and then

(6) $$\Lambda x G(xx) \rightarrow \Lambda x G(xB) \quad .$$

But this formula is not a theorem (according to exercise 74 of this chapter).

As a second example, consider the theorem

$$\Lambda x Fx \rightarrow Fy \quad ,$$

and substitute for 'F' the formula

$$\mathrm{V} y G(ay) \quad .$$

We obtain first

$$\Lambda x\{\mathrm{V} y G(ay)\}\, x \rightarrow \{\mathrm{V} y G(ay)\}\, y \quad ,$$

and then

(7) $$\Lambda x \mathrm{V} y G(xy) \rightarrow \mathrm{V} y G(yy) \quad .$$

But this formula too is not a theorem (according to exercise 75).

The difficulties uncovered by these two examples result from a clash of variables and can be avoided by requiring that the substituend and the formula in which the substitution is to take place contain no common variables. Accordingly, we say that a symbolic formula $\psi$ comes from a symbolic formula $\phi$ by *proper substitution of* a symbolic formula $\chi$ *for* a $k$-place *predicate letter* $\delta$ just in case $\psi$ can be obtained from $\phi$ by steps I and II above, and in addition there is no variable occurring in both $\phi$ and $\chi$.

We turn now to substitution on operation letters. We wish to say that the formula

(8) $$\bigwedge x\mathrm{F}x \rightarrow \mathrm{FB}(zy)$$

comes from (1) by *substitution for* the *operation letter* 'A'. As in the case of (2), we may think of (8) as obtained from (1) in two steps.

We consider a certain term, the *substituend*,

(9) $$\mathrm{B}(ay) \quad ,$$

and in (1) replace all occurrences of 'A' by (9) enclosed in braces, obtaining

(10) $$\bigwedge x\mathrm{F}x \rightarrow \mathrm{F}\{\mathrm{B}(ay)\}z$$

(which is not a formula).

Next we replace each part

$$\{\mathrm{B}(ay)\}\zeta$$

of (10), where $\zeta$ is a term, by

$$\mathrm{B}(\zeta y) \quad ,$$

that is, the result of replacing in (9) all occurrences of '$a$' by $\zeta$. Thus we obtain the formula (8).

Again we may ascend to the general situation. Let $\delta$ be a $k$-place operation letter occurring in a symbolic formula $\phi$, and let $\eta$ be a symbolic term that is to be substituted for $\delta$. (In the example, $\delta$ is 'A', $\phi$ is again (1), and $\eta$ is (9).) The *substitution* takes place in two steps, exactly analogous to those involved in substitution on predicate letters.

I. Throughout $\phi$, replace $\delta$ by $\eta$ enclosed in braces.

II. In the expression resulting from I, successively consider each part of the form

$$\{\eta\}\,\zeta_1 \ldots \zeta_k \quad ,$$

where $\zeta_1, \ldots, \zeta_k$ are terms. Replace each such part by a certain term—in fact, by the term obtained from $\eta$ by replacing all occurrences of '$a$' by $\zeta_1$, '$b$' by $\zeta_2$, etc., up to the $k$th variable, whose occurrences are to be replaced by $\zeta_k$.

If substitution is to lead from theorems only to theorems, a restriction is again necessary, as the following example indicates.

Let $\phi$, the formula in which the substitution is to take place, be

$$(11) \qquad \bigwedge y[Fy \to GA(y)] \to \bigwedge y \bigwedge z[Fz \to GA(z)] \quad .$$

This is a theorem (according to exercise 58). By substituting the term 'B($ya$)' for the operation letter 'A', we secure first

$$\bigwedge y[Fy \to G\{B(ya)\}(y)] \to \bigwedge y \bigwedge z[Fz \to G\{B(ya)\}(z)] \quad ,$$

and then

$$(12) \qquad \bigwedge y[Fy \to GB(yy)] \to \bigwedge y \bigwedge z[Fz \to GB(yz)] \quad .$$

But (12) is not a theorem (according to exercise 76).

The difficulty here can be avoided by exactly that restriction which was introduced in connection with substitution on predicate letters. Accordingly, we say that a symbolic formula $\psi$ comes from a symbolic formula $\phi$ by *proper substitution of* a symbolic term $\eta$ *for* a $k$-place *operation letter* $\delta$ just in case $\psi$ can be obtained from $\phi$ by the steps I and II above, and in addition there is no variable occurring in both $\phi$ and $\eta$.

Let us now consider substitution on variables. For free occurrences of variables the relevant notion has already been introduced on page 148; that is, the notion of *proper substitution of a term for a variable.* For bound occurrences of variables no notion of substitution is required, for the role it would play can be handled by alphabetic variance.

Thus we have three kinds of proper substitution—for predicate letters, for operation letters, and for (free) variables. An *instance* of a symbolic formula is obtained by iterated proper substitution; also, we shall consider each symbolic formula as an instance of itself. Thus a symbolic formula $\psi$ is said to be an *instance* of a symbolic formula $\phi$ just in case $\psi$ is $\phi$ or obtainable from $\phi$ by one or more operations of proper substitution.

For example, the theorem

$$(13) \qquad \bigwedge xFx \to Fy$$

has as an instance

$$\bigwedge x(Fx \vee Gx) \to (FA \vee GA) \quad .$$

In this case we have made two proper substitutions; for 'F' we have substituted '(F$a$ v G$a$)' and for '$y$' we have substituted 'A'. We may indicate this sequence of substitutions diagrammatically as follows:

$$\frac{F \qquad y}{(Fa \vee Ga) \qquad A} \quad .$$

Here the order in which the substitutions are made is unimportant, but this is not always true. For example,

$$\bigwedge x(\bigvee y Fy \vee Gx) \rightarrow (\bigvee y Fy \vee GA)$$

is also an instance of (13), in view of the sequence of substitutions

$$\frac{\begin{array}{cc} y & F \end{array}}{\begin{array}{cc} A & (\bigvee y Fy \vee Ga) \end{array}} \quad .$$

The reverse sequence,

$$\frac{\begin{array}{cc} F & y \end{array}}{\begin{array}{cc} (\bigvee y Fy \vee Ga) & A \end{array}} \quad ,$$

would lead, however, to an improper substitution on 'F'.

Our new notion of an instance subsumes the earlier notion of an instance of a sentential theorem. To see this, it is sufficient to consider substitution on 0-place predicate letters.

As before, the relation of instance is intended to hold only between *symbolic* formulas. This restriction is now to some extent forced upon us; to introduce a satisfactory notion of instance for formulas involving phrases of quantity and their stylistic variants would be extremely laborious.

## EXERCISES

53. Substitute (according to steps I and II) each of the formulas (i) – (x) for 'F' in the theorem

$$F(xy) \rightarrow \bigvee z F(zy) \quad .$$

In which cases is the substitution proper? Parts (i) and (ii) are solved for illustration.

(i) $G(bba)$

(Steps I and II lead to

$$G(yyx) \rightarrow \bigvee z G(yyz) \quad ,$$

and the substitution is proper.)

(ii) $G(bxa)$

(In this case substitution leads to

$$G(yxx) \rightarrow \bigvee z G(yxz) \quad .$$

The substitution is improper, for '$x$' is common to (ii) and the theorem above.)

(iii) $G(baa)$
(iv) $\bigvee w G(awb)$

(v) $\bigvee y G y$
(vi) $\bigvee w G w$
(vii) $G(abc)$
(viii) $G(bw)$
(ix) $G(azb)$
(x) $Ga$

54. For each of the formulas (i) – (x) below, state whether it comes from the theorem

$$(14) \qquad Fx \rightarrow \bigvee y F y$$

by proper substitution of a formula for 'F'; if so, indicate the relevant substituend. Parts (i) and (ii) are solved for illustration.

(i) $\sim G(xz) \rightarrow \bigvee y \sim G(yz)$

(The formula (i) comes from (14) by proper substitution of

$$\sim G(az)$$

for 'F'.)

(ii) $\sim G(xy) \rightarrow \bigvee y \sim G(yy)$

(The formula (ii) does not come from (14) by proper substitution on 'F'. As the reader can verify, '$\sim G(ay)$' is the only substituend that will permit the passage from (14) to (ii); but a clash of variables makes this substitution improper.)

(iii) $G(xx) \rightarrow \bigvee y G(yy)$
(iv) $G(xx) \rightarrow \bigvee y G(xy)$
(v) $G(xx) \rightarrow \bigvee y G(yx)$
(vi) $\bigwedge w G(wx) \rightarrow \bigvee y \bigwedge w G(wy)$
(vii) $\bigwedge w G(ww) \rightarrow \bigvee y \bigwedge w G(wy)$
(viii) $\bigwedge w G(ww) \rightarrow \bigvee y \bigwedge w G(ww)$
(ix) $G(xx) \rightarrow \bigvee y G(yz)$
(x) $P \rightarrow \bigvee y P$

55. Substitute each of the terms (i) – (x) below for 'A' in the theorem

$$\bigwedge x F A(xz) \rightarrow \bigvee y F A(yz) \quad .$$

In which cases is the substitution proper?

(i) $B(C(b)\, a)$
(ii) $B(ayb)$
(iii) $B(a)$
(iv) $B(b)$
(v) $B$
(vi) $B(acb)$

(vii) $B(C(a))$
(viii) $w$
(ix) $a$
(x) $B(A(bw)\ C(a))$

56. For each of the formulas (i) – (v) below, state whether it comes from the theorem

$$F(A(zy)\ y) \rightarrow \bigvee x\ F(A(xy)\ y)$$

by proper substitution of a term for 'A'; if so, indicate the relevant substituend.

(i) $F(B[C(z)]\ y) \rightarrow \bigvee x F(B[C(x)]\ y)$
(ii) $F\ B(zyy) \rightarrow \bigvee x F B(xyy)$
(iii) $F(A(yz)\ y) \rightarrow \bigvee x F(A(yx)\ y)$
(iv) $F(zy) \rightarrow \bigvee x F(xy)$
(v) $F(yy) \rightarrow \bigvee x F(yy)$

57. For each of the following pairs of formulas state whether the second is an instance of the first obtainable by the indicated sequence of substitutions. Cases (i) and (ii) are solved for illustration.

(i) $Fx \rightarrow \bigvee y Fy$ ; $G(xx) \rightarrow \bigvee y G(xy)$

$$\frac{F \qquad z}{G(za) \qquad x}$$

The second formula is an instance of the first obtainable by the indicated sequence of substitutions. (The reader should compare this case with part (iv) of exercise 54.)

(ii) $F(xy) \rightarrow \bigvee y F(xy)$ ; $[\bigvee x Hx \lor G(yz)] \rightarrow \bigvee y[\bigvee x Hx \lor G(yz)]$

$$\frac{F}{[\bigvee x Hx \lor G(bz)]}$$

Here the second formula is not an instance of the first obtained by the indicated substitution. (The second can, however, be obtained as an instance of the first by the following sequence of substitutions:

$$\frac{x \qquad F}{z \qquad [\bigvee x Hx \lor G(ba)]} \text{ .)}$$

(iii) $\bigwedge x(Fx \rightarrow P) \leftrightarrow (\bigvee x Fx \rightarrow P)$ ; $\bigwedge x[F(xy) \rightarrow \bigwedge z G(yz)] \leftrightarrow [\bigvee x F(xy) \rightarrow \bigwedge z G(yz)]$

$$\frac{P \qquad F \qquad w}{\bigwedge z G(yz) \qquad F(aw) \qquad y}$$

(iv) $\sim \bigvee x Fx \rightarrow \bigwedge x(Fx \rightarrow Gx)$ ; $\sim \bigvee x \bigvee y F(xy) \rightarrow \bigwedge x[\bigvee y F(xy) \rightarrow \bigwedge z G(A(x)\ z\ B(x))]$

$$\frac{\mathrm{F} \qquad \mathrm{G}}{\mathrm{V}y\mathrm{F}(ay) \qquad \Lambda z\mathrm{G}(\mathrm{A}(a)\ z\ \mathrm{B}(a))}$$

(v) $\Lambda x\mathrm{FA}(x\ \mathrm{B}) \rightarrow \mathrm{V}x\mathrm{FA}(xx)$ ; $\Lambda x\mathrm{FD}(\mathrm{B}(xz)) \rightarrow \mathrm{V}x\mathrm{FD}(\mathrm{B}(xx))$

$$\frac{\mathrm{B} \qquad \mathrm{A}}{z \qquad \mathrm{D}(\mathrm{B}(ab))}$$

(vi) $\mathrm{F}y \rightarrow \mathrm{V}x\mathrm{F}x$ ; $\mathrm{F}(xy) \rightarrow \mathrm{V}x\mathrm{F}(xx)$

$$\frac{\mathrm{F} \qquad z}{\mathrm{F}(za) \qquad x}$$

58. Show that formula (11) of page 160 is a theorem.

**9. Abbreviated derivations reconsidered.** Using the notions of the last two sections, we can now introduce more sweeping methods of abbreviating derivations than those carried over from chapter III. To begin with, we can drop from clause (7) the restriction to sentential theorems, since the general notion of an instance is now available.

Thus we replace (7) as formulated in chapter III by the following:

(7) *If $\phi$ is an instance of an alphabetic variant of a theorem that has already been proved, then $\phi$ may occur as a line.* (*Annotation: the number of the theorem in question, sometimes together with a diagrammatic indication of the sequence of substitutions involved.*)

Clause (8) of chapter III is no longer needed; any application of QN can be obtained by the revised clause (7) (applied to T203 or T204), together with one or two sentential steps. We replace the old clause (8) by a new clause involving the notion of alphabetic variance:

(8) *A symbolic formula may occur as a line if it is an alphabetic variant of an antecedent line.* (*Annotation: 'AV' and the number of the antecedent line.*)

Symbolic formulas differing only in equivalent parts are themselves equivalent, and this fact provides us with another useful device for abbreviating derivations. To incorporate this device into our system we replace the former clause (9) (which will reappear as clause (10)) by the following:

(9) *If $\phi$, $\phi'$ are symbolic formulas such that*

$$\phi \leftrightarrow \phi'$$

*is an instance of an alphabetic variant of a previously proved theorem, and $\psi$, $\psi'$ are symbolic formulas such that $\psi'$ is like $\psi$ except for having one or more occurrences of $\phi'$ where $\psi$ has occurrences of $\phi$, then*

$$\psi \leftrightarrow \psi'$$

*may occur as a line.* (*In such a case we may call*

$$\psi \leftrightarrow \psi'$$

*an* interchange of equivalents *on the basis of the previously proved theorem; as annotation we use 'IE' with a parenthetical reference to this theorem.*)

As before, we permit the compression of several steps into one, and the following clause makes provision for this.

(*10*) *A symbolic formula may occur as a line if it follows from antecedent lines by a succession of steps, and each intermediate step can be justified by one of clauses* (*2*), (*5a*), (*7*), (*8*), *or* (*9*). (*The annotation should determine the omitted succession of steps by indicating, in order of application, the antecedent lines, the premises, the inference rules, and the previously proved theorems employed. We should also make such indication of terms and formulas involved in the intermediate inferences as may be required to dispel remaining ambiguity.*)

The special annotations 'SC' and 'CD', which were introduced earlier to indicate special kinds of compression, will continue to be used. We shall now use the annotation 'QN' as an abbreviation for 'T203, BC, MP' or 'T204, BC, MP'; it will therefore comprehend all inferences that its previous use would have justified.

Clauses (7) – (10), like the earlier (7) – (9), satisfy the requirements given on page 58; that is, they are theoretically dispensable, and derivations constructed on their basis can be automatically checked for correctness (at least when membership in the class of premises is automatically decidable and annotations are present). It should be mentioned that an exact demonstration of the theoretical dispensability of clauses (7) – (10) would be more involved than in the case of previous abbreviations. To obtain a proof of an instance of a theorem from the proof of the theorem itself it is not sufficient to make simple replacements on the original proof: the restrictions imposed on variables in our basic clauses will sometimes necessitate a systematic change of variables in the original proof before the replacements can be made. The rule IE (and also AV in some of its applications) raises a special problem. IE cannot be regarded as an abbreviation of the same kind as QN, SC, and CD, for there is no one theorem to whose instances all applications of IE correspond. Theorems particularly useful in eliminating applications of IE are:

T96 $(P \leftrightarrow Q) \leftrightarrow (\sim P \leftrightarrow \sim Q)$

T97 $(P \leftrightarrow R) \wedge (Q \leftrightarrow S) \rightarrow [(P \rightarrow Q) \leftrightarrow (R \rightarrow S)]$

T98 $(P \leftrightarrow R) \wedge (Q \leftrightarrow S) \rightarrow (P \wedge Q \leftrightarrow R \wedge S)$

T99 $(P \leftrightarrow R) \wedge (Q \leftrightarrow S) \rightarrow (P \vee Q \leftrightarrow R \vee S)$

T100 $(P \leftrightarrow R) \wedge (Q \leftrightarrow S) \rightarrow [(P \leftrightarrow Q) \leftrightarrow (R \leftrightarrow S)]$

T213 $\bigwedge x(Fx \leftrightarrow Gx) \rightarrow (\bigwedge xFx \leftrightarrow \bigwedge xGx)$

T214 $\bigwedge x(Fx \leftrightarrow Gx) \rightarrow (\bigvee xFx \leftrightarrow \bigvee xGx)$

Elimination of an application of AV would require (in addition to the procedure necessary with IE) proving a suitable instance of T231 or T232.

The economy achieved by clauses (7) – (10) is illustrated by the following proofs:

1. ~~Show~~ $\bigvee x(\bigvee xFx \rightarrow Fx)$ (T229)
2. $\bigvee x(\bigvee yFy \rightarrow Fx) \leftrightarrow (\bigvee yFy \rightarrow \bigvee xFx)$ — T220$\left(\frac{P}{\bigvee yFy}\right)$
3. $\bigvee x(\bigvee xFx \rightarrow Fx) \leftrightarrow (\bigvee xFx \rightarrow \bigvee xFx)$ — 2, AV
4. $\bigvee x(\bigvee xFx \rightarrow Fx)$ — 3, BC, T1, MP

(Compare with the proof of T229 on page 112.)

1. ~~Show~~ $\bigwedge x\bigwedge yF(xy) \rightarrow \bigwedge y\bigwedge xF(yx)$ (T257)
2. $\bigwedge x\bigwedge yF(xy)$
3. $\bigwedge x\bigwedge zF(xz)$ — 2, AV
4. $\bigwedge y\bigwedge zF(yz)$ — 3, AV
5. $\bigwedge y\bigwedge xF(yx)$ — 4, AV

(Compare with the proof of T257 constructed in solving exercise 48 of chapter IV, p. 155.)

1. ~~Show~~ $\sim\bigvee x(Fx \wedge Gx) \leftrightarrow \bigwedge x(Fx \rightarrow \sim Gx)$ (T243)
2. $\sim\bigvee x(Fx \wedge Gx) \leftrightarrow \bigwedge x\sim(Fx \wedge Gx)$ — T204$\left(\frac{F}{(Fa \wedge Ga)}\right)$
3. $\sim\bigvee x(Fx \wedge Gx) \leftrightarrow \bigwedge x(Fx \rightarrow \sim Gx)$ — 2, IE(T39), BC, MP

The passage from line 2 to line 3 in this proof would look, if expanded, as follows:

2. $\sim \bigvee x(Fx \wedge Gx) \leftrightarrow \bigwedge x \sim (Fx \wedge Gx)$
2a. $[\sim \bigvee x(Fx \wedge Gx) \leftrightarrow \bigwedge x \sim (Fx \wedge Gx)] \leftrightarrow$ $[\sim \bigvee x(Fx \wedge Gx) \leftrightarrow \bigwedge x(Fx \rightarrow \sim Gx)]$ — IE$\left(T39 \frac{P \quad Q}{Fx \quad Gx}\right)$
2b. $[\sim \bigvee x(Fx \wedge Gx) \leftrightarrow \bigwedge x \sim (Fx \wedge Gx)] \rightarrow$ $[\sim \bigvee x(Fx \wedge Gx) \leftrightarrow \bigwedge x(Fx \rightarrow \sim Gx)]$ — 2a, BC
3. $\sim \bigvee x(Fx \wedge Gx) \leftrightarrow \bigwedge x(Fx \rightarrow \sim Gx)$ — 2, 2b, MP

This sort of inference will occur frequently. We shall therefore often omit 'BC' and 'MP' from an annotation like '2, IE(T39), BC, MP'.

(Compare the proof given here of T243 with the one given on page 115.)

Another example:

1. ~~*Show*~~ $\sim \bigvee x \bigvee y F(xy) \rightarrow \bigwedge x \bigwedge y \sim F(xy)$

| | | |
|---|---|---|
| 2. | $\sim \bigvee x \bigvee y F(xy)$ | |
| 3. | $\bigwedge x \sim \bigvee y F(xy)$ | 2, QN |
| 4. | $\bigwedge x \bigwedge y \sim F(xy)$ | 3, IE(T204) |

In this proof the reader should observe that he could not pass directly from line 3 to line 4, as he passed from line 2 to line 3, by means of QN. The passage from line 3 to line 4 would look, if expanded, as follows:

3. $\bigwedge x \sim \bigvee y F(xy)$
3a. $\bigwedge x \sim \bigvee y F(xy) \leftrightarrow \bigwedge x \bigwedge y \sim F(xy)$ — IE$\left(T204 \frac{F}{F(xa)}\right)$
3b. $\bigwedge x \sim \bigvee y F(xy) \rightarrow \bigwedge x \bigwedge y \sim F(xy)$ — 3a, BC
4. $\bigwedge x \bigwedge y \sim F(xy)$ — 3, 3b, MP

(The interchange of equivalents that occurs in line 3a results from the substitution of '$F(xa)$' for '$F$' in '$\sim \bigvee y Fy \leftrightarrow \bigwedge y \sim Fy$', which is an alphabetic variant of T204.)

Variants of the confinement laws, useful for the exercises to follow, can now be obtained without the tedious proofs they would have required in chapter III.

T272 1. ~~*Show*~~ $\bigwedge x(Fx \wedge P) \leftrightarrow \bigwedge x Fx \wedge P$

| | | |
|---|---|---|
| 2. | $\bigwedge x(P \wedge Fx) \leftrightarrow P \wedge \bigwedge x Fx$ | T215 |
| 3. | $\bigwedge x(Fx \wedge P) \leftrightarrow P \wedge \bigwedge x Fx$ | 2, IE(T24) |
| 4. | $\bigwedge x(Fx \wedge P) \leftrightarrow \bigwedge x Fx \wedge P$ | 3, IE(T24) |

T273 $\bigvee x(Fx \wedge P) \leftrightarrow \bigvee x Fx \wedge P$

T274 $\bigwedge x(Fx \vee P) \leftrightarrow \bigwedge x Fx \vee P$

T275 $\bigvee x(Fx \vee P) \leftrightarrow \bigvee x Fx \vee P$

## EXERCISES, GROUP I

Using the abbreviations of the present section give brief proofs of the following theorems. Exercise 59 is solved for illustration.

59. $(\bigwedge x\, Fx \leftrightarrow \bigvee x\, Gx) \leftrightarrow \bigvee x \bigvee y \bigwedge z \bigwedge w([Fx \rightarrow Gy] \wedge [Gz \rightarrow Fw])$ (T263)

| | | |
|---|---|---|
| 1. | ~~Show~~ $(\bigwedge x Fx \leftrightarrow \bigvee x Gx) \leftrightarrow \bigvee x \bigvee y \bigwedge z \bigwedge w([Fx \rightarrow Gy] \wedge [Gz \rightarrow Fw])$ | |
| 2. | $(\bigwedge x Fx \leftrightarrow \bigvee x Gx)$ $\leftrightarrow ([\bigwedge x Fx \rightarrow \bigvee x Gx] \wedge [\bigvee x Gx \rightarrow \bigwedge x Fx])$ | T81 |
| 3. | $\leftrightarrow (\bigvee x[Fx \rightarrow \bigvee x Gx] \wedge [\bigvee x Gx \rightarrow \bigwedge x Fx])$ | 2, AV, IE(T222), AV |
| 4. | $\leftrightarrow \bigvee x([Fx \rightarrow \bigvee x Gx] \wedge [\bigvee x Gx \rightarrow \bigwedge x Fx])$ | 3, AV, IE(T273), AV |
| 5. | $\leftrightarrow \bigvee x(\bigvee y[Fx \rightarrow Gy] \wedge [\bigvee x Gx \rightarrow \bigwedge x Fx])$ | 4, AV, IE(T220) |
| 6. | $\leftrightarrow \bigvee x \bigvee y([Fx \rightarrow Gy] \wedge [\bigvee x Gx \rightarrow \bigwedge x Fx])$ | 5, IE(T273) |
| 7. | $\leftrightarrow \bigvee x \bigvee y([Fx \rightarrow Gy] \wedge \bigwedge z[Gz \rightarrow \bigwedge x Fx])$ | 6, AV, IE(T221) |
| 8. | $\leftrightarrow \bigvee x \bigvee y \bigwedge z([Fx \rightarrow Gy] \wedge [Gz \rightarrow \bigwedge x Fx])$ | 7, IE(T215) |
| 9. | $\leftrightarrow \bigvee x \bigvee y \bigwedge z([Fx \rightarrow Gy] \wedge \bigwedge w[Gz \rightarrow Fw])$ | 8, AV, IE(T219) |
| 10. | $\leftrightarrow \bigvee x \bigvee y \bigwedge z \bigwedge w([Fx \rightarrow Gy] \wedge [Gz \rightarrow Fw])$ | 9, IE(T215) |

(It is understood, in the proof above, that the left side of line 2 is to be carried over to the blanks of lines 3 – 10.) The passage from line 2 to line 3 would look, if partially expanded, as follows:

| | | |
|---|---|---|
| 2. | $(\bigwedge x Fx \leftrightarrow \bigvee x Gx)$ $\leftrightarrow ([\bigwedge x Fx \rightarrow \bigvee x Gx] \wedge [\bigvee x Gx \rightarrow \bigwedge x Fx])$ | |
| 2a. | $\leftrightarrow ([\bigwedge x Fx \rightarrow \bigvee y Gy] \wedge [\bigvee x Gx \rightarrow \bigwedge x Fx])$ | 2, AV |
| 2b. | $\leftrightarrow (\bigvee x[Fx \rightarrow \bigvee y Gy] \wedge [\bigvee x Gx \rightarrow \bigwedge x Fx])$ | 2a, IE(T222) |
| 3. | $\leftrightarrow (\bigvee x[Fx \rightarrow \bigvee x Gx] \wedge [\bigvee x Gx \rightarrow \bigwedge x Fx])$ | 2b, AV |

60. $\bigvee x(Fx \rightarrow \bigwedge x Fx)$ (T230)

61. $\bigwedge x(\bigvee y[H(xy) \wedge Fy] \rightarrow \bigvee y[H(xy) \wedge Gy]) \leftrightarrow$
$\bigwedge x \bigwedge y \bigvee z(H(xy) \wedge Fy \rightarrow H(xz) \wedge Gz)$ (T265)

62. $\bigwedge x(Fx \wedge \bigvee y G(xy) \rightarrow \bigvee y[H(xy) \wedge \bigwedge z\, J(xyz)]) \leftrightarrow$
$\bigwedge x \bigwedge y \bigvee w \bigwedge z[Fx \wedge G(xy) \rightarrow H(xw) \wedge J(xwz)]$ (T266)

63. $\bigvee y \bigwedge x(Fx \rightarrow Gy) \rightarrow \bigwedge x \bigvee y(Fx \rightarrow Gy)$ (T268)

64. $\bigwedge x \bigwedge y \bigvee z(Fx \wedge Gy \rightarrow Hz) \leftrightarrow$
$\bigwedge y \bigvee z \bigwedge x(Fx \wedge Gy \rightarrow Hz)$ (T271)

## EXERCISES, GROUP II

Symbolize each of the following arguments, indicating the scheme of abbreviation used, and derive the conclusion of the symbolization from its premises. Exercise 65 is solved for illustration.

65. Alfred shaves all and only those inhabitants of Berkeley who do not shave themselves. Alfred is an inhabitant of Berkeley. Therefore Alfred does not shave himself.

On the basis of the scheme of abbreviation

$F^2$ : *a* is an inhabitant of *b*
$S^2$ : *a* shaves *b*
$A^0$ : Berkeley
$B^0$ : Alfred ,

the argument becomes

$$\Lambda x(F(xA) \rightarrow [S(Bx) \leftrightarrow \sim S(xx)]) \quad . \qquad F(BA) \quad \therefore \sim S(BB) \quad .$$

Its validity is established by the following derivation:

| | | |
|---|---|---|
| 1. | ~~*Show*~~ $\sim S(BB)$ | |
| 2. | $S(BB)$ | |
| 3. | $F(BA) \rightarrow [S(BB) \leftrightarrow \sim S(BB)]$ | 1st premise, UI |
| 4. | $\sim S(BB)$ | 2nd premise, 3 MP, BC, 2, MP |

66. Alfred shaves all and only those inhabitants of Berkeley who do not shave themselves. Alfred is an inhabitant of Berkeley. Therefore Alfred shaves himself.

(Exercises 65 and 66 justify the assertions under (2) of page 3.)

67. Every student is able to solve some problems and not able to solve some problems. Some teacher is able to solve all problems. ∴ Some teacher is not a student.

68. All members of the Board of Regents distrust every member of the Communist party. Every proponent of Marxism is a member of the Communist party. Some member of the Board of Regents does not distrust some proponent of Marxism. ∴ There is a Communist on the faculty.

69. No student likes every course he takes unless he takes no courses but philosophical studies. No mathematical study is a philosophical study. Some student takes nothing but mathematical studies and likes every course he takes. ∴ Some student does not take any courses.

70. No member of the club owes a debt to the treasurer of the club. A member of the club has not paid the entrance fee only if he owes a debt to the treasurer of the club. ∴ If the treasurer of the club is a member of the club, then he has paid the entrance fee.

## *EXERCISES, GROUP III

71. For each of the following monadic formulas, find an equivalent monadic formula without overlay (see p. 153).

$$\bigvee y \Lambda x(Fx \rightarrow \bigvee z[Gz \rightarrow Hy])$$
$$\bigvee x \bigvee y(Fx \leftrightarrow \sim Fy)$$
$$\Lambda x \Lambda y(Fx \leftrightarrow Fy)$$

72. Formulate a general procedure for transforming any monadic formula into an equivalent monadic formula without overlay.

**10. *Invalidity.** As Church has shown in [2], there is no automatic test for the validity of arguments formulated within the present symbolic language. There are, however, several ways of showing arguments invalid.

In the first place, the method of chapter III, that of *truth-functional expansions*, remains available. It is, however, applicable only to a restricted class of arguments. It must be used only in connection with symbolic arguments whose premises and conclusion are sentences containing no terms other than variables; incorrect results might otherwise be obtained. Further, even among arguments of this sort, cases will now arise of invalid arguments whose invalidity cannot be established by this method; an example is argument (1) below. In the present section, we therefore introduce methods of wider applicability.

By an *English translation of a symbolic argument* on the basis of a given scheme of abbreviation we understand another argument whose premises and conclusion are English translations, on the basis of that scheme, of the respective premises and conclusion of the symbolic argument—in other words, an English argument having the symbolic argument as a symbolization on the basis of the given scheme. As before, we say that an English *argument* is *false* if its premises are true sentences of English and its conclusion is a false sentence of English.

It happens that if a symbolic argument whose premises and conclusion are sentences has, on the basis of some scheme of abbreviation, an English translation that is false, then the symbolic argument cannot be valid. (This fact will not be proved here; it follows from Montague and Kalish [1].) Thus, to establish the invalidity (in the quantifier calculus) of a symbolic argument, it is sufficient to show that it has a false English translation.

For example, the argument

$$\bigvee x\mathrm{F}x \quad . \quad \bigvee x\mathrm{G}x \quad \therefore \bigvee x(\mathrm{F}x \wedge \mathrm{G}x)$$

is invalid, because it has the following, clearly false, English translation:

Something is an odd number. Something is an even number.
∴ Something is both an odd number and an even number.

In showing arguments invalid it is often necessary to determine the truth or falsehood of English sentences of a rather complex structure. Intuition is not in all cases an unequivocal guide. For instance, the sentence

If snow is red, then blood is green

may seem true to some, false to others. Our deductive system, however, provides assistance in the following way.

We say that an *English sentence* is *derivable* from other *English sentences* just in case there is a valid argument with the latter as premises and the former as conclusion. If an English sentence is derivable from true English sentences, then it will be regarded as true, and if a false English sentence is derivable from it, then it will be regarded as false. Thus, for example, the sentence above is to be considered true because it is derivable from the indisputably true sentence 'Snow is not red'. Also, if we were to doubt the falsehood of

For each $x$, $x$ is red,

it would be sufficient to consider the obviously false sentence

The White House is red.

As a further example, consider the sentence

There is an object $x$ such that if $x$ is identical with 2, then $x$ is not identical with $x$.

To show its truth it is sufficient to consider the true sentence

If 1 is identical with 2, then 1 is not identical with 1.

The method of showing invalidity given above is directly applicable only to those symbolic arguments whose premises and conclusion are sentences. We can, however, extend the method to arbitrary symbolic arguments.

As before, it happens that a symbolic argument is valid just in case each of its closures is valid. (For a definition of *closure*, see p. 126.) Thus to establish the invalidity of a symbolic argument it is sufficient to find a false English translation of one of its closures.

As a matter of fact, we may make a stronger statement: a symbolic argument is invalid (in the quantifier calculus) *just in case* a closure of it has a false English translation. Thus, for any symbolic argument, either a derivation may be supplied or else a false English translation of one of its closures will exist. (Yet there is no automatic way of deciding which alternative holds.)

The task of establishing invalidity can in some cases be simplified. The argument

(1) $$\bigwedge x \bigwedge y \bigwedge z[F(xy) \wedge F(yz) \rightarrow F(xz)] \quad . \quad \bigwedge x \bigvee y\ F(xy) \quad \therefore \bigvee x F(xx) \quad ,$$

has no obvious false translations. At first inspection, the argument

(2) For each $x$, for each $y$, for each $z$, if $x < y$ and $y < z$, then $x < z$. For each $x$, there is an object $y$ such that $x < y$. Therefore there is an object $x$ such that $x < x$

might present the appearance of a false English translation of (1). But this is not clear, for the second premise of (2) might reasonably be regarded as false. It does *not* assert that

for every *number* there is a greater,

which is true, but that

for every *thing* there is a greater,

which is doubtful. (Though such exceptions as the universe are amusing the doubt attaches to any object that is not a number; the meaning of '$<$' is not clearly determined for such objects.) The difficulty would seem to be resolved if we could in some way restrict the phrases of quantity in (2) so as to apply only to numbers. This is made possible by considering, instead of (1), another argument, called its *relativization*.

If $\phi$ is a symbolic sentence and $\pi$ a 1-place predicate letter not occurring in $\phi$, then by the *relativization of* $\phi$ *to* $\pi$ we understand that sentence which is obtained from $\phi$ by replacing each part of the form

$$\Lambda\alpha\psi$$

or

$$\mathrm{V}\alpha\psi \quad ,$$

where $\alpha$ is a variable and $\psi$ a formula, by

$$\Lambda\alpha(\pi\alpha \rightarrow \psi)$$

or

$$\mathrm{V}\alpha\,(\pi\alpha \wedge \psi)$$

respectively. For example, the relativization of the second premise of (1) to '$G^1$' is

$$\Lambda x(\mathrm{G}x \rightarrow \mathrm{V}y[\mathrm{G}y \wedge \mathrm{F}(xy)]) \quad .$$

If $\delta$ is a $k$-place operation letter and $\pi$ a 1-place predicate letter, then by the *closure axiom* for $\delta$ with respect to $\pi$ we understand the sentence

$$\pi\,\delta$$

if $k = 0$, or the sentence

$$\Lambda x_1 \ldots \Lambda x_k[\pi x_1 \wedge \ldots \wedge \pi x_k \rightarrow \pi\delta(x_1 \ldots x_k)]$$

if $k \geqslant 1$. For example, the closure axiom for '$A^0$' with respect to '$G^1$' is

$$\mathrm{GA} \quad ,$$

and for '$A^2$' with respect to '$G^1$' is

$$\Lambda x_1 \Lambda x_2[\mathrm{G}x_1 \wedge \mathrm{G}x_2 \rightarrow \mathrm{G\,A}(x_1 x_2)] \quad .$$

Let $A$ be a symbolic argument whose premises and conclusion are sentences, let $\pi$ be a 1-place predicate occurring in none of these, and let $\delta_1, \ldots, \delta_n$ be all the operation letters occurring in $A$. Then a *relativization of A to* $\pi$ is an argument obtained from $A$ by replacing its premises and conclusion by their relativizations to $\pi$ and adjoining, as additional premises, the closure axioms for $\delta_1, \ldots, \delta_n$ with respect to $\pi$, together with the sentence

$$\bigvee x\, \pi\, x \quad .$$

For example, a relativization of (1) to 'G¹' is the argument

$$(3)\quad \bigwedge x[Gx \to \bigwedge y(Gy \to \bigwedge z[Gz \to (F(xy) \wedge F(yz) \to F(xz))])] \quad . \quad \bigwedge x(Gx \to \bigvee y[Gy \wedge F(xy)]) \quad . \quad \bigvee xGx \quad \therefore \bigvee x(Gx \wedge F(xx)) \quad .$$

(Other relativizations to 'G¹' could be obtained by changing the order of the premises.)

If $A$, $B$ are symbolic arguments whose premises and conclusions are sentences, then $A$ is said to be a *relativization* of $B$ if there is a 1-place predicate letter $\pi$ not occurring in $B$ such that $A$ is a relativization of $B$ to $\pi$.

Now it happens that if an argument whose premises and conclusion are sentences is valid, then so is its relativization to any 1-place predicate letter not occurring in it. Thus to show such an argument invalid it is sufficient to find a false English translation of one of its relativizations. For example, the invalidity of (1) is established by the following false translation of (3):

> For all numbers $x$, $y$, $z$, if $x < y$ and $y < z$, then $x < z$. For each number $x$, there is a number $y$ such that $x < y$. There is a number. Therefore there is a number $x$ such that $x < x$.

As another example consider the argument

$$\bigwedge xF[A(x)\, x] \quad \therefore \bigvee xF[A(x)\, A(x)] \quad .$$

To establish the invalidity of this argument we consider the relativization

$$\bigwedge x(Gx \to F[A(x)\, x]) \quad . \quad \bigwedge x_1(Gx_1 \to GA(x_1)) \quad . \quad \bigvee xGx \quad \therefore \bigvee x(Gx \wedge F[A(x)\, A(x)]) \quad ,$$

together with the following false translation of the latter:

> The successor of a number is greater than that number. The successor of a number is a number. There is a number. Therefore the successor of some number is greater than itself.

The methods given above for showing invalidity suffer from two drawbacks. First, two of the notions involved, that of an *English translation* of a symbolic sentence, and that of the *truth* of an English sentence, have not been precisely characterized. This difficulty could be removed as follows. We could restrict ourselves to *literal* translations into English, and, further, limit consideration to that part of English which might be called *elementary arithmetic* (that is, the arithmetic of nonnegative integers). The first restriction would avoid the looseness involved in the notion of stylistic variance, and the second would provide the possibility of a precise characterization of truth; for the latter see Tarski [2]. (In practice we shall not make these restrictions but proceed informally in showing invalidity.)

The second drawback concerns the methods of showing sentences true or false. If we deal with sentences of ordinary English, empirical knowledge will in general be needed. This need can be obviated by the device, mentioned above, of replacing English by the language of elementary arithmetic. But even then the mathematical methods required for showing truth and falsehood will in some cases be very deep. Indeed, there are invalid arguments whose invalidity cannot be established within the mathematical system presently used. (This is one version of the incompleteness theorem of Gödel [2].) For such arguments there will be translations into elementary arithmetic that are in fact false but which we shall be unable to prove false.

## EXERCISES

The reader will recall that a symbolic formula $\phi$ is a theorem if and only if the argument with no premises and with $\phi$ as its conclusion is valid. One can thus obtain from the considerations of the present section methods of showing formulas not to be theorems.

Show that each of the following is not a theorem.

73. The converse of T253 (p. 152).
74. The formula (6) of page 158.
75. The formula (7) of page 158.
76. The formula (12) of page 160.

For each of the following arguments show that it is valid or show that it is invalid. Exercise 77 is solved for illustration.

77. $\bigvee x(Fx \wedge \bigwedge y[Gy \wedge \bigvee z(Fz \wedge H(yz)) \rightarrow H(yx)])$ .
$\bigvee y(Gy \wedge \bigvee z[Fz \wedge H(yz)])$ $\therefore$ $\bigvee x(Fx \wedge \bigwedge y[Gy \rightarrow H(yx)])$

A false English translation of this argument is the argument

> There is an even prime such that every prime less than or equal to some even prime is less than or equal to it.

> Some prime is less than or equal to an even prime.
> ∴ There is an even prime such that every prime is less than or equal to it.

(By a *prime* is understood any integer $x$ other than 1 or $-1$ that is divisible (without remainder) only by 1, $-1$, $x$, and $-x$.) The falsehood of this argument becomes apparent once it is recognized that there is no largest prime but that there is a largest even prime, in particular, 2.

78. $F(xy) \rightarrow \sim F(yx) \quad \therefore \sim F(xx)$

79. $F(xy) \wedge F(yz) \rightarrow F(xz) \quad . \quad F(xy) \rightarrow F(yx) \quad \therefore F(xx)$

80. $F(xy) \wedge F(yz) \rightarrow F(xz) \quad . \quad F(xy) \rightarrow F(yx)$
$\therefore \bigwedge x[\bigvee y F(xy) \rightarrow F(xx)]$

81. $\bigwedge x \bigwedge y(Fx \wedge Fy \rightarrow FA(xy)) \quad . \quad \bigvee x(Fx \wedge \bigwedge y[Fy \rightarrow GA(yx)])$
$\therefore \bigvee x(Fx \wedge Gx)$

82. $\bigwedge x(\bigwedge y[Fy \rightarrow H(xy)] \rightarrow \bigvee y[\sim Gy \wedge H(xy)]) \quad .$
$\therefore \bigwedge x(\bigwedge y[H(xy) \rightarrow Gy] \rightarrow \bigwedge y[Fy \rightarrow \sim H(xy)])$

83. $\bigwedge y[Fy \wedge \bigvee x\, G(xyA) \rightarrow H(yB)] \quad . \quad \bigwedge y(\bigvee x[Ix \wedge H(yx)] \rightarrow Jy) \quad .$
$\bigwedge x \bigwedge y[\bigvee z G(xyz) \wedge Jy \rightarrow K(yx)] \quad . \quad \sim \bigwedge x \bigwedge y[G(xyA) \wedge Fy \rightarrow K(yx)]$
$\therefore \bigvee x \sim Ix$

84. $FA \quad \therefore \bigwedge x(Fx \rightarrow Gx) \leftrightarrow \bigwedge x([Fx \wedge Gx] \vee [\sim Fx \wedge GA])$

85. $\bigwedge x(\bigvee y[F(yA) \wedge G(xy)] \rightarrow \bigvee y[F(yA) \rightarrow H(xB)]$
$\therefore \bigwedge y[F(yA) \wedge G(By) \rightarrow H(BB)]$

Give an interesting symbolization (see p. 77) of each of the following arguments, and for each symbolization either show that it is valid or show that it is invalid. Exercise 86 is solved for illustration.

86. No member of the club owes a debt to the treasurer of the club. A member of the club has not paid the entrance fee only if he owes a debt to the treasurer of the club. ∴ The treasurer of the club has paid the entrance fee.

On the basis of the scheme of abbreviation

$F^2$ : *a* is a member of *b*
$G^2$ : *a* owes a debt to *b*
$H^2$ : *a* has paid *b*
$A^0$ : the club
$B^1$ : the treasurer of *a*
$C^0$ : the entrance fee ,

No. 86 has the symbolization

$$\sim \bigvee x[F(xA) \wedge G(xB(A))] \quad .$$
$$\bigwedge x[F(xA) \wedge \sim H(xC) \rightarrow G(xB(A))] \quad \therefore H(B(A)C) \quad .$$

A relativization of this symbolization to '$J^1$',

$$\sim \bigvee x(Jx \wedge [F(x\ A) \wedge G(x\ B(A))]) \quad .$$
$$\bigwedge x(Jx \to [F(x\ A) \wedge \sim H(x\ C) \ \to\ G(x\ B(A))]) \quad .$$
$$JA \quad . \qquad \bigwedge x_1[Jx_1 \to J(B(x_1))] \quad .$$
$$JC \quad . \qquad \bigvee x Jx \quad \therefore\ H(B(A)\ C) \quad ,$$

has the following translation:

> No integer equals both 2 and $2^2$. For each integer $x$, if $x$ both equals and does not equal 2, then $x$ equals $2^2$. 2 is an integer. The square of an integer is an integer. 2 is an integer. There is an integer. Therefore $2^2$ equals 2.

On the basis of elementary arithmetical facts it is clear that the premises of the last argument are true and the conclusion false; hence the symbolization above is invalid.

87. There is a number such that the product of it and any number is even. ∴ There is an even number.

88. The distance between A and B is less than the distance between B and C. The distance between B and C is less than the distance between A and C. ∴ The distance between A and B is less than the distance between A and C.

89. Students who solve every problem on the final examination also solve every problem on the midterm examination. Some student does not solve the hardest problem on the midterm examination. ∴ There is a problem on the final examination that some student does not solve.

90. No student who fails some course that Rudolf teaches fails all the courses that Alfred teaches. Some student fails all the courses Alfred teaches and also fails all the courses that Rudolf teaches. ∴ If Rudolf teaches any courses, then there is a course that every student fails.

91. No member of the club owes a debt to the treasurer of the club. A member of the club has not paid the entrance fee only if he owes a debt to some member of the club. ∴ If the treasurer of the club is a member of the club, then he has paid the entrance fee.

For each of the following formulas, either show that it is a theorem or show that it is not a theorem.

92. $\sim \bigvee z \bigvee x\ F(xz) \leftrightarrow \bigwedge z \bigwedge w \bigwedge y(F(yw) \leftrightarrow \bigvee x[F(xz) \wedge \sim F(yy)])$

93. $\bigwedge z \bigvee y \bigwedge x(F(xy) \leftrightarrow F(xz) \wedge Gx) \to \bigvee y \bigvee x(F(xy) \leftrightarrow \bigvee z F(xz) \wedge Gx)$

94. $\bigvee y \bigwedge x(F(xy) \leftrightarrow \bigvee z F(xz) \wedge Gx) \to \bigwedge z \bigvee y \bigvee x(F(xy) \leftrightarrow F(xz) \wedge Gx)$

95. $\sim \bigvee y \bigwedge x \bigwedge z[F(xyz) \leftrightarrow \sim F(xzx)]$

**11. * Paradoxical inferences.** We have shown earlier that if the restrictions on variables are ignored, it is possible to validate false English arguments. Certain examples suggest that this can be done even when all rules are strictly observed. One such case was discussed on page 122 but was ruled out by the requirement concerning apparent variables imposed on schemes of abbreviation.

Another example will indicate the necessity of excluding from schemes of abbreviation those apparent variables which might be called 'intuitively bound'. Consider the theorem

(1) $$\bigwedge x \bigwedge y F(xy) \rightarrow \bigvee x F(xx)$$

and the scheme

(2) $F^2$ : there is an object *a* such that *a* differs from *b* .

In the process of obtaining an English translation on the basis of (2), (1) becomes first

> $\bigwedge x \bigwedge y$ {there is an object *a* such that *a* differs from *b*} $(xy) \rightarrow$
> $\bigvee x$ {there is an object *a* such that *a* differs from *b*} $(xx)$ ,

next

> $\bigwedge x \bigwedge y$ there is an object *x* such that *x* differs from *y* $\rightarrow$
> $\bigvee x$ there is an object *x* such that *x* differs from *x* ,

and finally

> If, for each *x*, for each *y*, there is an object *x* such that *x* differs from *y*, then there is an object *x* such that there is an object *x* such that *x* differs from *x* ,

which has as a stylistic variant

> If each thing differs from something, then something differs from itself.

We should thus have found a false English translation of a theorem, which contradicts the claim on page 171. The trouble is that (2) is not a scheme of abbreviation: its right-hand member has the apparent variable '*a*'.

Now consider the argument

(3) Phidias' chryselephantine statue of Athene had a removable golden garment. ∴ There is an object *x* such that Phidias' chryselephantine statue of *x* had a removable golden garment.

On the basis of the scheme

(4) $F^1$ : Phidias' chryselephantine statue of *a* had a removable golden garment

$A^0$ : Athene ,

(3) seems to have the symbolization

$$FA \quad \therefore \bigvee x Fx \quad ,$$

which is clearly valid. But the premise of (3) is, as a matter of history, true; and the conclusion of (3) is false. If there is something whose statue had a removable golden garment, what is it? Certainly not Athene, for there is no such thing as Athene.

The paradox is solved by observing that if (4) is to be a scheme of abbreviation, the word 'Athene' must be a name. But is 'Athene' a name? In its ordinary sense, no; for in this sense it designates nothing, and a name according to our characterization must always have a designation.

Another treatment is possible. We may distort the meaning of English so as to *make* a name of every expression having the structural characteristics of a name. To do this, we must assign a designation, which may be quite arbitrary, to each expression that purports to designate but in its accepted meaning does not. Let us accordingly choose the number 0 as the common designation of all such expressions. The argument (3), then, is valid. But its premise, which asserts that Phidias' chryselephantine statue of 0 had a removable golden garment, is false. To express the intended meaning of this premise, we must now resort to circumlocution such as,

> The chryselephantine statue executed by Phidias and called 'Phidias' statue of Athene' had a removable golden garment.

**12. *Historical remarks.** The observation that every monadic formula is equivalent to one in which no quantifier occurs within the scope of another quantifier is found in Behmann [1]. The idea of prenex normal form is due to Peirce (see Peirce [2] and [3]). The process of reduction to prenex normal form is given in Whitehead and Russell [1].

The notion of an *instance* of a theorem of the quantifier calculus has appeared in many forms. Inadequate versions occur in Hilbert and Ackermann [1], Carnap [1], and Quine [1]. Correct, though highly complex, forms are to be found in Hilbert and Bernays [1], Hilbert and Ackermann [2] and [3], and Church [3]. Our formulation is the result of simplifying and extending the version of Quine [3] and profits from a suggestion in Pager [1].

The rule IE was demonstrated for the sentential calculus in Post [1] and for the quantifier calculus in Hilbert and Ackermann [1].

That there is no automatic test for validity in the quantifier calculus was derived in Church [2] from the results of Church [1].

The assertion that a closure of an invalid argument of the quantifier calculus always has a false translation can be understood as expressing the *completeness*, in a certain sense, of this branch of logic. Completeness, in a somewhat different sense, was demonstrated in Gödel [1] for an axiomatic quantification theory, and the present assertion can be established by Gödel's methods.

## 13. Appendix: summary of the system of logic developed in chapters I – IV.

### INFERENCE RULES

(Here $\phi$, $\psi$, $\chi$ are to be symbolic formulas and $\alpha$ a variable.)

PRIMITIVE SENTENTIAL RULES:

$$\begin{array}{l} \phi \rightarrow \psi \\ \phi \\ \hline \psi \end{array}$$

*Modus ponens* (MP)

$$\begin{array}{l} \phi \rightarrow \psi \\ \sim\psi \\ \hline \sim\phi \end{array}$$

*Modus tollens* (MT)

$$\begin{array}{c} \phi \\ \hline \sim\sim\phi \end{array} \qquad \begin{array}{c} \sim\sim\phi \\ \hline \phi \end{array}$$

Double negation (DN)

$$\begin{array}{c} \phi \\ \hline \phi \end{array}$$

Repetition (R)

$$\begin{array}{c} \phi \wedge \psi \\ \hline \phi \end{array} \qquad \begin{array}{c} \phi \wedge \psi \\ \hline \psi \end{array}$$

Simplification (S)

$$\begin{array}{l} \phi \\ \psi \\ \hline \phi \wedge \psi \end{array}$$

Adjunction (Adj)

$$\begin{array}{c} \phi \\ \hline \phi \vee \psi \end{array} \qquad \begin{array}{c} \psi \\ \hline \phi \vee \psi \end{array}$$

Addition (Add)

$$\begin{array}{c} \phi \vee \psi \\ \sim\phi \\ \hline \psi \end{array} \qquad \begin{array}{c} \phi \vee \psi \\ \sim\psi \\ \hline \phi \end{array}$$

*Modus tollendo ponens* (MTP)

$$\frac{\phi \leftrightarrow \psi}{\phi \rightarrow \psi} \qquad \frac{\phi \leftrightarrow \psi}{\psi \rightarrow \phi}$$ Biconditional-conditional (BC)

$$\begin{array}{c} \phi \rightarrow \psi \\ \psi \rightarrow \phi \\ \hline \phi \leftrightarrow \psi \end{array}$$ Conditional-biconditional (CB)

DERIVED SENTENTIAL RULES:

$$\begin{array}{c} \phi \rightarrow \psi \\ \sim\phi \rightarrow \psi \\ \hline \psi \end{array} \qquad \begin{array}{c} \phi \vee \psi \\ \phi \rightarrow \chi \\ \psi \rightarrow \chi \\ \hline \chi \end{array} \qquad \begin{array}{c} \phi \rightarrow \chi \\ \psi \rightarrow \chi \\ \hline \phi \vee \psi \rightarrow \chi \end{array}$$ Separation of cases (SC)

$$\frac{\sim\phi \rightarrow \psi}{\phi \vee \psi}$$ Conditional-disjunction (CD)

PRIMITIVE QUANTIFICATIONAL RULES:

$$\frac{\bigwedge\alpha\phi}{\psi},$$ Universal instantiation (UI)

$$\frac{\psi}{\bigvee\alpha\phi},$$ Existential generalization (EG)

where $\psi$ comes from $\phi$ by proper substitution of a term for $\alpha$;

$$\frac{\bigvee\alpha\phi}{\psi},$$ Existential instantiation (EI)

where $\psi$ comes from $\phi$ by proper substitution of a variable for $\alpha$. (See p. 148 for a definition of 'proper substitution'.)

DERIVED QUANTIFICATIONAL RULES:

$$\frac{\sim\bigwedge\alpha\phi}{\bigvee\alpha\sim\phi} \qquad \frac{\bigvee\alpha\sim\phi}{\sim\bigwedge\alpha\phi}$$

$$\frac{\sim\bigvee\alpha\phi}{\bigwedge\alpha\sim\phi} \qquad \frac{\bigwedge\alpha\sim\phi}{\sim\bigvee\alpha\phi}$$ Quantifier negation (QN)

## DIRECTIONS FOR CONSTRUCTING A DERIVATION FROM A CLASS *K* OF SYMBOLIC FORMULAS

(1) If $\phi$ is any symbolic formula, then

*Show* $\phi$

may occur as a line. (Annotation: 'Assertion'.)

(2) Any member of *K* may occur as a line. (Annotation: 'Premise'.)

(3) If $\phi$, $\psi$ are symbolic formulas such that

$$\textit{Show}\ (\phi \rightarrow \psi)$$

occurs as a line, then $\phi$ may occur as the next line. (Annotation: 'Assumption'.)

(4) If $\phi$ is a symbolic formula such that

$$\textit{Show}\ \phi$$

occurs as a line, then

$$\sim \phi$$

may occur as the next line; if $\phi$ is a symbolic formula such that

$$\textit{Show}\ \sim \phi$$

occurs as a line, then $\phi$ may occur as the next line. (Annotation: 'Assumption'.)

(5a) A symbolic formula may occur as a line if it follows from antecedent lines (see p. 21) by a primitive inference rule other than EI.

(5b) A symbolic formula may occur as a line if it follows from an antecedent line by the inference rule EI, provided that the variable of instantiation (see p. 100) does not occur in any preceding line. (The annotation for (5a) and (5b) should refer to the inference rule employed and the numbers of the antecedent lines involved.)

(6) When the following arrangement of lines has appeared:

$$\begin{array}{l} \textit{Show}\ \phi \\ \chi_1 \\ \cdot \\ \cdot \\ \cdot \\ \chi_m \quad , \end{array}$$

where none of $\chi_1$ through $\chi_m$ contain uncancelled '*Show*' and either

(i) $\phi$ occurs unboxed among $\chi_1$ through $\chi_m$;

(ii) $\phi$ is of the form

$$(\psi_1 \rightarrow \psi_2)$$

and $\psi_2$ occurs unboxed among $\chi_1$ through $\chi_m$;

(iii) for some formula $\chi$, both $\chi$ and its negation occur unboxed among $\chi_1$ through $\chi_m$; or

(iv) $\phi$ is of the form

$$\Lambda\alpha_1 \ldots \Lambda\alpha_k\psi \quad ,$$

$\psi$ occurs unboxed among the lines $\chi_1$ through $\chi_m$, and the variables $\alpha_1$ through $\alpha_k$ are not free in lines antecedent to the displayed occurrence of

$$\textit{Show}\ \phi\quad ,$$

then one may simultaneously cancel the displayed occurrence of '*Show*' and box all subsequent lines.

The remaining clauses are abbreviatory (in the sense of page 58).

(7) If $\phi$ is an instance of an alphabetic variant of a theorem that has already been proved, then $\phi$ may occur as a line. (Annotation: the number of the theorem of which $\phi$ is an instance, sometimes together with a diagrammatic indication of the sequence of substitutions involved.) (For the notion of *instance*, see chapter IV, section 8.)

(8) A symbolic formula may occur as a line if it is an alphabetic variant of an antecedent line. (Annotation: 'AV' and the number of the antecedent line.) (For the notion of *alphabetic variance* see chapter IV, section 7.)

(9) If $\phi$, $\phi'$ are symbolic formulas such that

$$\phi \leftrightarrow \phi'$$

is an instance of an alphabetic variant of a previously proved theorem, and $\psi$, $\psi'$ are symbolic formulas such that $\psi'$ is like $\psi$ except for having one or more occurrences of $\phi'$ where $\psi$ has occurrences of $\phi$, then

$$\psi \leftrightarrow \psi'$$

may occur as a line. (Annotation: 'IE' ('interchange of equivalents'), together with a parenthetical reference to the theorem involved.)

(10) A symbolic formula may occur as a line if it follows from antecedent lines by a succession of steps, and each intermediate step can be justified by one of clauses (2), (5a), (7), (8), or (9). (The annotation should determine the omitted succession of steps by indicating, in order of application, the antecedent lines, the premises, the inference rules, and the previously proved theorems employed.) (The use of derived inference rules is comprehended under this clause.)

## DERIVABILITY

A derivation is *complete* just in case every line either is boxed or contains cancelled '*Show*'. A symbolic formula $\phi$ is *derivable* from a class $K$ of symbolic formula just in case a complete derivation from $K$ can be constructed in which

$$\text{~~\textit{Show}~~}\ \phi$$

occurs as an unboxed line.

## 14. Appendix: list of theorems of chapter IV.

T249 $\bigwedge x \bigwedge y F(xy) \rightarrow \bigvee x \bigvee y F(xy)$

T250 $\bigwedge x \bigwedge y \bigwedge z F(xyz) \leftrightarrow \sim \bigvee x \bigvee y \bigvee z \sim F(xyz)$

T251 $\bigwedge x \bigwedge y F(xy) \leftrightarrow \bigwedge y \bigwedge x F(xy)$

T252 $\bigvee x \bigvee y F(xy) \leftrightarrow \bigvee y \bigvee x F(xy)$

T253 $\bigvee x \bigwedge y F(xy) \rightarrow \bigwedge y \bigvee x F(xy)$

T254. $\bigvee x \bigvee y F(xy) \leftrightarrow \bigvee x \bigvee y [F(xy) \vee F(yx)]$

T255 $\sim \bigvee y \bigwedge x [F(xy) \leftrightarrow \sim F(xx)]$

T256 $\bigwedge z \bigvee y \bigwedge x [F(xy) \leftrightarrow F(xz) \wedge \sim F(xx)] \rightarrow \sim \bigvee z \bigwedge x F(xz)$

T257 $\bigwedge x \bigwedge y F(xy) \rightarrow \bigwedge y \bigwedge x F(yx)$

T258 $F(xA(x)) \leftrightarrow \bigvee y [\bigwedge z [F(zy) \rightarrow F(zA(x))] \wedge F(xy)]$

T259 $\bigwedge y \bigvee x (Fx \leftrightarrow \sim Fy) \leftrightarrow \bigvee x Fx \wedge \bigvee x \sim Fx$

T260 $\bigvee x \bigwedge y (Fx \leftrightarrow Fy) \leftrightarrow \sim \bigvee x Fx \vee \bigwedge x Fx$

T261 $\bigwedge x (Fx \rightarrow \bigvee y [Gy \wedge (Hy \vee Hx)]) \leftrightarrow$
$\bigvee x (Gx \wedge Hx) \vee \sim \bigvee x Fx \vee (\bigvee x Gx \wedge \bigwedge x [Fx \rightarrow Hx])$

T262 $(\bigwedge x Fx \rightarrow \bigwedge x Gx) \leftrightarrow \bigvee x \bigwedge y (Fx \rightarrow Gy)$

T263 $(\bigwedge x Fx \leftrightarrow \bigvee x Gx) \leftrightarrow \bigvee x \bigvee y \bigwedge z \bigwedge w ([Fx \rightarrow Gy) \wedge (Gz \rightarrow Fw])$

T264 $(\bigvee x Fx \rightarrow [\bigvee x Gx \rightarrow \bigwedge x Hx]) \leftrightarrow \bigwedge x \bigwedge y \bigwedge z (Fx \wedge Gy \rightarrow Hz)$

T265 $\bigwedge x (\bigvee y [H(xy) \wedge Fy] \rightarrow \bigvee y [H(xy) \wedge Gy]) \leftrightarrow$
$\bigwedge x \bigwedge y \bigvee z (H(xy) \wedge Fy \rightarrow H(xz) \wedge Gz)$

T266 $\bigwedge x (Fx \wedge \bigvee y G(xy) \rightarrow \bigvee y [H(xy) \wedge \bigwedge z J(xyz)]) \leftrightarrow$
$\bigwedge x \bigwedge y \bigvee w \bigwedge z [Fx \wedge G(xy) \rightarrow H(xw) \wedge J(xwz)]$

T267 $\bigwedge x \bigvee y (Fx \rightarrow Gy) \rightarrow \bigvee y \bigwedge x (Fx \rightarrow Gy)$

T268 $\bigvee y \bigwedge x (Fx \rightarrow Gy) \rightarrow \bigwedge x \bigvee y (Fx \rightarrow Gy)$

T269 $\bigvee y \bigwedge x (Fx \vee Gy) \leftrightarrow \bigwedge x \bigvee y (Fx \vee Gy)$

T270 $\bigvee y \bigwedge x (Fx \wedge Gy) \leftrightarrow \bigwedge x \bigvee y (Fx \wedge Gy)$

T271 $\bigwedge x \bigwedge y \bigvee z (Fx \wedge Gy \rightarrow Hz) \leftrightarrow \bigwedge y \bigvee z \bigwedge x (Fx \wedge Gy \rightarrow Hz)$

T272 $\bigwedge x(Fx \wedge P) \leftrightarrow \bigwedge xFx \wedge P$

T273 $\bigvee x(Fx \wedge P) \leftrightarrow \bigvee xFx \wedge P$

T274 $\bigwedge x(Fx \vee P) \leftrightarrow \bigwedge xFx \vee P$

T275 $\bigvee x(Fx \vee P) \leftrightarrow \bigvee xFx \vee P$

# *Chapter V
# Automatic procedures

**1. Introduction.** In earlier chapters we have given informal suggestions concerning the construction of derivations. For instance, the reader was advised on page 69 to derive each conjunct of a conjunction before attempting to derive the conjunction, and to derive a disjunction by first deriving a corresponding conditional. It was emphasized on several occasions that these suggestions are not infallible—that they will not lead to a derivation in every case in which one is possible.

The main purpose of this chapter is to present some infallible procedures of derivation. The new procedures, like the old suggestions, will consist of directions that can be followed automatically, without the exercise of ingenuity. They will differ from the earlier suggestions in two respects. The new procedures will lead to a derivation whenever one is possible; they are, however, less intuitive than the suggestions previously offered, and in some cases lead to much lengthier derivations.

Before presenting these automatic procedures, we must add three abbreviatory clauses to the directions for constructing a derivation. The new clauses, which will occupy the two following sections, are intended for use only in the present chapter. (In fact, none of the present chapter will be presupposed later except in optional sections indicated by asterisks.)

**2. Tautologies reconsidered.** The tautologies of chapter II are always symbolic sentences, but the notion of a tautology can easily be extended so as to apply to arbitrary symbolic formulas. Accordingly, we define first a *molecular* formula as one that has the form

$$\sim \psi \;,$$
$$(\psi \rightarrow \chi) \;,$$
$$(\psi \wedge \chi) \;,$$
$$(\psi \vee \chi) \;,$$

or

$$(\psi \leftrightarrow \chi) \;,$$

for some symbolic formulas $\psi$ and $\chi$. An *assignment* now correlates with each nonmolecular symbolic *formula* one of the truth values *T* or *F*. The *truth value* of an arbitrary symbolic formula, on the basis of a given assignment, can be computed just as in chapter II (pp. 73 – 74). A *tautology* is a symbolic formula whose truth value is *T* with respect to every possible assignment. A symbolic argument is *tautologically valid* if the truth value of its conclusion is *T* under every assignment for which all its premises have the value *T*; under the same conditions, the premises of the argument are said to *imply tautologically* the conclusion. Just as before, we may use *truth tables* to check whether a symbolic formula is a tautology or whether a symbolic argument with finitely many premises is tautologically valid.

A *subformula* of a formula $\phi$ is any formula that occurs within $\phi$.

In the present section we shall add the following abbreviatory clause to the directions for constructing a derivation:

(*11*) *Any tautology may occur as a line.*

The theoretical superfluity of clause (11) depends on the fact that every tautology is a theorem; this was asserted in chapter II and will be substantiated here. We shall, in fact, give directions on the basis of which we can construct a proof of any given tautology.

Suppose that we are confronted with a tautology, for example,

(1) $$(P \rightarrow Q) \rightarrow (P \wedge Q \leftrightarrow P) \quad .$$

We must first construct a truth table whose last column is headed by the tautology in question. The last column, then, will not contain '*F*'. In connection with (1) we have

| P | Q | $P \rightarrow Q$ | $P \wedge Q$ | $P \wedge Q \leftrightarrow P$ | $(P \rightarrow Q) \rightarrow (P \wedge Q \leftrightarrow P)$ |
|---|---|---|---|---|---|
| *T* | *T* | *T* | *T* | *T* | *T* |
| *T* | *F* | *F* | *F* | *F* | *T* |
| *F* | *T* | *T* | *F* | *T* | *T* |
| *F* | *F* | *T* | *F* | *T* | *T* |

On the basis of the truth table, now, we construct a proof of the tautology; intuitively, the proof can be regarded as a separation of cases, which correspond to the rows of the truth table. For simplicity, we give the construction only in connection with (1); the reader will observe, however, that the procedure is quite general.

If $\chi$ is the formula (1), then the proof of $\chi$ will have the following over-all structure:

1. ~~*Show*~~ $\chi$
2. ~~*Show*~~ $P \rightarrow \chi$

3. P
4. ~~*Show*~~ $Q \rightarrow \chi$
10. ~~*Show*~~ $\sim Q \rightarrow \chi$
16. $\chi$ 4, 10, SC
17. ~~*Show*~~ $\sim P \rightarrow \chi$
18. $\sim P$
19. ~~*Show*~~ $Q \rightarrow \chi$
25. ~~*Show*~~ $\sim Q \rightarrow \chi$
31. $\chi$ 19, 25, SC
32. $\chi$ 2, 17, SC

The argument by which line 4 will be established (in which 'P' and 'Q' serve as assumptions) will correspond to the first row of the truth table. The argument for line 10 (under the assumptions 'P' and '~Q') will correspond to the second row. Similarly, the arguments for lines 19 and 25 will correspond to the third and fourth rows respectively.

In greater detail, the general directions for proving (1) are as follows. We begin the proof by indicating that the tautology is to be shown:

1. *Show* $(P \rightarrow Q) \rightarrow (P \wedge Q \leftrightarrow P)$

We indicate next that the tautology holds when its first nonmolecular part is true:

1. *Show* $(P \rightarrow Q) \rightarrow (P \wedge Q \leftrightarrow P)$
2. *Show* $P \rightarrow [(P \rightarrow Q) \rightarrow (P \wedge Q \leftrightarrow P)]$

We assume the antecedent of line 2:

1. *Show* $(P \rightarrow Q) \rightarrow (P \wedge Q \leftrightarrow P)$
2. *Show* $P \rightarrow [(P \rightarrow Q) \rightarrow (P \wedge Q \leftrightarrow P)]$
3. P

We indicate now that the tautology holds when its second nonmolecular part is true:

1. *Show* $(P \rightarrow Q) \rightarrow (P \wedge Q \leftrightarrow P)$
2. *Show* $P \rightarrow [(P \rightarrow Q) \rightarrow (P \wedge Q \leftrightarrow P)]$

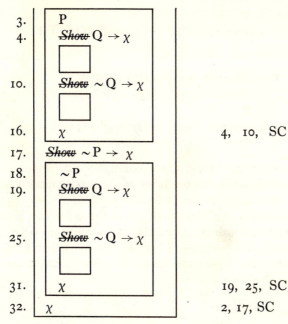

3. P
4. *Show* $Q \to [(P \to Q) \to (P \wedge Q \leftrightarrow P)]$

Again we assume the antecedent:

1. *Show* $(P \to Q) \to (P \wedge Q \leftrightarrow P)$
2. *Show* $P \to [(P \to Q) \to (P \wedge Q \leftrightarrow P)]$
3. P
4. *Show* $Q \to [(P \to Q) \to (P \wedge Q \leftrightarrow P)]$
5. Q

We have now, in lines 3 and 5, the assumptions corresponding to the first row of the truth table. On the basis of these assumptions, we proceed to 'compute' the values of the molecular subformulas of (1). That is, we treat the molecular subformulas of (1) in the order in which they appear at the head of our truth table, and for each subformula $\phi$ we derive either $\phi$ or its negation, according as '*T*' or '*F*' appears below $\phi$ in the first row of the truth table. (This can always be done because we have at our disposal the theorems

| | |
|---|---|
| T18 | $\sim P \to (P \to Q)$ , |
| T2 | $Q \to (P \to Q)$ , |
| T42 | $P \wedge \sim Q \to \sim(P \to Q)$ , |
| T43 | $\sim P \to \sim(P \wedge Q)$ , |
| T44 | $\sim Q \to \sim(P \wedge Q)$ , |
| T67 | $\sim P \wedge \sim Q \to \sim(P \vee Q)$ , |
| T84 | $P \wedge Q \to (P \leftrightarrow Q)$ , |
| T85 | $\sim P \wedge \sim Q \to (P \leftrightarrow Q)$ , |
| T88 | $P \wedge \sim Q \to \sim(P \leftrightarrow Q)$ , |
| T89 | $\sim P \wedge Q \to \sim(P \leftrightarrow Q)$ , |

which, together with the inference rules DN, Adj, and Add, correspond completely to the rules for assigning truth values.) In the truth table for (1), '*F*' does not appear in the first row; thus we obtain

1. *Show* $(P \to Q) \to (P \wedge Q \leftrightarrow P)$
2. *Show* $P \to [(P \to Q) \to (P \wedge Q \leftrightarrow P)]$
3. P
4. *Show* $Q \to [(P \to Q) \to (P \wedge Q \leftrightarrow P)]$
5. Q
6. $P \to Q$ — 5, T2, MP
7. $P \wedge Q$ — 3, 5, Adj

8. $P \land Q \leftrightarrow P$ — 3, 7, Adj, T84, MP
9. $(P \to Q) \to (P \land Q \leftrightarrow P)$ — 8, T2, MP

The conditional proof of line 4 is complete. We box and cancel, and pass to the case in which 'Q' is false.

1. *Show* $(P \to Q) \to (P \land Q \leftrightarrow P)$
2. *Show* $P \to [(P \to Q) \to (P \land Q \leftrightarrow P)]$
3. $P$
4. ~~*Show*~~ $Q \to [(P \to Q) \to (P \land Q \leftrightarrow P)]$
5. | $Q$
6. | $P \to Q$
7. | $P \land Q$
8. | $P \land Q \leftrightarrow P$
9. | $(P \to Q) \to (P \land Q \leftrightarrow P)$
10. *Show* $\sim Q \to [(P \to Q) \to (P \land Q \leftrightarrow P)]$
11. $\sim Q$

Again we derive 'computed' values, this time in accordance with the second row of the truth table:

1. *Show* $(P \to Q) \to (P \land Q \leftrightarrow P)$
2. *Show* $P \to [(P \to Q) \to (P \land Q \leftrightarrow P)]$
3. $P$
4. ~~*Show*~~ $Q \to [(P \to Q) \to (P \land Q \leftrightarrow P)]$
5. | $Q$
6. | $P \to Q$
7. | $P \land Q$
8. | $P \land Q \leftrightarrow P$
9. | $(P \to Q) \to (P \land Q \leftrightarrow P)$
10. *Show* $\sim Q \to [(P \to Q) \to (P \land Q \leftrightarrow P)]$
11. $\sim Q$
12. $\sim (P \to Q)$ — 3, 11, Adj, T42, MP
13. $\sim (P \land Q)$ — 11, T44, MP
14. $\sim (P \land Q \leftrightarrow P)$ — 3, 13, Adj, T89, MP
15. $(P \to Q) \to (P \land Q \leftrightarrow P)$ — 12, T18, MP

The conditional proof of line 10 is complete. Again, we box and cancel:

1. *Show* $(P \to Q) \to (P \land Q \leftrightarrow P)$
2. *Show* $P \to [(P \to Q) \to (P \land Q \leftrightarrow P)]$
3. $P$
4. ~~*Show*~~ $Q \to [(P \to Q) \to (P \land Q \leftrightarrow P)]$
5. | $Q$
6. | $P \to Q$

7. $P \wedge Q$
8. $P \wedge Q \leftrightarrow P$
9. $(P \rightarrow Q) \rightarrow (P \wedge Q \leftrightarrow P)$
10. ~~*Show*~~ $\sim Q \rightarrow [(P \rightarrow Q) \rightarrow (P \wedge Q \leftrightarrow P)]$
11. $\sim Q$
12. $\sim (P \rightarrow Q)$
13. $\sim (P \wedge Q)$
14. $\sim (P \wedge Q \leftrightarrow P)$
15. $(P \rightarrow Q) \rightarrow (P \wedge Q \leftrightarrow P)$

But the tautology in question now follows from lines 4 and 10 by separation of cases; thus we can complete the conditional proof of line 2:

1. *Show* $(P \rightarrow Q) \rightarrow (P \wedge Q \leftrightarrow P)$
2. ~~*Show*~~ $P \rightarrow [(P \rightarrow Q) \rightarrow (P \wedge Q \leftrightarrow P)]$
3. $P$
4. ~~*Show*~~ $Q \rightarrow [(P \rightarrow Q) \rightarrow (P \wedge Q \leftrightarrow P)]$
5. $Q$
6. $P \rightarrow Q$
7. $P \wedge Q$
8. $P \wedge Q \leftrightarrow P$
9. $(P \rightarrow Q) \rightarrow (P \wedge Q \leftrightarrow P)$
10. ~~*Show*~~ $\sim Q \rightarrow [(P \rightarrow Q) \rightarrow (P \wedge Q \leftrightarrow P)]$
11. $\sim Q$
12. $\sim (P \rightarrow Q)$
13. $\sim (P \wedge Q)$
14. $\sim (P \wedge Q \leftrightarrow P)$
15. $(P \rightarrow Q) \rightarrow (P \wedge Q \leftrightarrow P)$
16. $(P \rightarrow Q) \rightarrow (P \wedge Q \leftrightarrow P)$ 4, 10, SC

We continue in the same way, constructing 'cases' for the remaining rows of the truth table and 'computing' corresponding values, until finally all possibilities are exhausted:

1. ~~*Show*~~ $(P \rightarrow Q) \rightarrow (P \wedge Q \leftrightarrow P)$
2. ~~*Show*~~ $P \rightarrow [(P \rightarrow Q) \rightarrow (P \wedge Q \leftrightarrow P)]$
3. $P$
4. ~~*Show*~~ $Q \rightarrow [(P \rightarrow Q) \rightarrow (P \wedge Q \leftrightarrow P)]$
5. $Q$
6. $P \rightarrow Q$ 5, T2, MP

| | | |
|---|---|---|
| 7. | $P \wedge Q$ | 3, 5, Adj |
| 8. | $P \wedge Q \leftrightarrow P$ | 3, 7, Adj, T84, MP |
| 9. | $(P \rightarrow Q) \rightarrow (P \wedge Q \leftrightarrow P)$ | 8, T2, MP |
| 10. | ~~*Show*~~ $\sim Q \rightarrow [(P \rightarrow Q) \rightarrow (P \wedge Q \leftrightarrow P)]$ | |
| 11. | $\sim Q$ | |
| 12. | $\sim (P \rightarrow Q)$ | 3, 11, Adj, T42, MP |
| 13. | $\sim (P \wedge Q)$ | 11,T44,MP |
| 14. | $\sim (P \wedge Q \leftrightarrow P)$ | 3, 13, Adj, T89, MP |
| 15. | $(P \rightarrow Q) \rightarrow (P \wedge Q \leftrightarrow P)$ | 12,T18,MP |
| 16. | $(P \rightarrow Q) \rightarrow (P \wedge Q \leftrightarrow P)$ | 4, 10, SC |
| 17. | ~~*Show*~~ $\sim P \rightarrow [(P \rightarrow Q) \rightarrow (P \wedge Q \leftrightarrow P)]$ | |
| 18. | $\sim P$ | |
| 19. | ~~*Show*~~ $Q \rightarrow [(P \rightarrow Q) \rightarrow (P \wedge Q \leftrightarrow P)]$ | |
| 20. | $Q$ | |
| 21. | $P \rightarrow Q$ | 20, T2, MP |
| 22. | $\sim (P \wedge Q)$ | 18,T43,MP |
| 23. | $P \wedge Q \leftrightarrow P$ | 18, 22, Adj, T85, MP |
| 24. | $(P \rightarrow Q) \rightarrow (P \wedge Q \leftrightarrow P)$ | 23, T2, MP |
| 25. | ~~*Show*~~ $\sim Q \rightarrow [(P \rightarrow Q) \rightarrow (P \wedge Q \leftrightarrow P)]$ | |
| 26. | $\sim Q$ | |
| 27. | $P \rightarrow Q$ | 18,T18,MP |
| 28. | $\sim (P \wedge Q)$ | 26,T44,MP |
| 29. | $P \wedge Q \leftrightarrow P$ | 18, 28, Adj, T85, MP |
| 30. | $(P \rightarrow Q) \rightarrow (P \wedge Q \leftrightarrow P)$ | 29, T2, MP |
| 31. | $(P \rightarrow Q) \rightarrow (P \wedge Q \leftrightarrow P)$ | 19, 25, SC |
| 32. | $(P \rightarrow Q) \rightarrow (P \wedge Q \leftrightarrow P)$ | 2, 17, SC |

Thus we secure a complete proof of (1)—overly long, to be sure, but obtained by a simple general method.

## EXERCISES

Prove the following tautologies by the procedure of this section.

1. $(P \leftrightarrow [P \rightarrow Q]) \rightarrow Q$
2. $(Q \leftrightarrow P \wedge \sim P) \leftrightarrow \sim Q$

3. $([P \to Q] \to P) \to P$
4. $([P \to Q] \to Q) \leftrightarrow ([Q \to P] \to P)$

**3. Tautological implication; generalized indirect derivation.** The preceding discussion suggests two additional methods of abbreviating derivations, which, like clause (11), will be used only in the present chapter. The first of these consists in employing the following clause in the construction of derivations:

(*12*) *A symbolic formula may occur as a line if it is tautologically implied by antecedent lines.* (*Annotation: 'TI', together with the numbers of the antecedent lines involved.*)

The second of the new abbreviations concerns indirect derivation. In order to complete such a derivation, we must, according to earlier instructions, derive an *explicit contradiction*, that is, a pair of formulas of which one is the negation of the other. Our next abbreviation liberalizes these instructions: to complete an indirect derivation, we shall henceforth need only to derive a *truth-functional contradiction*, that is, a combination of formulas that cannot all simultaneously receive the truth value *T*. To be explicit, we say that the symbolic formulas $\phi_1 \ldots, \phi_n$ are *truth-functionally incompatible* just in case there is no assignment of truth values to nonmolecular symbolic formulas under which all of $\phi_1, \ldots, \phi_n$ receive the value *T*. Then the following clause introduces a new form of derivation, which we shall call *generalized indirect derivation.*

(*13*) *When the following arrangement of lines has appeared:*

*Show* $\phi$
$\chi_1$
.
.
.
$\chi_k$ ,

*where none of* $\chi_1$ *through* $\chi_k$ *contains uncancelled 'Show', and among* $\chi_1$ *through* $\chi_k$ *there occur unboxed truth-functionally incompatible lines, then one may simultaneously cancel the displayed occurrence of 'Show', and box all subsequent lines.*

Thus, for example, in the derivation

1. ~~*Show*~~ $\bigvee x(Fx \wedge \bigvee yGy)$
2. | $\sim \bigvee x(Fx \wedge \bigvee yGy)$
3. | $\sim \bigwedge x(Fx \to Hx)$ Premise

| | | |
|---|---|---|
| 4. | $\sim(Fz \rightarrow Hz)$ | 3, QN, EI |
| 5. | $\bigwedge x(\bigvee yGy \leftrightarrow Fx)$ | Premise |
| 6. | $\bigvee yGy \leftrightarrow Fz$ | 5, UI |
| 7. | $\sim(Fz \wedge \bigvee yGy)$ | 2, QN, UI |

the cancellation of '*Show*' and boxing of lines 2 through 7 are legitimized by clause (13); for, as the reader can verify by means of a truth table, lines 4, 6, and 7 are truth-functionally incompatible.

Clauses (12) and (13) are reducible to (11), and hence theoretically superfluous.

**4. A proof procedure for prenex formulas.** By a *prenex formula* we shall understand a formula in prenex normal form or, equivalently, a symbolic formula in which no quantifier occurs to the right of a parenthesis or negation sign. (This criterion is equivalent to the one on page 154 for formulas in official notation.) We are now in a position to describe an automatic procedure, applicable to any prenex formula, which will lead to a proof of that formula, if indeed the formula is a theorem. We shall content ourselves with presenting the procedure, without attempting to justify the claim made for it. (See, however, the historical remarks accompanying this chapter.) The reader will notice that the proofs generated by the procedure have a particularly simple structure. Only one form of derivation is employed—generalized indirect derivation—and no subsidiary derivations appear; further, only three inference rules are used—IE (used in connection with the laws of quantifier negation), UI, and EI.

Suppose that $\chi$ is a prenex formula. We begin an indirect derivation, as follows:

*Show* $\chi$

$\sim \chi$ (Assumption)

We now convert

$$\sim \chi$$

into prenex normal form by successive applications of IE, used in connection with the theorems

$$\sim \bigwedge xFx \leftrightarrow \bigvee x \sim Fx \quad , \qquad \text{(T203)}$$

$$\sim \bigvee xFx \leftrightarrow \bigwedge x \sim Fx \quad . \qquad \text{(T204)}$$

Thus we obtain

*Show* $\chi$

$\sim \chi$

.

.

.

$$\psi \quad ,$$

where $\psi$ is obtained from

$$\sim \chi$$

by 'moving in' the negation sign.

The remainder of the derivation, starting with $\psi$, constitutes an *elimination column*, constructed roughly as follows. We drop initial quantifiers one at a time by means of UI and EI. In the process we make sure that if the column contains a universal generalization

$$\Lambda \alpha \phi$$

and a free occurrence, in one of its lines, of a term $\zeta$, then it will also contain the formula that comes from $\phi$ by proper substitution of $\zeta$ for $\alpha$.

To be more explicit, we begin by examining $\psi$. If $\psi$ has the form

$$\Lambda \alpha \phi$$

and $\zeta_1, \ldots, \zeta_n$ are all the terms that are free in $\psi$, then we add to the column all the formulas that come from $\phi$ by proper substitution of one of $\zeta_1, \ldots, \zeta_n$ for $\alpha$; if there are no terms free in $\psi$, we add a formula that comes from $\phi$ by proper substitution for $\alpha$ of some variable not bound in any line of the derivation. If $\psi$ has the form

$$\mathrm{V} \alpha \phi \quad ,$$

we add to the column a formula that comes from $\phi$ by proper substitution for $\alpha$ of some variable new to the derivation. These additions are legitimate; they follow from $\psi$ by either UI or EI. If $\psi$ is *quantifier-free* (that is, contains no quantifiers), we make no additions; the elimination column then consists of $\psi$ alone.

We now examine the second line of the elimination column (if a second line has been obtained) and add lines to the column as follows.

(1) If the second line has the form

$$\Lambda \alpha \phi$$

and $\zeta_1, \ldots, \zeta_n$ are all the terms that are free in the first two lines of the elimination column, then we extend the column by subjoining all formulas that come from $\phi$ by proper substitution of one of $\zeta_1, \ldots, \zeta_n$ for $\alpha$. (The addition is justified by UI.)

(2) If the second line has the form

$$\mathrm{V} \alpha \phi \quad ,$$

then we subjoin to the column a formula that comes from $\phi$ by proper substitution for $\alpha$ of some variable new to the derivation. (The addition is justified by EI.)

(3) Whatever the form of the second line, we determine whether any terms are free in it that were not already free in the first line. If so, and if the first line is a universal generalization

$$\bigwedge \alpha \phi \quad ,$$

then we subjoin to the column all formulas that come from $\phi$ by proper substitution of one of these terms for $\alpha$. (The addition is justified by UI.)

We pass now to the third line (if indeed a third line has been obtained) and subject it to the same sort of treatment as the second line; and so on, for all following lines. In general, when we examine any line of the elimination column other than the first, we add lines at the bottom of the column in accordance with the following instructions. (It is clear that each addition is justified by either UI or EI.)

(*1*) *If the line in question has the form*

$$\bigwedge \alpha \phi$$

*and* $\zeta_1, \ldots, \zeta_n$ *are all the terms that are free in lines of the elimination column up to and including this line, then we add as lines all formulas that come from* $\phi$ *by proper substitution of one of* $\zeta_1, \ldots, \zeta_n$ *for* $\alpha$.

(*2*) *If the line in question has the form*

$$\bigvee \alpha \phi \quad ,$$

*then we add as a line a formula that comes from* $\phi$ *by proper substitution for* $\alpha$ *of some variable new to the derivation.*

(*3*) *Whatever the form of the line in question, we determine whether any terms are free in it that were not already free in the preceding lines. If so, let these terms be* $\zeta_1, \ldots, \zeta_n$. *For each universal generalization*

$$\bigwedge \alpha \phi$$

*that precedes the line in question in the elimination column, we add as lines all formulas that come from* $\phi$ *by proper substitution of one of* $\zeta_1, \ldots, \zeta_n$ *for* $\alpha$.

Now if the original formula $\chi$ is a theorem, this procedure will eventually lead to a truth-functional contradiction. It may be convenient, at various stages of the derivation, to test by a truth table the truth-functional compatibility of the formulas so far obtained; in fact, it will be sufficient to test the quantifier-free formulas of the elimination column, for among these the truth-functional contradiction, if there is one, must always appear. When we obtain a truth-functional contradiction, we may box and cancel in accordance with generalized indirect derivation. We shall then have a complete proof of $\chi$.

Let us prove, by means of the general procedure, the formula '$\bigwedge y \bigvee x(Fx \rightarrow Fy)$'. We begin an indirect derivation:

1. *Show* $\bigwedge y \bigvee x(Fx \rightarrow Fy)$
2. $\sim \bigwedge y \bigvee x(Fx \rightarrow Fy)$

Next we 'move in' the negation sign:

| | | |
|---|---|---|
| 3. | $\bigvee y \sim \bigvee x(Fx \rightarrow Fy)$ | 2, T203 |
| 4. | $\bigvee y \bigwedge x \sim (Fx \rightarrow Fy)$ | 3, IE(T204) |

Now we are to construct an elimination column, beginning with line 4. We notice that line 4 is an existential generalization and that the variable '*a*' has not yet occurred. Thus we add a fifth line, as follows:

| | | |
|---|---|---|
| 5. | $\bigwedge x \sim (Fx \rightarrow Fa)$ | 4, EI |

We now examine the fifth line of the derivation (the second of the elimination column), and observe that it is a universal generalization and that '*a*' is the only term free in this line or the preceding. Thus instruction (1) for adding lines leads us to

| | | |
|---|---|---|
| 6. | $\sim (Fa \rightarrow Fa)$ | 5, UI |

Instruction (2) is not applicable to line 5, because line 5 is not an existential generalization; nor is instruction (3), because the preceding line is not a universal generalization. Thus we pass to line 6. But none of the instructions is applicable here. Hence the procedure for extending the column has come to an end. But we easily see that line 6, the only quantifier-free line, forms by itself a truth-functional contradiction; that is, its truth value is *F* under any assignment of truth values to nonmolecular formulas. Thus, by generalized indirect derivation, we box and cancel, obtaining

| | | |
|---|---|---|
| 1. | ~~*Show*~~ $\bigwedge y \bigvee x(Fx \rightarrow Fy)$ | |
| 2. | $\sim \bigwedge y \bigvee x(Fx \rightarrow Fy)$ | |
| 3. | $\bigvee y \sim \bigvee x(Fx \rightarrow Fy)$ | 2, T203 |
| 4. | $\bigvee y \bigwedge x \sim (Fx \rightarrow Fy)$ | 3, IE(T204) |
| 5. | $\bigwedge x \sim (Fx \rightarrow Fa)$ | 4, EI |
| 6. | $\sim (Fa \rightarrow Fa)$ | 5, UI |

In the example just considered, the general procedure for extending the elimination column comes to an end. This is not always the case. For example, let us apply the procedure to the formula

$$\bigvee x \bigvee y \bigwedge z([Fx \rightarrow Gx] \rightarrow [Fy \rightarrow Gz]) \quad .$$

We begin as before:

1. *Show* $\bigvee x \bigvee y \bigwedge z([Fx \rightarrow Gx] \rightarrow [Fy \rightarrow Gz])$
2. $\sim \bigvee x \bigvee y \bigwedge z([Fx \rightarrow Gx] \rightarrow [Fy \rightarrow Gz])$

3. $\bigwedge x \sim \bigvee y \bigwedge z([Fx \rightarrow Gx] \rightarrow [Fy \rightarrow Gz])$ 2, T204
4. $\bigwedge x \bigwedge y \sim \bigwedge z([Fx \rightarrow Gx] \rightarrow [Fy \rightarrow Gz])$ 3, IE(T204)
5. $\bigwedge x \bigwedge y \bigvee z \sim ([Fx \rightarrow Gx] \rightarrow [Fy \rightarrow Gz])$ 4, IE(T203)

Since no terms are free in the first line of the elimination column (line 5), we choose a variable so far not bound, say '*a*', and add a corresponding sixth line:

6. $\bigwedge y \bigvee z \sim ([Fa \rightarrow Ga] \rightarrow [Fy \rightarrow Gz])$ 5, UI

We turn to line 6, find it to be a universal generalization, and observe that '*a*' is the only term that is free so far. Thus, following instruction (1) of page 195, we add a seventh line:

7. $\bigvee z \sim (Fa \rightarrow Ga] \rightarrow [Fa \rightarrow Gz])$ 6, UI

Instruction (3) is also applicable to line 6, for that line contains freely for the first time the variable '*a*', and the preceding line is a universal generalization. Thus instruction (3) would lead us to add

$\bigwedge y \bigvee z \sim ([Fa \rightarrow Ga] \rightarrow [Fy \rightarrow Gz])$ .

We do not, however, make this addition, for it would amount only to repeating line 6. (In general, we shall not add lines that repeat earlier lines.) We now consider line 7. (2) is the only applicable instruction. We obtain

8. $\sim([Fa \rightarrow Ga] \rightarrow [Fa \rightarrow Gb])$ 7, EI

Instruction (3) is applicable to line 8; it contains freely for the first time the variable '*b*', and there are two earlier universal generalizations (lines 5 and 6). Thus we obtain

9. $\bigwedge y \bigvee z \sim ([Fb \rightarrow Gb] \rightarrow [Fy \rightarrow Gz])$ 5, UI
10. $\bigvee z \sim ([Fa \rightarrow Ga] \rightarrow [Fb \rightarrow Gz])$ 6, UI

To line 9, only instruction (1) is applicable. There are two terms free up to this point, '*a*' and '*b*'; thus we obtain

11. $\bigvee z \sim ([Fb \rightarrow Gb] \rightarrow [Fa \rightarrow Gz])$ 9, UI
12. $\bigvee z \sim ([Fb \rightarrow Gb] \rightarrow [Fb \rightarrow Gz])$ 9, UI

Only instruction (2) is applicable to line 10. We obtain

13. $\sim([Fa \rightarrow Ga] \rightarrow [Fb \rightarrow Gc])$ 10, EI

Again, in connection with line 11, only instruction (2) is applicable. We obtain

14. $\sim([Fb \rightarrow Gb] \rightarrow [Fa \rightarrow Gd)]$ 11, EI

But now we observe that lines 8, 13, and 14 form a truth-functional contradiction; the reader may check this fact by constructing a truth table. Thus we may box and cancel, to obtain the complete (abbreviated) proof

| | | |
|---|---|---|
| 1. | ~~*Show*~~ $\bigvee x \bigvee y \bigwedge z([\mathrm{F}x \to \mathrm{G}x] \to [\mathrm{F}y \to \mathrm{G}z])$ | |
| 2. | $\sim \bigvee x \bigvee y \bigwedge z([\mathrm{F}x \to \mathrm{G}x] \to [\mathrm{F}y \to \mathrm{G}z])$ | |
| 3. | $\bigwedge x \sim \bigvee y \bigwedge z([\mathrm{F}x \to \mathrm{G}x] \to [\mathrm{F}y \to \mathrm{G}z])$ | 2, T204 |
| 4. | $\bigwedge x \bigwedge y \sim \bigwedge z([\mathrm{F}x \to \mathrm{G}x] \to [\mathrm{F}y \to \mathrm{G}z])$ | 3, IE(T204) |
| 5. | $\bigwedge x \bigwedge y \bigvee z \sim ([\mathrm{F}x \to \mathrm{G}x] \to [\mathrm{F}y \to \mathrm{G}z])$ | 4, IE(T203) |
| 6. | $\bigwedge y \bigvee z \sim ([\mathrm{F}a \to \mathrm{G}a] \to [\mathrm{F}y \to \mathrm{G}z])$ | 5, UI |
| 7. | $\bigvee z \sim ([\mathrm{F}a \to \mathrm{G}a] \to [\mathrm{F}a \to \mathrm{G}z])$ | 6, UI |
| 8. | $\sim ([\mathrm{F}a \to \mathrm{G}a] \to [\mathrm{F}a \to \mathrm{G}b])$ | 7, EI |
| 9. | $\bigwedge y \bigvee z \sim ([\mathrm{F}b \to \mathrm{G}b] \to [\mathrm{F}y \to \mathrm{G}z])$ | 5, UI |
| 10. | $\bigvee z \sim ([\mathrm{F}a \to \mathrm{G}a] \to [\mathrm{F}b \to \mathrm{G}z])$ | 6, UI |
| 11. | $\bigvee z \sim ([\mathrm{F}b \to \mathrm{G}b] \to [\mathrm{F}a \to \mathrm{G}z])$ | 9, UI |
| 12. | $\bigvee z \sim ([\mathrm{F}b \to \mathrm{G}b] \to [\mathrm{F}b \to \mathrm{G}z])$ | 9, UI |
| 13. | $\sim ([\mathrm{F}a \to \mathrm{G}a] \to [\mathrm{F}b \to \mathrm{G}c])$ | 10, EI |
| 14. | $\sim ([\mathrm{F}b \to \mathrm{G}b] \to [\mathrm{F}a \to \mathrm{G}d])$ | 11, EI |

Instead of boxing and cancelling, we might well have added lines to the column. Consideration of lines 12, 13, and 14 would have led to new lines, and these to further lines. Indeed, the general procedure would never, in this case, terminate naturally. It is of course pointless to continue adding lines after a contradiction has been obtained; there is, however, no general method for computing when, if ever, this stage will be reached.

## EXERCISES

Apply the general procedure just described to obtain proofs of the following theorems. Exercise 5 is solved for illustration.

5. $\bigvee x \bigvee y[\mathrm{FA}(x\mathrm{B}) \to \mathrm{F}y]$

| | | |
|---|---|---|
| 1. | ~~*Show*~~ $\bigvee x \bigvee y[\mathrm{FA}(x\mathrm{B}) \to \mathrm{F}y]$ | |
| 2. | $\sim \bigvee x \bigvee y[\mathrm{FA}(x\mathrm{B}) \to \mathrm{F}y]$ | |
| 3. | $\bigwedge x \sim \bigvee y[\mathrm{FA}(x\mathrm{B}) \to \mathrm{F}y]$ | 2, T204 |
| 4. | $\bigwedge x \bigwedge y \sim [\mathrm{FA}(x\mathrm{B}) \to \mathrm{F}y]$ | 3, IE(T204) |
| 5. | $\bigwedge y \sim [\mathrm{FA}(\mathrm{BB}) \to \mathrm{F}y]$ | 4, UI |
| 6. | $\sim [\mathrm{FA}(\mathrm{BB}) \to \mathrm{FB}]$ | 5, UI |
| 7. | $\sim [\mathrm{FA}(\mathrm{BB}) \to \mathrm{FA}(\mathrm{BB})]$ | 5, UI |

6. $\bigvee x \bigvee y[\mathrm{FA}(x\mathrm{B}) \to \mathrm{FA}(yy)]$
7. $\bigvee x \bigwedge y(\mathrm{F}x \vee [\mathrm{F}y \to \mathrm{G}y])$
8. $\bigwedge y \bigvee x \bigwedge z \bigvee w[\mathrm{F}(xy) \to \mathrm{F}(zw)]$
9. $\bigwedge x \bigwedge y \bigwedge z \bigvee w(\mathrm{F}x \wedge \sim \mathrm{F}z \leftrightarrow [\mathrm{F}w \leftrightarrow \sim \mathrm{F}y])$

**5. A derivation procedure for prenex arguments.** The procedure just described in connection with individual prenex formulas can be extended quite naturally so as to apply to arguments. In particular, we must consider symbolic arguments with finitely many premises, all of whose formulas are prenex, and whose premises are sentences.

Suppose that such an argument has $\phi_1, \ldots, \phi_n$ as its premises and $\psi$ as its conclusion. Then the following general procedure will lead, if the argument is valid, to a complete derivation of $\psi$ from $\phi_1, \ldots, \phi_n$.

We begin an indirect proof of $\psi$:

*Show* $\psi$
$\sim \psi$

As before, we convert

$\sim \psi$

into prenex normal form, obtaining

*Show* $\psi$
$\sim \psi$
.
.
.
$\chi$ ,

where $\chi$ is inferred from

$\sim \psi$

by 'moving in' the negation sign. Now we add the premises of the argument:

*Show* $\psi$
$\sim \psi$
.
.
.
$\chi$
$\phi_1$
.
.
.
$\phi_n$ .

The elimination column is again regarded as beginning with $\chi$. Lines are added in accordance with the instructions given on pages 194 – 95. If the argument is valid, a truth-functional contradiction will be obtained, and the generalized indirect proof may be completed.

To illustrate the present procedure, let us construct a derivation corresponding to the argument

$$\bigwedge x \bigwedge y[Fx \wedge Gy \rightarrow H(xy)] \quad . \quad \bigvee x \bigwedge y(Fx \wedge [Jy \rightarrow \sim H(xy)])$$
$$\therefore \bigwedge x(Gx \rightarrow \sim Jx)$$

Derivation:

| | | |
|---|---|---|
| 1. | ~~*Show*~~ $\bigwedge x(Gx \rightarrow \sim Jx)$ | |
| 2. | $\sim \bigwedge x(Gx \rightarrow \sim Jx)$ | |
| 3. | $\bigvee x \sim (Gx \rightarrow \sim Jx)$ | |
| 4. | $\bigwedge x \bigwedge y[Fx \wedge Gy \rightarrow H(xy)]$ | Premise |
| 5. | $\bigvee x \bigwedge y(Fx \wedge [Jy \rightarrow \sim H(xy)])$ | Premise |
| 6. | $\sim (Ga \rightarrow \sim Ja)$ | 3, EI |
| 7. | $\bigwedge y(Fb \wedge [Jy \rightarrow \sim H(by)])$ | 5, EI |
| 8. | $\bigwedge y[Fa \wedge Gy \rightarrow H(ay)]$ | 4, UI |
| 9. | $Fb \wedge [Ja \rightarrow \sim H(ba)]$ | 7, UI |
| 10. | $Fb \wedge [Jb \rightarrow \sim H(bb)]$ | 7, UI |
| 11. | $\bigwedge y[Fb \wedge Gy \rightarrow H(by)]$ | 4, UI |
| 12. | $Fa \wedge Ga \rightarrow H(aa)$ | 8, UI |
| 13. | $Fa \wedge Gb \rightarrow H(ab)$ | 8, UI |
| 14. | $Fb \wedge Ga \rightarrow H(ba)$ | 11, UI |
| 15. | $Fb \wedge Gb \rightarrow H(bb)$ | 11, UI |

The reader can easily verify that lines 6, 9, and 14 are truth-functionally incompatible.

Here and in subsequent sections we consider only arguments with finitely many premises. The restriction is not essential. A slight modification of the procedures of this chapter would provide derivation procedures for arguments with infinitely many premises, provided only that the set of premises be *recursively enumerable*. (Loosely speaking, a set of expressions is *recursively enumerable* just in case there is an enumeration of the members of the set according to which one can find automatically, for each number $n$, the $n$th expression in the enumeration. For an exact definition, see, for instance, Kleene [1].)

## EXERCISES

Establish the validity of the following arguments by the derivation procedure just described.

10. $\bigwedge y \bigvee x([Fy \rightarrow Gy] \vee [Fx \wedge Hx])$ . $\bigwedge x(Fx \rightarrow \sim Hx)$ . $\bigvee x Fx \quad \therefore \bigvee x(Fx \wedge Gx)$

11. $\bigwedge x \bigwedge y \bigwedge z[F(xy) \wedge F(yz) \rightarrow F(xz)]$ . $\bigwedge x \bigwedge y[F(xy) \rightarrow F(yx)]$ $\therefore \bigwedge x \bigwedge y[F(xy) \rightarrow F(xx)]$

12. $\bigwedge x \bigwedge y(Fx \wedge Fy \rightarrow FA(xy))$ . $\bigvee x \bigwedge y(Fx \wedge [Fy \rightarrow GA(xy)])$ $\therefore \bigvee x(Fx \wedge Gx)$

**6. Conversion to prenex form.** In the foregoing sections we considered only those arguments whose formulas are prenex. Before extending our considerations further, we must attend to the task of *converting a formula into prenex form.* It was mentioned in chapter IV that every symbolic formula is equivalent to a formula in prenex normal form. This assertion can be strengthened. Every symbolic formula $\phi$ is equivalent to a prenex formula with the same free variables as $\phi$; such a formula is called a *prenex form* of $\phi$. We must now give directions for finding a prenex form of an arbitrary formula $\phi$ and for proving its equivalence with $\phi$.

Suppose we are given a symbolic formula $\phi$ that is not yet in prenex normal form. First we eliminate '$\leftrightarrow$'; that is, we replace, one at a time, each subformula

$$\psi \leftrightarrow \chi$$

by an equivalent formula, on the basis of T81:

$$(\mathrm{P} \leftrightarrow \mathrm{Q}) \leftrightarrow (\mathrm{P} \rightarrow \mathrm{Q}) \wedge (\mathrm{Q} \rightarrow \mathrm{P}) \quad .$$

This transformation requires several applications of IE, corresponding to the occurrences of '$\leftrightarrow$'. Let the formulas produced by these steps, which lead from $\phi$ to a formula without '$\leftrightarrow$', be, in order, $\phi_1, \ldots, \phi_n$.

The formula $\phi_n$, like the original formula $\phi$, is not yet in prenex normal form. Thus $\phi_n$ contains at least one *noninitial* occurrence of a quantifier phrase, that is, an occurrence standing to the right of a parenthesis or negation sign. (We have in mind our official notation, so that in

$$\bigvee x \mathrm{F} x \rightarrow \bigvee x \mathrm{G} x$$

both occurrences of '$\bigvee x$' are regarded as noninitial.)

We fix our attention on the first such occurrence, and convert it to an initial occurrence. The occurrence will begin a subformula of $\phi_n$ of the form

(1) $$\bigwedge \alpha \psi$$

or

(2) $$\bigvee \alpha \psi \quad ,$$

where $\alpha$ is a variable and $\psi$ a formula. We first, on the basis of AV, change $\alpha$ to a variable $\alpha'$ that is new to $\phi_n$; that is, we replace (1) or (2) by

$$\bigwedge \alpha' \psi'$$

or

$$\bigvee \alpha' \psi'$$

respectively, where $\psi'$ comes from $\psi$ by proper substitution of $\alpha'$ for $\alpha$. Thus we obtain from $\phi_n$ a formula $\phi_{n+1}$.

Now we move the occurrence of

$$\bigwedge\alpha'$$

or

$$\bigvee\alpha'$$

to the left of all parentheses and negation signs, proceeding step by step in accordance with the laws of quantifier negation and confinement:

| | |
|---|---|
| T203 | $\sim \bigwedge x\mathrm{F}x \leftrightarrow \bigvee x \sim \mathrm{F}x$ |
| T204 | $\sim \bigvee x\mathrm{F}x \leftrightarrow \bigwedge x \sim \mathrm{F}x$ |
| T219 | $\bigwedge x(\mathrm{P} \rightarrow \mathrm{F}x) \leftrightarrow (\mathrm{P} \rightarrow \bigwedge x\mathrm{F}x)$ |
| T220 | $\bigvee x(\mathrm{P} \rightarrow \mathrm{F}x) \leftrightarrow (\mathrm{P} \rightarrow \bigvee x\mathrm{F}x)$ |
| T221 | $\bigwedge x(\mathrm{F}x \rightarrow \mathrm{P}) \leftrightarrow (\bigvee x\mathrm{F}x \rightarrow \mathrm{P})$ |
| T222 | $\bigvee x(\mathrm{F}x \rightarrow \mathrm{P}) \leftrightarrow (\bigwedge x\mathrm{F}x \rightarrow \mathrm{P})$ |
| T215 | $\bigwedge x(\mathrm{P} \wedge \mathrm{F}x) \leftrightarrow \mathrm{P} \wedge \bigwedge x\mathrm{F}x$ |
| T272 | $\bigwedge x(\mathrm{F}x \wedge \mathrm{P}) \leftrightarrow \bigwedge x\mathrm{F}x \wedge \mathrm{P}$ |
| T216 | $\bigvee x(\mathrm{P} \wedge \mathrm{F}x) \leftrightarrow \mathrm{P} \wedge \bigvee x\mathrm{F}x$ |
| T273 | $\bigvee x(\mathrm{F}x \wedge \mathrm{P}) \leftrightarrow \bigvee x\mathrm{F}x \wedge \mathrm{P}$ |
| T217 | $\bigwedge x(\mathrm{P} \vee \mathrm{F}x) \leftrightarrow \mathrm{P} \vee \bigwedge x\mathrm{F}x$ |
| T274 | $\bigwedge x(\mathrm{F}x \vee \mathrm{P}) \leftrightarrow \bigwedge x\mathrm{F}x \vee \mathrm{P}$ |
| T218 | $\bigvee x(\mathrm{P} \vee \mathrm{F}x) \leftrightarrow \mathrm{P} \vee \bigvee x\mathrm{F}x$ |
| T275 | $\bigvee x(\mathrm{F}x \vee \mathrm{P}) \leftrightarrow \bigvee x\mathrm{F}x \vee \mathrm{P}$ |

Let the formulas corresponding to the various steps in this transformation be $\phi_{n+2}, \ldots, \phi_{n+k}$. In the last of these formulas the occurrence of

$$\bigwedge\alpha'$$

or

$$\bigvee\alpha'$$

has become an initial occurrence. We now consider the first noninitial occurrence of a quantifier phrase in $\phi_{n+k}$ and transform this into an initial occurrence just as before. We repeat the process until all noninitial occurrences of quantifier phrases have been removed. Thus, when we combine all stages of the process, we obtain a sequence of formulas $\phi, \phi_1, \ldots, \phi_j, \psi$, in which each formula is equivalent to the preceding by IE (applied in connection with the theorems just listed), and such that the final formula $\psi$ has no noninitial occurrences of quantifier phrases.

The formula $\psi$ is the prenex form that was sought: it is equivalent to $\phi$, it is in prenex normal form, and it contains the same free variables as $\phi$. Further, a proof of the biconditional

$$\phi \leftrightarrow \psi$$

can easily be constructed, in terms of the intermediate formulas $\phi_1, \ldots, \phi_j$:

1. ~~*Show*~~ $\phi \leftrightarrow \psi$

| | |
|---|---|
| 2. | $\phi \leftrightarrow \phi_1$ |
| 3. | $\phi_1 \leftrightarrow \phi_2$ |
| | . |
| | . |
| | . |
| $j + 1$. | $\phi_{j-1} \leftrightarrow \phi_j$ |
| $j + 2$. | $\phi_j \leftrightarrow \psi$ |
| $j + 3$. | $\phi \leftrightarrow \psi$ |

Each of the lines (2) – ($j$ + 2) is justified by AV (assisted by the theorem '$P \leftrightarrow P$'), or by IE, used in connection with an appropriate theorem; and the line ($j$ + 3) is obtained from (2) – ($j$ + 2) by TI.

For example, let us convert the formula

(3) $$\bigwedge x Fx \rightarrow \bigvee y Fy \vee Gx$$

into prenex form. We fix our attention on the first noninitial occurrence of a quantifier phrase, change the variable occurring in it to one new to (3), and, on the basis of T222, move it to the beginning of (3), obtaining first

$$\bigwedge z Fz \rightarrow \bigvee y Fy \vee Gx$$

and then

(4) $$\bigvee z(Fz \rightarrow \bigvee y Fy \vee Gx) \quad .$$

We treat similarly the first noninitial quantifier phrase in (4), obtaining successively

$$\bigvee z(Fz \rightarrow \bigvee w Fw \vee Gx) \quad ,$$
$$\bigvee z(Fz \rightarrow \bigvee w(Fw \vee Gx)) \quad ,$$
(5) $$\bigvee z \bigvee w(Fz \rightarrow Fw \vee Gx) \quad .$$

The formula (5) is a prenex form of (3), and the following is a proof of its equivalence to (3):

1. ~~*Show*~~ $(\bigwedge x Fx \rightarrow \bigvee y Fy \vee Gx) \leftrightarrow \bigvee z \bigvee w(Fz \rightarrow Fw \vee Gx)$

| | | |
|---|---|---|
| 2. | $(\bigwedge x Fx \rightarrow \bigvee y Fy \vee Gx) \leftrightarrow (\bigwedge z Fz \rightarrow \bigvee y Fy \vee Gx)$ | T91, AV |

| | | |
|---|---|---|
| 3. | $(\bigwedge z Fz \to \bigvee y Fy \vee Gx) \leftrightarrow \bigvee z(Fz \to \bigvee y Fy \vee Gx)$ | EI(T222) |
| 4. | $\bigvee z(Fz \to \bigvee y Fy \vee Gx) \leftrightarrow \bigvee z(Fz \to \bigvee w Fw \vee Gx)$ | T91, AV |
| 5. | $\bigvee z(Fz \to \bigvee w Fw \vee Gx) \leftrightarrow \bigvee z(Fz \to \bigvee w(Fw \vee Gx))$ | IE(T275) |
| 6. | $\bigvee z(Fz \to \bigvee w(Fw \vee Gx)) \leftrightarrow \bigvee z \bigvee w(Fz \to Fw \vee Gx)$ | IE(T220) |
| 7. | $(\bigwedge x Fx \to \bigvee y Fy \vee Gx) \leftrightarrow \bigvee z \bigvee w(Fz \to Fw \vee Gx)$ | 2 – 6, TI |

**EXERCISES**

Convert each of the following formulas into prenex form and prove the equivalence of the formula and its prenex form.

13. $\bigvee x \bigwedge y F(xy) \to \bigwedge y \bigvee x F(xy)$
14. $\bigvee x[Fx \wedge \bigwedge y G(xy)] \to \sim \bigwedge x Hx$
15. $\bigwedge x(Fx \to Gx) \to \bigwedge x(\bigvee y[H(xy) \wedge Fy] \to \bigvee y[H(xy) \wedge Gy])$
16. $\bigwedge x(Fx \leftrightarrow Gx) \to (\bigvee x Fx \leftrightarrow \bigvee x Gx)$

**7. A derivation procedure for arbitrary symbolic arguments.** We consider now an arbitrary symbolic argument with finitely many premises. Let these premises be $\phi_1, \ldots, \phi_n$, and let the conclusion be $\psi$. The derivation procedure is roughly this. We first derive from the premises their closures, then convert both the premises and the conclusion into prenex form, and finally apply the derivation procedure of section 5.

In greater detail, the derivation procedure is the following. Let

(1) $$\bigwedge \alpha_1 \ldots \bigwedge \alpha_k \, \phi_1$$

be a closure of $\phi_1$, the first premise. We begin the derivation with a derivation of (1):

~~*Show*~~ $\bigwedge \alpha_1 \ldots \bigwedge \alpha_k \, \phi_1$

| |
|---|
| $\phi_1$ |

By the procedure described in section 6, we find a prenex sentence $\psi_1$ that is equivalent to (1), and further a proof of the equivalence. We add this proof to the derivation to obtain

~~*Show*~~ $\bigwedge \alpha_1 \ldots \bigwedge \alpha_k \, \phi_1$

| |
|---|
| $\phi_1$ |

~~*Show*~~ $\bigwedge \alpha_1 \ldots \bigwedge \alpha_k \, \phi_1 \leftrightarrow \psi_1$

| |
|---|
| |

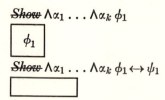

We repeat the procedure for each of the other premises, obtaining

~~Show~~ $\wedge\alpha_1 \ldots \wedge\alpha_k\ \phi_1$

$\boxed{\phi_1}$

~~Show~~ $\wedge\alpha_1 \ldots \wedge\alpha_k\ \phi_1 \leftrightarrow \psi_1$

$\boxed{\phantom{xxxxxxxxxx}}$

.
.
.

~~Show~~ $\wedge\beta_1 \ldots \wedge\beta_j\ \phi_n$

$\boxed{\phi_n}$

~~Show~~ $\wedge\beta_1 \ldots \wedge\beta_j\ \phi_n \leftrightarrow \psi_n$

$\boxed{\phantom{xxxxxxxxxx}}$

where $\beta_1, \ldots, \beta_j$ are the free variables of $\phi_n$. Again by the procedure of section 6, we find a prenex form $\chi$ of the conclusion $\psi$, together with a proof of the equivalence. This proof we add to the derivation:

~~Show~~ $\wedge\alpha_1 \ldots \wedge\alpha_k\ \phi_1$

$\boxed{\phi_1}$

~~Show~~ $\wedge\alpha_1 \ldots \wedge\alpha_k\ \phi_1 \leftrightarrow \psi_1$

$\boxed{\phantom{xxxxxxxxxx}}$

.
.
.

~~Show~~ $\wedge\beta_1 \ldots \wedge\beta_j\ \phi_n$

$\boxed{\phantom{xxxxxxxxxx}}$

~~Show~~ $\wedge\beta_1 \ldots \wedge\beta_j\ \phi_n \leftrightarrow \psi_n$

$\boxed{\phantom{xxxxxxxxxxxxxxx}}$

~~Show~~ $\psi \leftrightarrow \chi$

$\boxed{\phantom{xxxxxxxxxx}}$

At this point we begin an indirect proof of $\psi$, and from the assumption

$$\sim\psi \quad ,$$

together with an equivalence occurring earlier in the derivation, we infer

$$\sim\chi \quad .$$

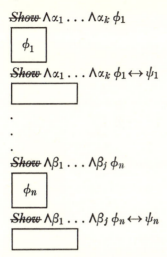

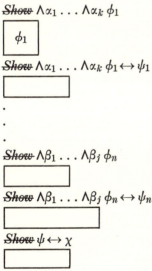

Next we 'move in' the negation sign, obtaining a prenex counterpart $\chi'$ of

$$\sim\chi \quad .$$

Thus the derivation becomes:

~~Show~~ $\wedge\alpha_1 \ldots \wedge\alpha_k\, \phi_1$

$\boxed{\phi_1}$

~~Show~~ $\wedge\alpha_1 \ldots \wedge\alpha_k\, \phi_1 \leftrightarrow \psi_1$

$\boxed{\qquad\qquad}$

.

.

.

~~Show~~ $\wedge\beta_1 \ldots \wedge\beta_j\, \phi_n$

$\boxed{\phi_n}$

~~Show~~ $\wedge\beta_1 \ldots \wedge\beta_j\, \phi_n \leftrightarrow \psi_n$

$\boxed{\qquad\qquad}$

~~Show~~ $\psi \leftrightarrow \chi$

$\boxed{\qquad\qquad}$

~~Show~~ $\psi$

$\sim\psi$

$\sim\chi$

.

.

.

$\chi'$

Next we infer the formulas $\psi_1$ through $\psi_n$, obtaining

~~Show~~ $\wedge\alpha_1 \ldots \wedge\alpha_k\, \phi_1$

$\boxed{\phi_1}$

~~Show~~ $\wedge\alpha_1 \ldots \wedge\alpha_k\, \phi_1 \leftrightarrow \psi_1$

$\boxed{\qquad\qquad}$

.

.

.

~~Show~~ $\wedge\beta_1 \ldots \wedge\beta_j\, \phi_n$

$\boxed{\phi_n}$

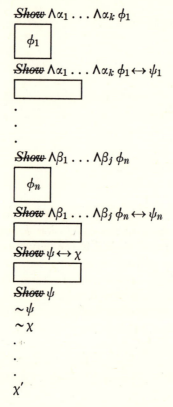

~~Show~~ $\bigwedge\beta_1 \ldots \bigwedge\beta_j\ \phi_n \leftrightarrow \psi_n$

~~Show~~ $\psi \leftrightarrow \chi$

~~Show~~ $\psi$

$\sim\psi$

$\sim\chi$

.

.

.

$\chi'$

$\psi_1$

.

.

.

$\psi_n$

We now proceed as in section 5, regarding the elimination column as beginning with the line $\chi'$. A truth-functional contradiction will ensue just in case the conclusion $\psi$ is derivable from the premises $\phi_1$ through $\phi_n$.

The present method, besides serving as a derivation procedure for symbolic arguments, provides also a proof procedure for arbitrary symbolic formulas. To apply the method to a formula $\phi$, we consider the argument without premises whose conclusion is $\phi$.

## EXERCISES

Establish the validity of each of the following arguments by the derivation procedure described in this section. Exercise 17 is solved for illustration.

17. $\mathrm{F}x \rightarrow \bigwedge x\mathrm{G}x \quad . \quad \bigvee x(\mathrm{F}x \wedge \sim\bigwedge x\mathrm{H}x) \quad \therefore \sim\bigwedge x(\mathrm{G}x \rightarrow \mathrm{H}x)$

| | | |
|---|---|---|
| 1. | ~~Show~~ $\bigwedge x(\mathrm{F}x \rightarrow \bigwedge x\mathrm{G}x)$ | |
| 2. | $\mathrm{F}x \rightarrow \bigwedge x\mathrm{G}x$ | Premise |
| 3. | ~~Show~~ $\bigwedge x(\mathrm{F}x \rightarrow \bigwedge x\mathrm{G}x) \leftrightarrow \bigwedge x\bigwedge y(\mathrm{F}x \rightarrow \mathrm{G}y)$ | |
| 4. | $\bigwedge x(\mathrm{F}x \rightarrow \bigwedge x\mathrm{G}x) \leftrightarrow \bigwedge x(\mathrm{F}x \rightarrow \bigwedge y\mathrm{G}y)$ | T91, AV |
| 5. | $\bigwedge x(\mathrm{F}x \rightarrow \bigwedge y\mathrm{G}y) \leftrightarrow \bigwedge x\bigwedge y(\mathrm{F}x \rightarrow \mathrm{G}y)$ | IE(T219) |
| 6. | $\bigwedge x(\mathrm{F}x \rightarrow \bigwedge x\mathrm{G}x) \leftrightarrow \bigwedge x\bigwedge y(\mathrm{F}x \rightarrow \mathrm{G}y)$ | 4, 5, TI |
| 7. | ~~Show~~ $\bigvee x(\mathrm{F}x \wedge \sim\bigwedge x\mathrm{H}x) \leftrightarrow \bigvee x\bigvee y(\mathrm{F}x \wedge \sim\mathrm{H}y)$ | |
| 8. | $\bigvee x(\mathrm{F}x \wedge \sim\bigwedge x\mathrm{H}x) \leftrightarrow \bigvee x(\mathrm{F}x \wedge \sim\bigwedge y\mathrm{H}y)$ | T91, AV |
| 9. | $\bigvee x(\mathrm{F}x \wedge \sim\bigwedge y\mathrm{H}y) \leftrightarrow \bigvee x(\mathrm{F}x \wedge \bigvee y \sim\mathrm{H}y)$ | IE(T203) |
| 10. | $\bigvee x(\mathrm{F}x \wedge \bigvee y \sim\mathrm{H}y) \leftrightarrow \bigvee x\bigvee y(\mathrm{F}x \wedge \sim\mathrm{H}y)$ | IE(T216) |
| 11. | $\bigvee x(\mathrm{F}x \wedge \sim\bigwedge x\mathrm{H}x) \leftrightarrow \bigvee x\bigvee y(\mathrm{F}x \wedge \sim\mathrm{H}y)$ | 8, 9, 10, TI |

| | | |
|---|---|---|
| 12. | ~~Show~~ $\sim\bigwedge x(Gx \to Hx) \leftrightarrow \bigvee x \sim (Gx \to Hx)$ | |
| 13. | $\sim\bigwedge x(Gx \to Hx) \leftrightarrow \bigvee x \sim (Gx \to Hx)$ | T203 |
| 14. | ~~Show~~ $\sim\bigwedge x(Gx \to Hx)$ | |
| 15. | $\bigwedge x(Gx \to Hx)$ | |
| 16. | $\sim\bigvee x \sim (Gx \to Hx)$ | 12, 15, TI |
| 17. | $\bigwedge x \sim\sim (Gx \to Hx)$ | 16, T204, TI |
| 18. | $\bigwedge x \bigwedge y(Fx \to Gy)$ | 1, 3, TI |
| 19. | $\bigvee x \bigvee y(Fx \wedge \sim Hy)$ | Premise, 7, TI |
| 20. | $\sim\sim(Ga \to Ha)$ | 17, UI |
| 21. | $\bigvee y(Fb \wedge \sim Hy)$ | 19, EI |
| 22. | $\bigwedge y(Fa \to Gy)$ | 18, UI |
| 23. | $Fb \wedge \sim Hc$ | 21, EI |
| 24. | $\sim\sim(Gb \to Hb)$ | 17, UI |
| 25. | $\bigwedge y(Fb \to Gy)$ | 18, UI |
| 26. | $Fa \to Ga$ | 22, UI |
| 27. | $Fa \to Gb$ | 22, UI |
| 28. | $\sim\sim(Gc \to Hc)$ | 17, UI |
| 29. | $\bigwedge y(Fc \to Gy)$ | 18, UI |
| 30. | $Fa \to Gc$ | 22, UI |
| 31. | $Fb \to Ga$ | 25, UI |
| 32. | $Fb \to Gb$ | 25, UI |
| 33. | $Fb \to Gc$ | 25, UI |

In line 1 of the derivation, a closure of the first premise is established. In each of lines 3, 7, and 12, a closure of a formula of the argument is shown to be equivalent to a prenex form of that formula. Line 14 begins the derivation of the conclusion of the argument, and the elimination column of this derivation begins with line 17. The reader can verify that lines 23, 28, and 33 are truth-functionally incompatible.

18. $Fx \to \bigvee y[Gy \wedge H(xy)] \wedge \bigvee y[Gy \wedge \sim H(xy)]$ .
$\bigvee x(Tx \wedge \bigwedge y[Gy \to H(xy)])$ $\therefore \bigvee x(Tx \wedge \sim Fx)$

19. $\bigvee y(\bigwedge z[F(zy) \to F(z\ \acute{A}(x))] \wedge F(xy))$ $\therefore \bigwedge x F(x\ A(x))$

**8. A decision procedure for certain prenex formulas.** A subsidiary purpose of this chapter is to present *decision procedures* for certain kinds of formulas and arguments. A *decision procedure* for a class of *formulas* is an automatic method for determining of each formula in the class whether or not it is a theorem. A *decision procedure* for a class of *arguments* is an automatic method for determining of each argument in the class whether or not it is valid.

We have observed earlier (p. 170) that for the class of all symbolic formulas (of the full quantifier calculus) and for the class of all symbolic arguments (of this calculus) no decision procedures are available. For some special classes, however, decision procedures may be found. Indeed, in some cases the automatic derivation procedures described earlier in this chapter lead quite simply to decision procedures.

The cases we wish to consider depend upon the following fact (which will not be proved here): for formulas and arguments of certain structures, the elimination column of our automatic derivation procedure will come to an end, and in such cases we may test for provability or validity by generating a complete elimination column and performing a truth-table check of the truth-functional compatibility of the quantifier-free formulas in it. It should be mentioned that we exclude from our considerations formulas containing operation letters.

The first case is that of prenex formulas without operation letters in which no universal quantifier follows an existential quantifier. To determine whether such a formula is a theorem we need only attempt to prove it by means of the derivation procedure described in section 4. The elimination column so obtained will come to an end. Thus it is possible to test the quantifier-free formulas of the elimination column for truth-functional compatibility by constructing a truth table. We shall obtain truth-functional incompatibility just in case the original formula is a theorem.

For instance, the formula

(1) $$\bigwedge y \bigvee x(\mathrm{F}x \rightarrow \mathrm{F}y)$$

has the required form. Its elimination column, constructed on page 196, was seen to terminate, and a truth-table check would have disclosed that (1) is a theorem.

As another example, consider the formula

(2) $$\bigwedge x \bigwedge y \bigvee z[\mathrm{F}(xz) \vee \mathrm{G}(yz) \rightarrow \mathrm{F}(xy)] \quad .$$

The automatic derivation procedure leads us to the following:

$$\begin{array}{l}
\textit{Show } \bigwedge x \bigwedge y \bigvee z[\mathrm{F}(xz) \vee \mathrm{G}(yz) \rightarrow \mathrm{F}(xy)] \\
\sim \bigwedge x \bigwedge y \bigvee z[\mathrm{F}(xz) \vee \mathrm{G}(yz) \rightarrow \mathrm{F}(xy)] \\
\bigvee x \sim \bigwedge y \bigvee z[\mathrm{F}(xz) \vee \mathrm{G}(yz) \rightarrow \mathrm{F}(xy)] \\
\bigvee x \bigvee y \sim \bigvee z[\mathrm{F}(xz) \vee \mathrm{G}(yz) \rightarrow \mathrm{F}(xy)] \\
\bigvee x \bigvee y \bigwedge z \sim [\mathrm{F}(xz) \vee \mathrm{G}(yz) \rightarrow \mathrm{F}(xy)] \\
\bigvee y \bigwedge z \sim [\mathrm{F}(az) \vee \mathrm{G}(yz) \rightarrow \mathrm{F}(ay)] \\
\bigwedge z \sim [\mathrm{F}(az) \vee \mathrm{G}(bz) \rightarrow \mathrm{F}(ab)] \\
\sim [\mathrm{F}(aa) \vee \mathrm{G}(ba) \rightarrow \mathrm{F}(ab)] \\
\sim [\mathrm{F}(ab) \vee \mathrm{G}(bb) \rightarrow \mathrm{F}(ab)]
\end{array}$$

The quantifier-free formulas to be considered are

$$\begin{array}{l}
\sim [\mathrm{F}(aa) \vee \mathrm{G}(ba) \rightarrow \mathrm{F}(ab)] \quad , \\
\sim [\mathrm{F}(ab) \vee \mathrm{G}(bb) \rightarrow \mathrm{F}(ab)] \quad .
\end{array}$$

A truth-table check reveals that these formulas are truth-functionally compatible. Thus, according to our decision procedure, the formula (2) is not a theorem.

### EXERCISES

Using the decision procedure just described, determine which of Nos. 20 – 28 are theorems.

20. $\Lambda x \Lambda y \mathrm{V} z \mathrm{V} w[F(xy) \vee F(yx) \rightarrow F(zw)]$
21. $\mathrm{V} x \mathrm{V} y \mathrm{V} z([Fx \rightarrow Gx] \rightarrow [Fy \rightarrow Gz])$
22. $\Lambda x \Lambda y \Lambda z \mathrm{V} w([Fx \leftrightarrow Fw] \rightarrow \sim Fy \vee Fz)$
23. $\Lambda z \Lambda y \Lambda x \Lambda u \Lambda v([F(xy) \leftrightarrow F(xz) \wedge \sim F(xx)] \rightarrow \sim F(uv))$
24. $\Lambda x \Lambda y \mathrm{V} z \mathrm{V} w([Fw \rightarrow Gw] \rightarrow [Fy \wedge H(xy) \rightarrow Gz \wedge H(xz)])$
25. $\Lambda z \Lambda x \Lambda w \Lambda u \mathrm{V} v \mathrm{V} y(F(xz) \rightarrow [F(yw) \leftrightarrow (F(uv) \rightarrow F(yy))])$
26. $\Lambda x \Lambda y \Lambda z \Lambda w([F(xz) \rightarrow G(yw)] \rightarrow G(yw))$
27. $\Lambda z \Lambda y \mathrm{V} w \mathrm{V} u \mathrm{V} x([F(xu) \rightarrow G(xu) \vee H(xu)] \rightarrow$
$[F(yz) \rightarrow G(yz)] \wedge [F(wz) \wedge H(wz)])$
28. $\mathrm{V} y \mathrm{V} x \mathrm{V} z[F(xyz) \leftrightarrow F(xzx)]$

29. * Give a decision procedure for symbolic formulas without operation letters and *without overlay* (see p. 153), that is, an automatic procedure whereby one can determine whether such a formula is a theorem. In view of the decision procedure just described, it is sufficient to give an automatic procedure for transforming any formula of the kind in question into an equivalent prenex formula, also without operation letters, in which no universal quantifier follows an existential quantifier. (One such procedure can be obtained with the aid of the normal forms mentioned on page 72, in exercises 40 – 43 of chapter II.) Notice that the solution of this problem, combined with that of exercise 72 of chapter IV (p. 170), will give a decision procedure for all monadic formulas. (An alternative decision procedure was given in section 9 of chapter III.)

**9. A decision procedure for certain prenex arguments.** The second decision procedure we consider is a natural extension of the first. In particular, we treat here arguments (again without operation letters) whose premises, finite in number, are prenex *sentences* in which no existential quantifier follows a universal, and whose conclusion is a prenex formula in which no universal quantifier follows an existential. To determine whether such an argument is valid we apply to the argument the derivation procedure of section 5. Again the elimination column will come to an end, and we may test its quantifier-free formulas for truth-functional compatibility. We shall obtain truth-functional incompatibility just in case the argument is valid.

For example, the argument

$$\Lambda x \Lambda y[Fx \wedge Gy \rightarrow H(xy)] \quad . \quad \mathrm{V} x \Lambda y(Fx \wedge [Jy \rightarrow H(xy)])$$
$$\therefore \Lambda x(Gx \rightarrow \sim Jx)$$

has the required form. Its elimination column, constructed on page 200, was seen to terminate and to contain a truth-functional contradiction.

Consider also the argument

(1) $\bigvee y \bigwedge x(Fx \rightarrow Gy)$ . $\bigvee y \bigvee z(Gy \rightarrow Hz)$ $\therefore$ $\bigwedge x \bigvee z(Fx \rightarrow Hz)$

Our derivation procedure leads us to the following:

*Show* $\bigwedge x \bigvee z(Fx \rightarrow Hz)$
$\sim \bigwedge x \bigvee z(Fx \rightarrow Hz)$
$\bigvee x \sim \bigvee z(Fx \rightarrow Hz)$
$\bigvee x \bigwedge z \sim (Fx \rightarrow Hz)$
$\bigvee y \bigwedge x(Fx \rightarrow Gy)$
$\bigvee y \bigvee z(Gy \rightarrow Hz)$
$\bigwedge z \sim (Fa \rightarrow Hz)$
$\bigwedge x(Fx \rightarrow Gb)$
$\bigvee z(Gc \rightarrow Hz)$
$\sim (Fa \rightarrow Ha)$
$Fa \rightarrow Gb$
$Fb \rightarrow Gb$
$\sim (Fa \rightarrow Hb)$
$Gc \rightarrow Hd$
$\sim (Fa \rightarrow Hc)$
$Fc \rightarrow Gb$
$\sim (Fa \rightarrow Hd)$
$Fd \rightarrow Fb$

A truth-table check of the quantifier-free formulas in this derivation would reveal that these formulas are truth-functionally compatible. Thus, according to our decision procedure, the argument (1) is invalid.

## EXERCISES

Using the decision procedure just described, determine which of the following arguments are valid.

30. $\bigwedge x \bigwedge y[Fx \wedge Gy \rightarrow H(xy)])$ . $\bigvee x \bigvee y[Fx \wedge Jy \wedge \sim H(xy)]$ $\therefore$ $\bigvee x(Jx \wedge \sim Gx)$

31. $\bigvee y \bigwedge x \bigwedge z(Fx \wedge [Gy \rightarrow H(xy)] \rightarrow [Gz \rightarrow H(xz)])$ . $\bigvee y \bigvee x[Fx \wedge \sim H(xy)]$ $\therefore$ $\bigvee x \bigvee y[Fx \wedge Gy \wedge \sim H(xy)]$

32. $\bigvee x \bigvee y \bigwedge z[Fz \rightarrow Gx \wedge H(zx) \wedge Gy \wedge \sim H(zy)]$ . $\bigvee x \bigwedge y(Jx \wedge [Gy \rightarrow H(xy)])$ $\therefore$ $\bigvee x(Jx \wedge \sim Fx)$

33. $\bigvee y \bigvee z \bigwedge x(F(xy) \leftrightarrow F(xz) \wedge Gx)$ $\therefore$ $\bigwedge z \bigwedge x \bigvee y(F(xy) \leftrightarrow F(xz) \wedge Gx)$

**10. Historical remarks.** The task of giving a derivation procedure, if no special conditions are imposed, is trivial. We need only settle upon an ordering of our inference rules and forms of derivation and give directions, based on that ordering, which will ensure that all possible inferences are drawn. If we are presented with a theorem, we may automatically obtain a proof for it by carrying out these directions until they lead us to one. The derivation procedures given in this chapter, however, lead to derivations of a particularly simple structure. For example, we obtain by the procedure of section 4 (our basic procedure for the quantifier calculus) derivations whose lines are always subformulas of the formula to be proved or else come from such subformulas by proper substitution on variables.

The procedure of section 2 (used to show the superfluity of clause (11)) is essentially Kalmár's proof of the completeness of the sentential calculus (see Kalmár [1]). The proof procedure of section 4 is closely related to a procedure given in Herbrand [2]. The claim made on page 193, that our procedure will always lead to a derivation if one is possible, follows from a theorem (known as the *Herbrand Theorem*) proved in Herbrand [2] or, alternatively, from Gentzen's *Extended Hauptsatz*, which is proved in Gentzen [1]. Here we must use the fact that our development of the quantifier calculus is equivalent to those of Herbrand and Gentzen; this follows from Montague and Kalish [1].

A decision procedure (differing considerably from ours) for prenex formulas in which no operation letters occur and no universal quantifier follows an existential quantifier was first given in Bernays and Schönfinkel [1]; a procedure quite similar to ours was given in Quine [5]. For a survey of other cases for which decision procedures have been found, see Church [3], pp. 245 - 94, and the comprehensive treatment in Ackermann [1].

# Chapter VI
# 'IS' (in one sense)

**1. Terms and formulas.** Not all valid arguments can be reached by the procedures of chapters I – IV. For example, the argument

(1) Mark Twain is identical with Samuel Clemens. Samuel Clemens wrote *Tom Sawyer*. ∴ Mark Twain wrote *Tom Sawyer* ,

unlike those discussed in the preceding chapters, depends for its validity on the meaning of the phrase 'is identical with'.

We shall abbreviate 'is identical with' by the familiar sign ' = '. The first premise of (1) then becomes

(2) Mark Twain = Samuel Clemens ,

which is regarded as true just in case the two names 'Mark Twain' and 'Samuel Clemens' designate the same object. ' = ' is known as the *identity sign*, and the result, like (2), of flanking it with terms is the *identity* formed from those terms.

The *language* with which we shall deal is that of chapter IV (see p. 133), with ' = ' added to its symbols. We obtain no new *terms*. To the characterization of the class of *formulas* given in chapter IV a fifth clause must be added, leading to the following exhaustive characterization:

(*1*) *All formulas of English* (*that is, sentences of English or expressions like sentences of English except for the occurrence of variables*) *are formulas.*

(*2*) *The result of writing a* k*-place predicate letter followed by* k *terms is a formula.* (*In particular, a* 0*-place predicate letter is itself a formula.*)

(*3*) *If* $\phi$ *and* $\psi$ *are formulas, so are*

$$\sim\phi \quad ,$$
$$(\phi \rightarrow \psi) \quad ,$$
$$(\phi \wedge \psi) \quad ,$$
$$(\phi \vee \psi) \quad ,$$
$$(\phi \leftrightarrow \psi) \quad .$$

(*4*) *If* $\phi$ *is a formula and* $\alpha$ *a variable, then*

$$\Lambda\alpha\phi \quad ,$$
$$\mathrm{V}\alpha\phi$$

*are formulas.*

(5) *If* $\zeta$ *and* $\eta$ *are terms, then*

$$\zeta = \eta$$

*is a formula.*

Thus, for example, each of the following is a formula of our present language:

| | |
|---|---|
| (3) | Scott = the author of *Waverley* |
| (4) | $7 + 5 = 12$ |
| (5) | $x = y$ |
| (6) | $A^0 = B^1x$ . |

The *symbolic terms* and *formulas* of our present language are those which can be constructed exclusively from variables, predicate letters, operation letters, parentheses, sentential connectives, quantifiers, and the identity sign. More precisely, the *symbolic terms* are those of chapter IV (see clauses (a) and (b) of page 134), and the class of *symbolic formulas* is exhaustively characterized as follows:

(*A*) *The result of writing a* k*-place predicate letter followed by* k *symbolic terms is a symbolic formula.* (*In particular, a* 0*-place predicate letter is itself a symbolic formula.*)

(*B*) *If* $\phi$ *and* $\psi$ *are symbolic formulas, so are*

$$\sim \phi \quad ,$$
$$(\phi \rightarrow \psi) \quad ,$$
$$(\phi \wedge \psi) \quad ,$$
$$(\phi \vee \psi) \quad ,$$
$$(\phi \leftrightarrow \psi) \quad .$$

(*C*) *If* $\phi$ *is a symbolic formula and* $\alpha$ *a variable, then*

$$\Lambda\alpha\phi \quad ,$$
$$\mathrm{V}\alpha\phi$$

*are symbolic formulas.*

(*D*) *If* $\zeta$ *and* $\eta$ *are symbolic terms, then*

$$\zeta = \eta$$

*is a symbolic formula.*

Of the preceding formulas, then, (3) and (4) are not symbolic, and (5) and (6) are symbolic.

We shall continue to use the informal notational conventions introduced in section 3 of chapter IV. Again we emphasize that official notation must be mentally restored before decisions concerning freedom and bondage and applications of inference rules can be made.

**2. Translation and symbolization.** At this point we must relinquish consideration of a large class of English formulas—those we shall call *nonextensional*. An English *sentence* $\phi$ is said to be *extensional* if, whenever a name occurring in it is replaced in one or more of its occurrences by another name designating the same object, the resulting sentence has the same truth value as $\phi$; that is, is true or false according as $\phi$ is true or false. Most sentences which come naturally to mind are extensional, for instance,

(1) Socrates is fat.

Replacement of 'Socrates' by another name designating Socrates will leave the truth value of (1) unchanged. For example,

The snub-nosed teacher of Plato is fat

is true if and only if (1) is true. However, the sentence

(2) Necessarily Socrates is Socrates

is not extensional. 'Socrates' and 'the snub-nosed teacher of Plato' designate the same object, and (2) is true; but

Necessarily the snub-nosed teacher of Plato is Socrates

is false. As an another example consider

Most students believe that Cicero denounced Catiline,

which is seen to be nonextensional by comparison with

Most students believe that Tully denounced Catiline.

The use of so-called modal terms, such as 'necessarily' and 'possibly', and the subordinate conjunction 'that' is in general likely to produce nonextensional sentences; and it is perhaps in this way that most nonextensional sentences originate. There are, however, other cases not so easy to detect, for example, the truth

Hamlet intended to kill the man behind the arras,

which, upon exchanging two names of the same person, becomes the falsehood

Hamlet intended to kill Polonius.

By an *extensional formula of English* we understand either an extensional sentence of English or a formula containing variables which, upon replacement of all its variables by English names, becomes an extensional sentence of English. Thus '$x$ is fat' is extensional, but 'necessarily $x$ is Socrates', 'most students believe that $x$ denounced $y$', and '$x$ intended to kill $y$' are nonextensional.

An English *name* $\zeta$ is said to be *extensional* if, whenever $\zeta'$ is a name obtained from $\zeta$ by replacing one or more occurrences of an English name within $\zeta$ by another English name designating the same object, then $\zeta$ and $\zeta'$ designate the same object. By an *extensional term of English* we understand either an extensional English name or a term containing variables which, upon replacement of all its variables by English names, becomes an extensional English name. For example,

the father of $a$

and

$$7 + a$$

are extensional terms of English, but

the person whom $a$ believed to be $b$

is not.

We shall now admit abbreviations only of those formulas and terms of English whose variables are the first $k$ variables, for some number $k$, which contain no apparent variables, and which are extensional. (The reason for the additional requirement of extensionality is discussed in section 5 below.)

Thus an *abbreviation* will now be either (1) an ordered pair whose first member is a $k$-place predicate, for some number $k \geqslant 0$, and whose second member is an extensional formula of English containing exactly the first $k$ variables and having no apparent variables, or (2) an ordered pair whose first member is a $k$-place operation letter, for some number $k \geqslant 0$, and whose second member is an extensional term of English containing exactly the first $k$ variables and having no apparent variables.

*Literal translation into English on the basis of a given scheme of abbreviation* is handled as in chapter IV, section 4, with one modification. In step (v) (p. 140), we replace ' = ' by 'is identical with', in addition to eliminating sentential connectives and quantifier phrases in favor of the corresponding phrases of connection and quantity.

As before, we say that an English formula $\psi$ is a *free translation* (or simply a *translation*) of a symbolic formula $\phi$ (or that $\phi$ is a *symbolization* of $\psi$) on the basis of a given scheme of abbreviation if $\psi$ is a stylistic variant of the literal English translation of $\phi$ based on that scheme.

For example, consider the scheme of abbreviation

(3) $A^0$ : Mark Twain
$B^0$ : Samuel Clemens

and the symbolic sentence

(4) $$A = B \quad .$$

The literal English translation of (4) on the basis of the scheme (3) is

(5) Mark Twain is identical with Samuel Clemens,

and a free translation of (4) would be

(6) Mark Twain is Samuel Clemens.

Thus we regard (6) as a stylistic variant of (5). 'Is' does not, however, always have the sense of *identity*. In some contexts, for example,

(7) Socrates is wise,

'is' has the sense of *predication*. (7) is true just in case Socrates has the characteristic, possibly common to many individuals, of wisdom; it is not asserted that the words 'Socrates' and 'wise' designate the same object (or even that 'wise' designates any object), as the sense of identity would require.

The logic of predication can best be regarded as part of the theory developed in chapters III and IV. The present chapter concerns the logic of identity.

(5) has other stylistic variants than (6); for example,

Mark Twain is the same as Samuel Clemens,
Mark Twain and Samuel Clemens are the same,
Mark Twain and Samuel Clemens are identical,
Mark Twain equals Samuel Clemens.

The negation of (4) has free translations other than those suggested by the free translations of (4). For example,

Mark Twain differs from Samuel Clemens,
Mark Twain is distinct from Samuel Clemens,

and

Mark Twain is other than Samuel Clemens

are all free translations of

$$\sim A = B$$

on the basis of the scheme (3).

To find a symbolization of a given formula of English on the basis of a given scheme of abbreviation, the reader will find it useful to proceed roughly as follows:

(*1*) *Introduce 'is identical with', phrases of quantity, and phrases of connection, the latter accompanied by parentheses and occurring canonically, in place of their stylistic variants.*

(*2*) *Reverse the steps leading from a symbolic formula to a literal English translation.*

For example, consider the scheme of abbreviation

$F^1$ : *a* is a citizen
$A^1$ : the wife of *a*

and the sentence

Anyone who is the wife of a citizen is also a citizen.

In step (1) we obtain

For each $x$ (if there is an object $y$ such that ($y$ is a citizen and $x$ is identical with the wife of $y$), then $x$ is a citizen) ,

and carrying through the successive parts of step (2), we obtain the symbolization

$$\bigwedge x(\bigvee y[Fy \wedge x = A(y)] \rightarrow Fx) \quad .$$

As another example, consider the scheme of abbreviation

$F^2$ : *a* is a member of *b*
$G^2$ : *a* has defeated *b*
$A^0$ : the team
$B^1$ : the captain of *a* ,

and the sentence

The captain of the team has defeated every other member of the team.

The symbolic translation in this case is

$$\bigwedge x[F(xA) \wedge \sim x = B(A) \rightarrow G(B(A)\, x)] \quad .$$

## EXERCISES

1. For each formula of group A, find the formula of group B that is a symbolization of it.

GROUP A

(1) Something differs from everything except itself.
(2) Everything is identical with something.
(3) Everything is distinct from something.

(4) Everything equal to a thing differs from everything not equal to that thing.
(5) Nothing differs from everything.
(6) Two things equal to the same thing are equal to each other.
(7) Nothing differs from itself.

GROUP B

(a) $\sim \bigvee x \bigwedge y \sim x = y$
(b) $\bigwedge x \bigvee y\ x = y$
(c) $\bigvee x \bigwedge y(\sim y = x \rightarrow \sim x = y)$
(d) $\bigwedge x \bigwedge y(\bigvee z[x = z \wedge y = z] \rightarrow x = y)$
(e) $\bigwedge x \bigvee y \sim y = x$
(f) $\sim \bigvee x \sim x = x$
(g) $\bigwedge x \bigwedge z[x = z \rightarrow \bigwedge y(\sim y = z \rightarrow \sim x = y)]$

Symbolize each of the following formulas on the basis of the scheme of abbreviation that accompanies it.

2. None but Alfred and his teacher are able to solve the problem. ($F^2$ : *a* is able to solve *b*; $A^0$ : Alfred; $B^0$ : the problem; $C^1$ : the teacher of *a*)

3. Alfred attended the conference and arrived at it before everyone else who attended it except Mary. ($F^2$ : *a* attended *b*; $G^3$ : *a* arrived at *b* before *c*; $A^0$ : Alfred; $B^0$ : the conference; $C^0$ : Mary)

4. Alfred is the only member of the class who can read Greek. ($F^2$ : *a* is a member of *b*; $G^2$ : *a* can read *b*; $A^0$ : Alfred; $B^0$ : the class; $C^0$ : Greek)

5. Greensleeves can jump farther than any other frog in Calaveras County. ($F^2$ : *a* can jump farther than *b*; $G^1$ : *a* is a frog; $H^2$ : *a* is in *b*; $A^0$ : Greensleeves; $B^0$ : Calaveras County)

6. Anyone whose mother is the wife of a citizen and whose father has no other children is the son of a citizen. ($F^1$ : *a* is a citizen; $G^2$ : *a* is a child of *b*; $A^1$ : the mother of *a*; $B^1$ : the wife of *a*; $C^1$ : the father of *a*; $D^1$ : the son of *a*)

7. If Alfred's sister is the wife of Alfred's brother-in-law, then Alfred and his sister have the same father-in-law. ($A^0$ : Alfred; $B^1$ : the wife of *a*; $C^1$ : the sister of *a*; $D^1$ : the brother-in-law of *a*; $E^1$ : the father-in-law of *a*)

8. None but the lonely heart can know my sadness. ($F^1$ : *a* is a lonely heart; $G^2$ : *a* can know *b*; $A^0$ : my sadness)

9. Bach is more ingenious than any other composer. ($F^2$ : *a* is more ingenious than *b*; $G^1$ : *a* is a composer; $B^0$ : Bach)

10. Pergolesi was more promising than any other composer of his time. ($F^2$ : *a* was more promising than *b*; $G^1$ : *a* is a composer; $H^2$ : *a* was contemporary with *b*; $A^0$ : Pergolesi)

**3. Inference rules; theorems.** Notions of bondage and freedom, as well as the notion of proper substitution of a term for a variable, are to be understood here as in chapter IV (see pp. 136–37, 148). For the logi of identity, two inference rules must be added to those already at ou disposal. The new rules lead, respectively, from a symbolic formula

$$\psi$$

to the formula

$$\Lambda\alpha(\alpha = \zeta \rightarrow \phi) \quad ,$$

and from a symbolic formula

$$\Lambda\alpha(\alpha = \zeta \rightarrow \phi)$$

to the formula

$$\psi \quad ,$$

where $\psi$ comes from $\phi$ by proper substitution of the symbolic term $\zeta$ fo the variable $\alpha$, and $\alpha$ does not occur in $\zeta$. Diagrammatically the tw rules present the following appearance:

Identity–1 (Id–1): $$\frac{\psi}{\Lambda\alpha(\alpha = \zeta \rightarrow \phi)}$$

Identity–2 (Id–2): $$\frac{\Lambda\alpha(\alpha = \zeta \rightarrow \phi)}{\psi} \quad ,$$

where $\alpha$, $\zeta$, $\phi$, $\psi$ satisfy the conditions above.

The rule Id-1 permits the inference from

$$\mathrm{FA}$$

to

$$\Lambda x(x = \mathrm{A} \rightarrow \mathrm{F}x) \quad ;$$

and Id-2 permits the converse inference. For a somewhat more complex illustration, consider the formulas

(1) $\mathrm{GB}(\mathrm{A}x)$

(2) $\Lambda y(y = \mathrm{B}(\mathrm{A}x) \rightarrow \mathrm{G}y)$ ;

from (1) we can infer (2) by Id-1, and from (2) we can infer (1) by Id-2. However, from (1) we cannot infer

$$\Lambda x(x = \mathrm{B}(\mathrm{A}x) \rightarrow \mathrm{G}x) \quad ,$$

for '$x$' occurs in '$\mathrm{B}(\mathrm{A}x)$'.

We may call that branch of logic which essentially involves the identity sign, as well as quantifiers and sentential connectives, the *identity calculus.*

The directions for constructing, within this calculus, an unabbreviated derivation from given symbolic premises are those of chapter IV (see section 13, clauses (1) – (6)), with the understanding that Id-1 and Id-2 are now to be included among the primitive inference rules, and the word 'formula' is to assume the broader sense of the present chapter.

The characterizations of a *complete* derivation, *derivability*, a *proof*, a *theorem*, an *argument*, a *symbolic argument*, a *valid symbolic argument*, an *English argument*, a *symbolization* of an English argument, and a *valid English argument* remain as before (see pp. 95, 103, and 117).

In the identity calculus it is convenient to have at our disposal the methods for abbreviating derivations that appear in chapter IV (see section 13, clauses (7) – (10)). We may adopt these clauses without modification, along with the characterizations of the pertinent notions (*alphabetic variant*, *instance*; see sections 7 and 8 of chapter IV). Here too we must of course understand the word 'formula' in the sense of the present chapter. In the context of the identity calculus clauses (7) – (10) continue to satisfy the requirements given on page 58.

We assign to the theorems of the identity calculus numbers beginning with 301. Theorems 301, 302, and 303 are the respective laws of *reflexivity*, *symmetry*, and *transitivity* of identity.

T301

1. ~~*Show*~~ $x = x$
2. ~~*Show*~~ $\wedge y(y = x \rightarrow y = x)$
3. $y = x \rightarrow y = x$ — T1
4. $x = x$ — 2, Id-2

The formula '$x = x$' comes from the formula '$y = x$' by proper substitution of '$x$' for '$y$'; further, the variable '$y$' does not occur in the term '$x$'. Thus the transition from line 2 to line 4 constitutes a correct application of Id-2.

T302

1. ~~*Show*~~ $x = y \rightarrow y = x$
2. $y = y$ — T301
3. $\wedge x(x = y \rightarrow y = x)$ — 2, Id-1
4. $x = y \rightarrow y = x$ — 3, UI

In checking the transition from line 2 to line 3, we must observe that '$y = y$' comes from '$y = x$' by proper substitution of '$y$' for '$x$', and that '$x$' does not occur in '$y$'.

T303

1. ~~*Show*~~ $x = y \wedge y = z \rightarrow x = z$
2. $x = y \wedge y = z$
3. $y = z$ — 2, S
4. $\wedge x(x = y \rightarrow x = z)$ — 3, Id-1
5. $x = z$ — 4, UI, 2, S, MP

In connection with line 4, we must observe that '$y = z$' comes from '$x = z$' by proper substitution of '$y$' for '$x$', and that '$x$' does not occur in '$y$'.

T304 can be derived easily from T302, and is often useful in connection with IE. T305 – T307 are formulations, alternative to T303, of the law of transitivity of identity.

T304 $\quad x = y \leftrightarrow y = x$

T305 $\quad x = y \wedge z = y \rightarrow x = z$

T306 $\quad y = x \wedge y = z \rightarrow x = z$

T307 $\quad y = x \wedge z = y \rightarrow x = z$

The pattern of inference corresponding to T302 and those corresponding to T303, T305 – T307 will appear frequently in the sequel; thus, as with SC, CD, and QN, we shall employ special annotations to indicate their use. The derived rule corresponding to T302—the rule of symmetry (annotation: 'Sm')—and those corresponding to T303, T305 – T307—the rules of transitivity (annotation for all forms: 'T')—appear as follows:

Symmetry (Sm): $\dfrac{\zeta = \eta}{\eta = \zeta}$

Transitivity (T): $\dfrac{\zeta = \eta \quad \eta = \theta}{\zeta = \theta} \qquad \dfrac{\zeta = \eta \quad \theta = \eta}{\zeta = \theta} \qquad \dfrac{\eta = \zeta \quad \eta = \theta}{\zeta = \theta} \qquad \dfrac{\eta = \zeta \quad \theta = \eta}{\zeta = \theta}$

Here $\zeta$, $\eta$, $\theta$ may be any symbolic terms.

Inferences justified by Sm and T are familiar to every student of mathematics; for example, T corresponds to the principle that things equal to the same thing are equal to each other.

T308

| | | |
|---|---|---|
| 1. | ~~Show~~ $x = y \rightarrow (Fx \leftrightarrow Fy)$ | |
| 2. | $x = y$ | |
| 3. | ~~Show~~ $Fx \rightarrow Fy$ | |
| 4. | $Fx$ | |
| 5. | $\bigwedge y(y = x \rightarrow Fy)$ | 4, Id-1 |
| 6. | $y = x \rightarrow Fy$ | 5, UI |
| 7. | $y = x$ | 2, Sm |
| 8. | $Fy$ | 6, 7, MP |
| 9. | ~~Show~~ $Fy \rightarrow Fx$ | |
| 10. | $Fy$ | |
| 11. | $\bigwedge x(x = y \rightarrow Fx)$ | 10, Id-1 |
| 12. | $Fx$ | 11, UI, 2, MP |
| 13. | $Fx \leftrightarrow Fy$ | 3, 9, CB |

The pattern of inference corresponding to T308 is essentially Leibniz' principle of the *indiscernibility of identicals* (the converse of his better-known principle of the identity of indiscernibles), which asserts that if two things are identical, then anything true of one is also true of the other. We introduce the abbreviated annotation 'LL' for this pattern of inference, which has the following form:

$$\frac{\zeta = \zeta'}{\phi \leftrightarrow \phi'} \quad .$$

Here $\zeta$ and $\zeta'$ are to be symbolic terms, $\phi$ is to be a symbolic formula, and $\phi'$ is to be like $\phi$ except for having one or more free occurrence of $\zeta'$ where $\phi$ has free occurrences of $\zeta$. (To show that all cases of LL correspond to alphabetic variants of instances of T308 would require a brief argument, which we do not intend to present here.)

In the following arguments the conclusion follows from the premises by LL:

$$x = y \quad \therefore\ \mathrm{F}(xx) \leftrightarrow \mathrm{F}(xy) \quad .$$
$$x = y \quad \therefore\ \bigwedge z \mathrm{G}(xz) \leftrightarrow \bigwedge z\, \mathrm{G}(yz) \quad .$$

T309 – T311 will prove useful in connection with the rule IE. The proof of T309 is a trivial application of Id-1 and Id-2; those of T310 and T311 are elementary, now that LL is available.

T309 $\quad \mathrm{F}x \leftrightarrow \bigwedge y(y = x \rightarrow \mathrm{F}y)$

T310

| | | |
|---|---|---|
| 1. | ~~Show~~ $\mathrm{F}x \leftrightarrow \bigvee y(y = x \wedge \mathrm{F}y)$ | |
| 2. | ~~Show~~ $\mathrm{F}x \rightarrow \bigvee y(y = x \wedge \mathrm{F}y)$ | |
| 3. | $\mathrm{F}x$ | |
| 4. | $x = x$ | T301 |
| 5. | $\bigvee y(y = x \wedge \mathrm{F}y)$ | 3, 4, Adj, EG |
| 6. | ~~Show~~ $\bigvee y(y = x \wedge \mathrm{F}y) \rightarrow \mathrm{F}x$ | |
| 7. | $\bigvee y(y = x \wedge \mathrm{F}y)$ | |
| 8. | $z = x \wedge \mathrm{F}z$ | 7, EI |
| 9. | $\mathrm{F}z$ | 8, S |
| 10. | $z = x$ | 8, S |
| 11. | $\mathrm{F}x$ | 9, 10, LL, BC, MP |
| 12. | $\mathrm{F}x \leftrightarrow \bigvee y(y = x \wedge \mathrm{F}y)$ | 2, 6, CB |

The passage from lines 9 and 10 to line 11 in the last proof would look, if expanded, as follows:

9. $\mathrm{F}z$
10. $z = x$

| | | |
|---|---|---|
| 10a. | $Fz \leftrightarrow Fx$ | 10, LL |
| 10b. | $Fz \rightarrow Fx$ | 10a, BC |
| 11. | $Fx$ | 9, 10b, MP |

This sort of inference will occur very frequently. We shall therefore often replace the annotation 'LL, BC, MP' by 'LL'.

T311 $\quad Fx \wedge x = y \leftrightarrow Fy \wedge x = y$

T312 and T313 are instances, and T314 and T315 are two-variable analogues, of T309 and T310 respectively.

T312 $\quad x = y \leftrightarrow \bigwedge z(z = x \rightarrow z = y)$

T313 $\quad x = y \leftrightarrow \bigvee z(z = x \wedge z = y)$

T314 $\quad F(xy) \leftrightarrow \bigwedge z \bigwedge w(z = x \wedge w = y \rightarrow F(zw))$

T315 $\quad F(xy) \leftrightarrow \bigvee z \bigvee w(z = x \wedge w = y \wedge F(zw))$

T316 $\quad$ 1. ~~*Show*~~ $x = y \rightarrow A(x) = A(y)$

| | | |
|---|---|---|
| 2. | $x = y$ | |
| 3. | $A(x) = A(x)$ | T301 |
| 4. | $A(x) = A(y)$ | 2, 3, LL |

The pattern of inference corresponding to T316 is Euclid's postulate, 'When equals are substituted for equals, the results are equal'. We add this pattern to our stock of derived rules; it presents the following appearance:

Euclid's law (EL): $\dfrac{\zeta = \zeta'}{\eta = \eta'}$

Here $\zeta$, $\zeta'$, $\eta$, $\eta'$ are to be symbolic terms, and $\eta'$ is to be like $\eta$ except for having one or more occurrences of $\zeta'$ where $\eta$ has occurrences of $\zeta$. (As with LL, a small argument would be required to show that all cases of EL can be obtained as instances of T316.) For example, from

$$x = y$$

we may pass to

$$A(xx) = A(xy)$$

by EL.

The addition of the identity sign to our symbolism enables us to express numerical conditions symbolically. For example, if we let '$F^1$' abbreviate '*a* is a god', then

(3) $\quad \bigvee x Fx$

asserts that there is at least one god, and

(4) $$\bigwedge x \bigwedge y(Fx \wedge Fy \to x = y)$$

that there is at most one god; hence the conjunction of (3) and (4) expresses the doctrine of monotheism. The position that there are at least two gods is expressed by

(5) $$\bigvee x \bigvee y(Fx \wedge Fy \wedge \sim x = y) \quad ,$$

and that there are no more than two by

(6) $$\bigwedge x \bigwedge y \bigwedge z(Fx \wedge Fy \wedge Fz \to x = y \vee x = z \vee y = z) \quad .$$

The conjunction of (5) and (6) asserts that there are exactly two gods. In this way we can assert the existence of exactly three gods, exactly four gods, exactly as many gods as we may wish to countenance. And what has been illustrated by reference to gods holds equally well for apostles, apples, Africans—any finite collection the number of whose members we wish to specify.

The fact that exactly one thing satisfies a given condition can be expressed in a variety of ways, some of them, like the right-hand constituents of T317 and T318, shorter than the conjunction of (3) and (4). T319 and T320 provide similar alternatives to the conjunction of (5) and (6).

T317 $\bigvee xFx \wedge \bigwedge x \bigwedge y(Fx \wedge Fy \to x = y) \leftrightarrow \bigvee x(Fx \wedge \bigwedge y[Fy \to x = y])$

T318 $\bigvee xFx \wedge \bigwedge x \bigwedge y(Fx \wedge Fy \to x = y) \leftrightarrow \bigvee y \bigwedge x(Fx \leftrightarrow x = y)$

T319 $\bigvee x \bigvee y(Fx \wedge Fy \wedge \sim x = y) \wedge \bigwedge x \bigwedge y \bigwedge z(Fx \wedge Fy \wedge Fz \to x = y \vee x = z \vee y = z) \leftrightarrow \bigvee x \bigvee y(Fx \wedge Fy \wedge \sim x = y \wedge \bigwedge z[Fz \to x = z \vee y = z])$

T320 $\bigvee x \bigvee y(Fx \wedge Fy \wedge \sim x = y) \wedge \bigwedge x \bigwedge y \bigwedge z(Fx \wedge Fy \wedge Fz \to x = y \vee x = z \vee y = z) \leftrightarrow \bigvee x \bigvee y(\sim x = y \wedge \bigwedge z[Fz \leftrightarrow z = x \vee z = y])$

T321 expresses the fact that each thing is identical with at least one thing (itself); T322, that each thing is identical with exactly one thing.

T321 $\bigvee x\, x = y$

T322 $\bigvee y \bigwedge x(x = z \leftrightarrow x = y)$

T323 – T329 will be particularly useful in the next chapter.

T323 $\bigvee y \bigwedge x(\bigwedge z[x = z \leftrightarrow z = u] \leftrightarrow x = y)$

T324 $\bigvee y(\bigwedge x[Fx \leftrightarrow x = y] \wedge Gy) \leftrightarrow \bigvee y \bigwedge x(Fx \leftrightarrow x = y) \wedge \bigwedge x(Fx \to Gx)$

T325 $\bigvee y \bigwedge x(Fx \leftrightarrow x = y) \to (\bigwedge x[Fx \to Gx] \leftrightarrow \bigvee x[Fx \wedge Gx])$

T326 $\sim \bigvee y \bigwedge x(\sim x = x \leftrightarrow x = y)$

T327 $\bigwedge x\, A(x) = B(x) \rightarrow [\bigvee x\, FA(x) \leftrightarrow \bigvee x\, FB(x)]$

T328 $\bigwedge a \bigwedge b \bigwedge c[(\bigvee z \bigwedge x[Fx \leftrightarrow x = z] \rightarrow Fa) \wedge (\sim \bigvee z \bigwedge x[Fx \leftrightarrow x = z] \rightarrow a = c) \wedge (\bigvee z \bigwedge y[Fy \leftrightarrow y = z] \rightarrow Fb) \wedge (\sim \bigvee z \bigwedge y[Fy \leftrightarrow y = z] \rightarrow b = c) \rightarrow a = b]$

T329 $\bigwedge a \bigwedge b \bigwedge c[(\bigvee y \bigwedge x[Fx \leftrightarrow x = y] \rightarrow Fa) \wedge (\sim \bigvee y \bigwedge x[Fx \leftrightarrow x = y] \rightarrow a = c) \wedge (\bigvee y \bigwedge x[Gx \leftrightarrow x = y] \rightarrow Gb) \wedge (\sim \bigvee y \bigwedge x[Gx \leftrightarrow x = y] \rightarrow b = c) \wedge \bigwedge x(Fx \leftrightarrow Gx) \rightarrow a = b]$

### EXERCISES

11. Prove T311, T314, T315.
12. Prove T317 - T320.
13. Prove T323 - T327.
14. Prove T328 and T329.

Show that each of the following arguments is valid.

15. Some student is feminine. Every feminine student is liked by the teacher. The teacher likes at most one student. ∴ Every student whom the teacher likes is feminine.

16. The wife of a citizen is also a citizen. Anyone whose mother is a citizen is also a citizen. ∴ Anyone whose mother is the wife of a citizen is a citizen.

17. No men except Alfred, Rudolf, and Alonzo attended the meeting. Of those who attended the meeting only men smoked. Alfred did not smoke. ∴ Of those who attended the meeting at most two smoked.

18. 'The boxed sentence is not true' is true if and only if the boxed sentence is not true. The boxed sentence is 'the boxed sentence is not true'. Therefore the boxed sentence is neither true nor not true. (The reference is to the boxed sentence on page 3; this is the *paradox of the liar* as formulated in Tarski [2].)

19. $\bigwedge y \bigvee z \bigwedge x[F(xz) \leftrightarrow x = y]$ ∴ $\sim \bigvee w \bigwedge x(F(xw) \leftrightarrow \bigwedge u[F(xu) \rightarrow \bigvee y(F(yu) \wedge \sim \bigvee z[F(zu) \wedge F(zy)])])$

(Exercise 19 corresponds to the *paradox of grounded classes* as described in Montague [1].)

**4. *Invalidity.** We employ again the notions of an *English translation* of a symbolic argument, a *relativization* of a symbolic *sentence*, a *closure axiom*, and a *relativization* of a symbolic *argument;* see section 10 of chapter IV.

Here as in the quantifier calculus it is impossible to devise an automatic test for validity. We can, however, establish invalidity by use of the various methods introduced in chapter IV. (The method of truth-functional expansions, introduced in chapter III, will no longer yield correct results, even in connection with symbolic arguments containing no operation letters; the reader will perhaps be interested in showing this, and in adapting the method to the identity calculus.) For instance, if a closure of a symbolic argument has a false English translation, then the argument will not be valid. We can, however, no longer claim the converse.

For example, consider the following symbolic argument:

(1) $\bigwedge x \bigwedge y \bigwedge z(x = y \vee y = z \vee z = x) \quad \therefore \bigvee x \bigvee y \sim x = y$ .

This argument is invalid, yet none of its closures has a false translation. Clearly, all translations of its closures are stylistic variants of the argument

> There are at most two things. Therefore there are at least two things;

and none of these is false, for there are more than two things.

We can assert, however, that a symbolic argument is invalid in the identity calculus just in case a relativization of one of its closures has a false English translation. Consider, for example, the relativization of (1) to '$G^1$':

$$\bigwedge x(Gx \rightarrow \bigwedge y[Gy \rightarrow \bigwedge z(Gz \rightarrow x = y \vee y = z \vee z = x)]) \ . \ \bigvee x Gx \quad \therefore \bigvee x(Gx \wedge \bigvee y[Gy \wedge \sim x = y]) \ .$$

A false translation of this argument can be obtained on the basis of the scheme

$G^1$ : *a* is Socrates .

## EXERCISES

For each of the following arguments, either show that it is valid or show that it is invalid.

20. $\bigvee x \bigvee y \sim x = y \quad \therefore \bigvee x Fx \wedge \bigvee x \sim Fx$
21. $\bigwedge x(Fx \rightarrow Gx \vee Hx) \ . \ \bigvee y \bigwedge x(Gx \leftrightarrow x = y) \ . \ \bigvee y \bigwedge x(Hx \leftrightarrow x = y) \quad \therefore \bigvee x \bigvee y(Fx \wedge Fy \wedge \sim x = y)$
22. $\bigwedge x(Fx \rightarrow Gx \vee Hx) \ . \ \bigvee y \bigwedge x(Gx \leftrightarrow x = y) \ . \ \bigvee y \bigwedge x(Hx \leftrightarrow x = y) \quad \therefore \bigwedge x \bigwedge y \bigwedge z(Fx \wedge Fy \wedge Fz \wedge \sim x = y \wedge \sim x = z \rightarrow y = z)$
23. $\therefore \bigvee y \bigwedge x(x = y \wedge Fy \leftrightarrow \bigvee z\, x = z \wedge Fx)$
24. $\therefore \bigwedge z \bigwedge y \bigwedge x(x = y \wedge Fy \leftrightarrow x = z \wedge Fx)$
25. $\therefore A(x) = B(x) \rightarrow [\bigvee x\, FA(x) \leftrightarrow \bigvee x\, FB(x)]$

Show that each of the following inference rules, if incorporated into our system of logic, would permit the validation of a symbolic argument that, by methods now available, can be shown invalid. (The restrictions on variables involved in Id-1 and Id-2 are thus seen to be necessary.) In all cases $\alpha$ is to be variable, $\zeta$ a symbolic term, and $\phi$, $\psi$ symbolic formulas. (Exercise 26 is solved for illustration.)

26. $$\frac{\bigwedge\alpha\,(\alpha = \zeta \rightarrow \phi)}{\psi}\ ,$$

where $\psi$ comes from $\phi$ by proper substitution of $\zeta$ for $\alpha$ (but $\alpha$ is permitted to occur in $\zeta$).

Argument:

$\bigwedge x[x = A(xB) \rightarrow (Fx \rightarrow x = B)]$ . $FA(BB)$ $\therefore$ $A(BB) = B$ .

Derivation:

1. ~~Show~~ $A(BB) = B$
2. ~~Show~~ $\bigwedge x(FA(xB) \rightarrow A(xB) = B)$
3. $\bigwedge x[x = A(xB) \rightarrow (Fx \rightarrow x{=}B)]$ — 1st premise
4. $FA(xB) \rightarrow A(xB) = B$ — 3, the variant given above of Id-2
5. $A(BB) = B$ — 2, UI, 2nd premise, MP

Translation:

> For each $x$, if $x$ equals $x + 1$, then if $x$ is a positive integer, then $x$ equals 1. $1 + 1$ is a positive integer. Therefore $1 + 1$ equals 1.

It becomes clear that the translation is a false argument (and hence that the symbolic argument is invalid) once it is recognized that there is no positive integer equal to itself plus 1.

27. $$\frac{\psi}{\bigwedge\alpha\,(\alpha = \zeta \rightarrow \phi)}\ ,$$

where $\psi$ comes from $\phi$ by proper substitution of $\zeta$ for $\alpha$ (but $\alpha$ is permitted to occur in $\zeta$).

28. $$\frac{\bigwedge\alpha\,(\alpha = \zeta \rightarrow \phi)}{\psi}\ ,$$

where $\psi$ is like $\phi$ except for having free occurrences of $\zeta$ at some (but not all) places where $\phi$ has free occurrences of $\alpha$.

29. $$\frac{\psi}{\bigwedge\alpha\,(\alpha = \zeta \rightarrow \phi)}\ ,$$

where $\psi$ is like $\phi$ except for having free occurrences of $\zeta$ at some (but not all) places where $\phi$ has free occurrences of $\alpha$.

30. $$\frac{\bigwedge\alpha\,(\alpha = \zeta \rightarrow \phi)}{\psi}\ ,$$

where $\psi$ is like $\phi$ except for having occurrences of $\zeta$ (free or bound) wherever $\phi$ has free occurrences of $\alpha$.

31. $$\frac{\psi}{\Lambda\alpha\,(\alpha = \zeta \rightarrow \phi)}\ ,$$

where $\psi$ is like $\phi$ except for having occurrences of $\zeta$ (free or bound) wherever $\phi$ has free occurrences of $\alpha$.

**5.* Paradoxical inferences.** We now consider the consequences of admitting nonextensional formulas into schemes of abbreviation. Consider the valid symbolic argument

(1) $A = B \quad . \quad FA \quad \therefore FB$

and the scheme

(2) $F^1$ : Kepler was unaware that *a* is greater than 6
$A^0$ : the number of planets
$B^0$ : 9 .

(This scheme violates the requirement of section 2 in that 'Kepler was unaware that *a* is greater than 6' is nonextensional.) On the basis of (2), the argument

(3) The number of planets is 9. Kepler was unaware that the number of planets is greater than 6. Therefore Kepler was unaware that 9 is greater than 6

would be a translation of (1), and hence, by the definition of validity for English arguments, would be valid. But the premises of (3) are true, and the conclusion of (3) false. Thus if (2) were admitted as a scheme of abbreviation, a claim on page 227 would be falsified: we should be able to find a valid symbolic argument with a false English translation.

It is not to be expected that all unsatisfactory examples are precluded by the requirement of extensionality. Indeed, there are English arguments that completely fulfill the requirements for validity given in sections 2 and 3 but that on intuitive grounds would tend to be regarded as invalid. Consider, for example, the argument

(4) Alcibiades is the most notorious Athenian traitor. Everyone believes that Alcibiades is honest. Therefore everyone believes that the most notorious Athenian traitor is honest,

which, under the scheme

(5) $F^1$ : everyone believes that *a* is honest
$A^0$ : Alcibiades
$B^0$ : the most notorious Athenian traitor ,

is a translation of the valid symbolic argument (1).

Now (5) is indeed a scheme of abbreviation; 'everyone believes that *a* is honest' is extensional because every sentence obtained from it by replacing '*a*' by an English name is false. We have not found a fallacy, for the second premise of (4) is false; yet (4) would seem invalid on intuitive grounds because it is *possible*, given a sufficiently gullible and misinformed world, for its premises to be true and its conclusion false. As far as we can determine, no such examples can be found which are actually false English arguments; thus the claim on page 227 can still be maintained.

The difficulties encountered here would arise in connection with any known development of the quantifier and identity calculi, and are by no means peculiar to our formulation. (See, for example, Frege [2], Quine [4], and Carnap [2].) Indeed, it is not to be expected that a completely precise analysis of validity for a natural language, faithfully reflecting all intuitions that have accrued during the historical development of the language, could be either simple, elegant, or of major philosophical interest. Two alternative courses are appropriate: one might apply formal procedures quite generally, with measured indifference to ordinary use; or one might restrict consideration to a definite part of language that is naturally accessible to these procedures. Our logical system can be enlisted to the service of either alternative, and the second alternative is explicitly pursued in chapters VIII and IX.

It is interesting that within the framework of chapters III and IV no clear-cut counterintuitive cases seem possible as long as apparent variables are excluded from schemes of abbreviation. (The reader may wish to assay the accuracy of this observation, perhaps by attempting to develop paradoxes involving nonextensional contexts within the logic of quantification without identity.)

**6. * Historical remarks.** The identity calculus is often called the first-order predicate calculus with identity. It has been discussed extensively in Whitehead and Russell [1], Hilbert and Bernays [1], and Scholz [1].

The claim that if no closure of a symbolic argument has a relativization with a false translation, then the argument is valid in the identity calculus, is essentially Gödel's completeness theorem for this calculus (see Gödel [1]).

The principle of the indiscernibility of identicals is first to be found in Leibniz [1].

## 7. Appendix: list of theorems of chapter VI.

T301 $x = x$

T302 $x = y \rightarrow y = x$

T303 $x = y \wedge y = z \rightarrow x = z$

T304 $x = y \leftrightarrow y = x$

T305 $x = y \wedge z = y \rightarrow x = z$

T306 $y = x \wedge y = z \rightarrow x = z$

T307 $y = x \wedge z = y \rightarrow x = z$

T308 $x = y \rightarrow (\mathrm{F}x \leftrightarrow \mathrm{F}y)$

T309 $\mathrm{F}x \leftrightarrow \bigwedge y(y = x \rightarrow \mathrm{F}y)$

T310 $\mathrm{F}x \leftrightarrow \bigvee y(y = x \wedge \mathrm{F}y)$

T311 $\mathrm{F}x \wedge x = y \leftrightarrow \mathrm{F}y \wedge x = y$

T312 $x = y \leftrightarrow \bigwedge z(z = x \rightarrow z = y)$

T313 $x = y \leftrightarrow \bigvee z(z = x \wedge z = y)$

T314 $\mathrm{F}(xy) \leftrightarrow \bigwedge z \bigwedge w[z = x \wedge w = y \rightarrow \mathrm{F}(zw)]$

T315 $\mathrm{F}(xy) \leftrightarrow \bigvee z \bigvee w[z = x \wedge w = y \wedge \mathrm{F}(zw)]$

T316 $x = y \rightarrow \mathrm{A}(x) = \mathrm{A}(y)$

T317 $\bigvee x\mathrm{F}x \wedge \bigwedge x \bigwedge y(\mathrm{F}x \wedge \mathrm{F}y \rightarrow x = y) \leftrightarrow \bigvee x(\mathrm{F}x \wedge \bigwedge y[\mathrm{F}y \rightarrow x = y])$

T318 $\bigvee x\mathrm{F}x \wedge \bigwedge x \bigwedge y(\mathrm{F}x \wedge \mathrm{F}y \rightarrow x = y) \leftrightarrow \bigvee y \bigwedge x(\mathrm{F}x \leftrightarrow x = y)$

T319 $\bigvee x \bigvee y(\mathrm{F}x \wedge \mathrm{F}y \wedge \sim x = y) \wedge \bigwedge x \bigwedge y \bigwedge z(\mathrm{F}x \wedge \mathrm{F}y \wedge \mathrm{F}z \rightarrow$
$x = y \vee x = z \vee y = z) \leftrightarrow \bigvee x \bigvee y(\mathrm{F}x \wedge \mathrm{F}y \wedge \sim x = y \wedge$
$\bigwedge z[\mathrm{F}z \rightarrow x = z \vee y = z])$

T320 $\bigvee x \bigvee y(\mathrm{F}x \wedge \mathrm{F}y \wedge \sim x = y) \wedge \bigwedge x \bigwedge y \bigwedge z(\mathrm{F}x \wedge \mathrm{F}y \wedge \mathrm{F}z \rightarrow x = y \vee$
$x = z \vee y = z) \leftrightarrow \bigvee x \bigvee y(\sim x = y \wedge \bigwedge z[\mathrm{F}z \leftrightarrow z = x \vee z = y])$

T321 $\bigvee x\, x = y$

T322 $\bigvee y \bigwedge x(x = z \leftrightarrow x = y)$

T323 $\bigvee y \bigwedge x(\bigwedge z[x = z \leftrightarrow z = u] \leftrightarrow x = y)$

T324 $\bigvee y(\bigwedge x[\mathrm{F}x \leftrightarrow x = y] \wedge \mathrm{G}y) \leftrightarrow \bigvee y \bigwedge x(\mathrm{F}x \leftrightarrow x = y) \wedge$
$\bigwedge x(\mathrm{F}x \rightarrow \mathrm{G}x)$

T325 $\bigvee y \bigwedge x(\mathrm{F}x \leftrightarrow x = y) \rightarrow (\bigwedge x[\mathrm{F}x \rightarrow \mathrm{G}x] \leftrightarrow \bigvee x[\mathrm{F}x \wedge \mathrm{G}x])$

T326 $\sim \bigvee y \bigwedge x(\sim x = x \leftrightarrow x = y)$

T327 $\bigwedge x\, \mathrm{A}(x) = \mathrm{B}(x) \rightarrow [\bigvee x\, \mathrm{FA}(x) \leftrightarrow \bigvee x\, \mathrm{FB}(x)]$

T328 $\bigwedge a \bigwedge b \bigwedge c[(\bigvee z \bigwedge x[Fx \leftrightarrow x = z] \rightarrow Fa) \wedge (\sim \bigvee z \bigwedge x[Fx \leftrightarrow x = z] \rightarrow a = c) \wedge (\bigvee z \bigwedge y[Fy \leftrightarrow y = z] \rightarrow Fb) \wedge (\sim \bigvee z \bigwedge y[Fy \leftrightarrow y = z] \rightarrow b = c) \rightarrow a = b]$

T329 $\bigwedge a \bigwedge b \bigwedge c[(\bigvee y \bigwedge x[Fx \leftrightarrow x = y] \rightarrow Fa) \wedge (\sim \bigvee y \bigwedge x[Fx \leftrightarrow x=y] \rightarrow a = c) \wedge (\bigvee y \bigwedge x[Gx \leftrightarrow x = y] \rightarrow Gb) \wedge (\sim \bigvee y \bigwedge x[Gx \leftrightarrow x = y] \rightarrow b = c) \wedge \bigwedge x(Fx \leftrightarrow Gx) \rightarrow a = b]$

# Chapter VII

# 'THE'

**1. Descriptive phrases, terms, formulas.** There are arguments whose validity depends on the meaning of the singular definite article. For example, the argument

(1) Exactly one student failed. Alfred is the student who failed.
Anyone who failed is illogical. $\therefore$ Alfred is illogical

is intuitively valid, yet the procedures of the identity calculus are insufficient to establish its validity.

The word 'the', in many of its occurrences, may be replaced by 'the object $x$ such that'. For instance, the second premise of (1) may be paraphrased as

(2) Alfred is identical with the object $x$ such that $x$ is a student and $x$ failed ,

and

the composer of *Don Giovanni*

as

(3) the object $x$ such that $x$ composed *Don Giovanni* .

We shall abbreviate 'the object . . . such that' by '$\iota$'. (2) and (3) then become

$$\text{Alfred} = \iota x(x \text{ is a student and } x \text{ failed})$$

and

$$\iota x\ x \text{ composed } \textit{Don Giovanni}$$

respectively. '$\iota$' is known as the *descriptive operator* and may be written before any variable to form a *descriptive phrase*.

The reader will recall that an *English name* is an expression of English that designates (at least within a given context) a single object, an *English sentence* is an expression of English that is either true or false, a *term of English* is either an English name or an expression containing occurrences of variables that becomes an English name when some or all of these occurrences are replaced by English names, and a *formula of English* is either an English sentence or an expression containing occurrences of variables that becomes an English sentence when some or all of these occurrences are replaced by English names.

We shall regard expressions of the form

the object $\alpha$ such that ,

where $\alpha$ is a variable, as belonging to English; such expressions will be called *phrases of description*. Accordingly, (2) will be construed as an English sentence, and (3) as an English name.

An expression of the form

the object $\alpha$ such that $\phi$ ,

where $\alpha$ is a variable and $\phi$ a formula of English whose only variable is $\alpha$, will be called a *definite description*. Examples are (3) above,

(4) the object $x$ such that $x$ is presently king of France ,

and

(5) the object $x$ such that $x$ wrote *Iolanthe* .

English usage is not always clear as to the designation of definite descriptions. Some clarification is therefore in order. A definite description

(6) the object $\alpha$ such that $\phi$

is said to be *proper* if exactly one object satisfies $\phi$; in this case, following ordinary usage, we understand (6) as designating that object. In the case of *improper* definite descriptions, that is, expressions of the form (6) for which either no object or more than one object satisfy $\phi$, ordinary usage provides no guidance; it therefore falls to us to specify their meaning. It is convenient to select a common designation for all improper definite descriptions. What object we choose for this purpose is unimportant, but for the sake of definiteness let us choose the number 0. Thus expressions of the form (6), if improper, will henceforth be understood as designating 0.

It follows that (4) and (5), as well as (3), are English names; (3) – (5) respectively designate Mozart, 0, and 0. (4) and (5) are improper because nothing is presently king of France and there are two things (Gilbert and Sullivan) that wrote *Iolanthe*.

More generally, it is now clear that every expression of the form

the object $\alpha$ such that $\phi$ ,

where $\alpha$ is a variable and $\phi$ an arbitrary formula of English, is a term of English.

It also follows from our conventions regarding the designation of definite descriptions that the English sentences

> The object $x$ such that $x$ is presently king of France is married

and

> The object $x$ such that $x$ is presently king of France is less than 1

are respectively false and true.

For the treatment of definite descriptions we now consider a *language* obtained from that of chapter VI (see p. 213) by adding the descriptive operator.

The characterization of *terms* and *formulas* must be somewhat more involved than before, for the following reason. The grammatical function of descriptive phrases (as of their English counterparts) is to produce terms from formulas. Consequently, not only may terms occur within formulas, but also formulas within terms. The best way of introducing the present notions of term and formula is, it seems, to characterize them simultaneously. Like earlier characterizations the one below is to be regarded as exhaustive; that is, nothing is to be regarded as a term or formula unless it can be reached by successive applications of the following clauses.

(1) *Terms of English are terms. (In particular, variables are terms.)*

(2) *Formulas of English are formulas.*

(3) *The result of writing a* k*-place operation letter followed by* k *terms is a term. (In particular, a 0-place operation letter is itself a term.)*

(4) *The result of writing a* k*-place predicate letter followed by* k *terms is a formula. (In particular, a 0-place predicate letter is itself a formula.)*

(5) *If* $\zeta$, $\eta$ *are terms, then*

$$\zeta = \eta$$

*is a formula.*

(6) *If* $\phi$, $\psi$ *are formulas, then so are*

$$\sim\phi \quad ,$$
$$(\phi \rightarrow \psi) \quad ,$$
$$(\phi \vee \psi) \quad ,$$
$$(\phi \wedge \psi) \quad ,$$
$$(\phi \leftrightarrow \psi) \quad .$$

(7) *If* $\phi$ *is a formula and* $\alpha$ *a variable, then*

$$\bigwedge\alpha\phi \quad ,$$
$$\bigvee\alpha\phi$$

*are formulas.*

(8) *If* $\phi$ *is a formula and* $\alpha$ *a variable, then*

$$\text{℩}\alpha\phi$$

*is a term.*

Clause (8) leads not only to

$$\text{℩}x\ x \text{ composed } \textit{Don Giovanni} \quad ,$$

but also to seemingly meaningless combinations such as

$$\text{℩}x \text{ Socrates is bald} \quad .$$

As with vacuous quantification, it would be artificial to exclude these possibilities; their interpretation will be explained in due course.

The *symbolic terms and formulas* of our language are those terms and formulas which can be constructed exclusively from variables, predicate letters, operation letters, parentheses, sentential connectives, quantifiers, the identity sign, and the descriptive operator. More precisely, our *symbolic terms and formulas* can be exhaustively characterized by clauses (3) – (8) above, reading everywhere 'symbolic term' for 'term' and 'symbolic formula' for 'formula', together with the clause:

(1′) All variables are symbolic terms.

Examples of symbolic terms are

$$\text{℩}xF^1x \quad ,$$
$$\text{℩}xG^2xA^0 \quad ,$$
$$\text{℩}y\bigwedge xH^2xy \quad ,$$
$$\text{℩}z\ z = \text{℩}wI^2wB^0 \quad ,$$
$$\text{℩}x\bigwedge y(\sim y = x \rightarrow J^2xy) \quad .$$

Symbolic terms produced by clause (8), that is, terms of the form

$$\text{℩}\alpha\phi \quad ,$$

where $\alpha$ is a variable and $\phi$ a symbolic formula, are called *descriptive terms*.

## EXERCISES

1. State whether each of the following expressions is a term, a formula, or neither:

(a) $\text{℩}xF^0$
(b) $\text{℩}xF^1x$
(c) $\text{℩}xF^1y$
(d) $\text{℩}x(x = y)$
(e) $\text{℩}x(F^1x \wedge G^1x)$

(f) $\iota x A^1 x$

(g) $A^1 \iota x F^1 x$

(h) $F^1 \iota x G^1 x$

(i) $F^2 \iota x G^1 x\, y$

(j) $A^1 \iota x A^1 x = x$

(k) $\iota x\, x = \iota x\, x = x$

(l) $F^2 A^0 \iota x F^2 x A^0$

**2. Bondage and freedom.** The descriptive operator, like the quantifiers, binds variables. For example, in the formula

$$G^1 \iota x F^1 x$$

we shall say that both occurrences of '$x$' are bound. Further, we shall speak of *bondage in a term* as well as bondage in a formula. For example, we shall say that '$x$' is bound in the term

$$\iota x F^1 x \quad .$$

Revisions are thus necessitated in our characterizations of bondage and freedom. It is convenient to treat bondage and freedom for variables separately before considering the case of arbitrary symbolic terms.

An *occurrence* of a *variable* $\alpha$ is now said to be *bound in* a symbolic term or formula $\phi$ just in case it stands within an occurrence in $\phi$ of an expression

$$\wedge \alpha \psi \quad ,$$
$$\vee \alpha \psi \quad ,$$

or

$$\iota \alpha \psi \quad ,$$

where $\psi$ is a symbolic formula. An *occurrence* of a *variable* is *free in* a symbolic term or formula $\phi$ just in case it stands within $\phi$ but is not bound in $\phi$. A *variable* itself is *bound* or *free in* a symbolic term or formula $\phi$ according as there is a bound or free occurrence of it in $\phi$.

For example, in the formula

$$G^2 \iota x F^1 x\, x \quad ,$$

the first and second occurrences of '$x$' are bound, but the third occurrence is free, and '$x$' itself is both bound and free.

An *occurrence* of an arbitrary symbolic *term* $\zeta$ is said to be *bound in* a symbolic term or formula $\phi$ if it stands within an occurrence in $\phi$ of some expression

$$\wedge \alpha \psi \quad ,$$
$$\vee \alpha \psi \quad ,$$

or

$$\iota \alpha \psi \quad ,$$

where $\psi$ is a symbolic formula and $\alpha$ a variable which is free in $\zeta$. An *occurrence* of a symbolic *term* is *free in* $\phi$ if and only if it stands within $\phi$ but is not bound in $\phi$. A symbolic *term* is itself *bound* or *free in* $\phi$ according as there is a bound or free occurrence of it in $\phi$.

For example, in the formula

(1) $$(\bigvee y F^1 ℩x H^2 xy \vee G^1 ℩x H^2 xy) \quad ,$$

the first occurrence of the term

(2) $$℩x H^2 xy$$

is bound, and the second free. Thus (2) is both bound and free in (1).

Observe that bondage and freedom as applied to variables (and as characterized on p. 237) are special cases of bondage and freedom as applied to arbitrary symbolic terms.

A *symbolic sentence* can be characterized as a symbolic formula in which no variable is free, and a *symbolic name* as a symbolic term in which no variable is free.

The characterization of *proper substitution of a term for a variable* remains as before, even though our language now includes descriptive terms. Thus we again say that a symbolic formula $\psi$ *comes from* a symbolic formula $\phi$ by *proper substitution of* a symbolic *term* $\zeta$ *for* a *variable* $\alpha$ if $\psi$ is like $\phi$ except for having free occurrences of $\zeta$ wherever $\phi$ has free occurrences of $\alpha$.

## EXERCISES

Consider the formula

$$\bigvee x\, ℩x\, x = B^1 y = ℩z(\bigwedge y\, A^1 x = B^1 y \vee H^2 A^0 z) \quad .$$

2. In this formula identify each occurrence of a term that is not a variable as bound or free.
3. Which terms (including variables) are bound in the formula?
4. Which terms (including variables) are free in the formula?

**3. Informal notational conventions.** Once again we shall employ the conventions of chapter II (pp. 40, 51, and 65) for omitting parentheses and replacing some of them by brackets.

As in chapter IV, we shall usually omit superscripts from operation letters and predicate letters, inserting parentheses and brackets to avoid ambiguity. In the present context, 1-place predicate letters, as well as predicate letters of two or more places and operation letters of one or more places, will require parentheses or brackets. Thus we may write

(1) $$A^1 ℩x F^1 x$$

informally as

$$A(℩xF[x]) \quad ,$$

but not as

$$(2) \qquad A(℩xFx) \quad .$$

The official counterpart of (2) is not (1) but

$$A^2℩xF^0\,x \quad .$$

Thus, in unofficial notation, all predicate and operation letters of one place or more will bring with them a pair of parentheses or brackets.

(In formulating inference rules and criteria of bondage and freedom, we always have in mind official notation.)

### EXERCISES

5. For each of the following formulas (in unofficial notation) delete inserted parentheses and brackets and restore omitted superscripts in accordance with the conventions of this section.

(a) $℩xF(x)$
(b) $A(℩xF(x))$
(c) $F(℩xG(Ax))$
(d) $A[℩x\,A(x) = x]$
(e) $F[A\,℩xF(xA)]$

**4. Translation and symbolization.** An *abbreviation* is to be understood in the same way as in the last chapter (p. 216). The process of *literal translation into English on the basis of a given scheme of abbreviation* may now begin with either a symbolic term or a symbolic formula, and if successful will end with a term or formula of English. The process consists of the steps given in chapter IV (pp. 139 – 40), with one modification. Step (v) should now read as follows:

(*v*) *Eliminate sentential connectives, quantifier phrases,* '$=$', *and descriptive phrases in favor of the corresponding phrases of English, preserving all parentheses.*

An English term or formula $\psi$ is said to be a *free translation* (or simply a *translation*) of a symbolic term or formula $\phi$ (and $\phi$ is said to be a *symbolization* of $\psi$) on the basis of a given scheme of abbreviation if $\psi$ is a stylistic variant of the literal English translation of $\phi$ based on that scheme.

For example, on the basis of the scheme

$F^1$ : *a* is a doctor
$G^2$ : *a* is father of *b*
$H^2$ : *a* loves *b*
$I^2$ : *a* wrote *b*

$J^2$ : *a* is larger than *b*
$A^0$ : Socrates
$B^0$ : *Waverley* ,

the symbolic terms

(1) $\text{ɿ}xF(x)$ ,
(2) $\text{ɿ}xG(xA)$ ,
(3) $\text{ɿ}y\Lambda xH(xy)$ ,
(4) $\text{ɿ}z\, z = \text{ɿ}wI(wB)$ ,
(5) $\text{ɿ}x\Lambda y(\sim y = x \rightarrow J(xy))$

have the following respective literal English translations:

the object *x* such that *x* is a doctor,

the object *x* such that *x* is father of Socrates,

the object *y* such that, for each *x*, *x* loves *y*,

the object *z* such that *z* is identical with the object *w* such that *w* wrote *Waverley*,

the object *x* such that for each *y* (if it is not the case that *y* is identical with *x*, then *x* is larger than *y*) .

(Observe that in step (i) of the process of translation, that is, the restoration of official notation, (1) – (5) become the examples on page 236.) The following terms of English are free translations of (1) – (5):

(6) the doctor,
(7) the father of Socrates,
what everyone loves,
he who is identical with the one who wrote *Waverley*,
(8) the largest thing.

Thus we regard 'the', 'what', 'he who', and 'the one who', in some of their uses, as stylistic variants of 'the object . . . such that'.

It should not be supposed that every occurrence of 'the' can be supplanted by a phrase of description. Consider, for example, the sentences

(9) The whale is a mammal,
(10) The members of the club are Republicans,
(11) The trespassers were prosecuted.

(9) asserts that every whale is a mammal, not that the object *x* such that *x* is a whale is a mammal; and attempts to express (10) and (11) with the help of phrases of description would lead to ungrammatical results. In (6) – (8), 'the' generates a name, but this is not the case in (9) – (11); even 'the whale' cannot reasonably be construed as designating in (9) a single object.

To find a symbolization of a given term or formula of English on the basis of a given scheme of abbreviation the reader will find it useful to proceed roughly as follows:

(1) *Introduce 'is identical with', phrases of description, phrases of quantity, and phrases of connection, the latter accompanied by parentheses and occurring canonically, in place of their stylistic variants.*

(2) *Reverse the steps leading from a symbolic term or formula to a literal English translation.*

For example, consider the scheme of abbreviation

$F^2$ : *a* is wife of *b*
$G^2$ : *a* is more salacious than *b*
$H^1$ : *a* is a woman
$J^2$ : *a* is mentioned by *b*
$A^0$ : Justinian
$B^0$ : Gibbon

and the sentence

> The wife of Justinian is the most salacious woman mentioned by Gibbon.

In step (1) we might obtain

> The object *x* such that *x* is wife of Justinian is identical with the object *x* such that ((*x* is a woman and *x* is mentioned by Gibbon) and for each *y* (if ((*y* is a woman and *y* is mentioned by Gibbon) and it is not the case that *y* is identical with *x*), then *x* is more salacious than *y*)) ,

and carrying through the successive parts of step (2) we obtain the symbolization

$$\text{℩}xF(xA) = \text{℩}x(H(x) \wedge J(xB) \wedge \bigwedge y[H(y) \wedge J(yB) \wedge \sim y = x \rightarrow G(xy)]) \quad .$$

## EXERCISES

Symbolize each of the following formulas on the basis of the scheme of abbreviation which accompanies it. Exercise 6 is solved for illustration.

6. The man who lives at the North Pole does not live there.

On the basis of the scheme of abbreviation

$F^1$ : *a* is a man
$G^2$ : *a* lives at *b*
$A^0$ : the North Pole ,

No. 6 becomes

$$\sim G(\text{℩}x[F(x) \wedge G(xA)]\ A) \quad .$$

7. The positive square root of 2 is a positive even prime. ($F^1$ : *a* is positive; $G^2$ : *a* is a square root of *b*; $H^1$ : *a* is even; $J^1$ : *a* is prime; $A^0$ : 2)

8. Mary loves the one who loves Mary. ($F^2$ : *a* loves *b*; $A^0$ : Mary)

9. If he who murdered Desdemona was murdered by Desdemona's murderer, then he committed suicide. ($F^2$ : *a* murdered *b*; $G^1$ : *a* committed suicide; $A^0$ : Desdemona)

10. The hardest problem on the examination was solved by no one. ($F^1$ : *a* is a problem on the examination; $G^2$ : *a* is harder than *b*; $H^2$ : *a* was solved by *b*)

11. If the author of *Waverley* is the author of *Ivanhoe*, then the author of *Ivanhoe* is an author of *Waverley*. ($F^2$ : *a* is an author of *b*; $A^0$ : *Waverley*; $B^0$ : *Ivanhoe*)

12. God is that than which nothing greater can be conceived. ($A^0$ : God; $G^2$ : *a* is greater than *b*; $F^1$ : *a* can be conceived)

13. The positive square root of the square of the even prime is irrational. ($F^1$ : *a* is positive; $G^2$ : *a* is a square root of *b*; $H^1$ : *a* is even; $I^1$ : *a* is prime; $J^1$ : *a* is irrational; $A^1$ : the square of *a*)

14. The smallest positive integer is that positive integer which when multiplied by itself equals itself. ($F^2$ : *a* is less than *b*; $G^1$ : *a* is positive; $H^1$ : *a* is an integer; $A^2$ : the product of *a* and *b*)

15. There is no greatest positive integer. ($F^2$ : *a* is greater than *b*; $G^1$ : *a* is positive; $H^1$ : *a* is an integer)

**5. Inference rules.** For the logic of the descriptive operator we must adopt two new inference rules, the first of which is the following.

Proper descriptions (PD): $$\frac{\bigvee\beta\bigwedge\alpha(\phi \leftrightarrow \alpha = \beta)}{\psi}$$

Here $\alpha$ is to be a variable and $\phi$ a symbolic formula, $\beta$ is to be a variable not free in $\phi$, and $\psi$ is to come from $\phi$ by proper substitution of

$$\text{℩}\alpha\phi$$

for $\alpha$. PD corresponds to the principle that if a given condition is satisfied by one and only one object, then the object satisfying that condition satisfies it. PD leads, for example, from the sentence

$$\bigvee y \bigwedge x(F(x) \leftrightarrow x = y)$$

(which could be translated as the assertion that exactly one thing wrote *Waverley*), to the sentence

$$F[\text{℩}xF(x)]$$

(which would then assert that the author of *Waverley* wrote *Waverley*). A somewhat more complex application of PD is the inference from the formula

$$\bigvee y \bigwedge x(F(x) \wedge G(xzw) \leftrightarrow x = y)$$

(which could be taken as asserting that there is exactly one integer between $z$ and $w$) to the formula

$$F(\text{℩}x[F(x) \wedge G(xzw)]) \wedge G(\text{℩}x[F(x) \wedge G(xzw)]\ z\ w)$$

(which would then assert that the integer between $z$ and $w$ is an integer between $z$ and $w$).

Upon first consideration of such examples, one might believe that any English sentence that is a translation of a conclusion of PD is true. But consideration of further examples would quickly dispel this belief. The sentence

> The man who lives at the center of the earth is a man living at the center of the earth

is false, for it implies the falsehood that some man lives at the center of the earth. The assumption of propriety that constitutes the premise of PD is, however, sufficient to exclude such examples.

Because we seldom have occasion to use improper descriptive terms (that is, symbolic terms corresponding to improper definite descriptions), we could develop a significant part of the logic of the descriptive operator on the basis of the single rule PD. There is, however, a strong reason for introducing along with PD a rule pertaining to improper descriptive terms. Without such a rule, we could not extend AV and IE to the context

$$\text{℩}\alpha\phi$$

(where $\phi$ is a symbolic formula and $\alpha$ a variable); for example, we could not justify the inference from

$$F(\text{℩}xG(x))$$

to

$$F(\text{℩}yG(y))$$

or

$$F(\text{℩}x \sim\sim G(x)) \quad .$$

There are other reasons for adding a rule concerned with improper descriptions, and one of these reasons will appear in section 11. The relevant new rule has the following form.

Improper descriptions (ID): $\dfrac{\sim \bigvee\beta \bigwedge\alpha(\phi \leftrightarrow \alpha = \beta)}{\text{℩}\alpha\phi = \text{℩}\gamma \sim \gamma = \gamma}$

Here $\alpha$, $\beta$, and $\gamma$ are to be variables, and $\phi$ a symbolic formula in which $\beta$ is not free.

The rule ID has no intuitive counterpart, simply because ordinary language shuns improper definite descriptions. The rule corresponds, however, to the resolution that every improper definite description is to designate the same object as the particular improper definite description

the object $x$ such that $x$ differs from $x$ ,

and this resolution accords with the interpretation given on page 234.

The adoption of ID will enable us later to extend AV and IE to contexts involving descriptive phrases.

### EXERCISES

16. For each of the following pairs of formulas, state whether the second follows from the first by PD, by ID, or by neither.

(a) $\bigvee y \bigwedge x(\mathrm{F}(x) \wedge x = z \leftrightarrow x = y)$
$\mathrm{F}(\text{℩}x[\mathrm{F}(x) \wedge x = z]) \wedge \text{℩}x[\mathrm{F}(x) \wedge x = z] = z$

(b) $\sim \bigvee y \bigwedge x(\mathrm{H}(z) \leftrightarrow x = y)$
$\text{℩}z\mathrm{H}(z) = \text{℩}w \sim w = w$

(c) $\bigvee y \bigwedge x(\mathrm{F}(xy) \leftrightarrow x = y)$
$\mathrm{F}(\text{℩}x\mathrm{F}(xy)y)$

(d) $\bigvee y \bigwedge x(\mathrm{F}(z) \vee \bigwedge z\mathrm{G}(xz) \leftrightarrow x = y)$
$\mathrm{F}(z) \vee \bigwedge z\mathrm{G}(\text{℩}x[\mathrm{F}(z) \vee \bigwedge z\mathrm{G}(xz)]z)$

(e) $\sim \bigvee y \bigwedge x(\bigvee y\mathrm{F}(xy) \leftrightarrow x = y)$
$\text{℩}x\bigvee y\mathrm{F}(xy) = \text{℩}w \sim w = w$

(f) $\bigvee y \bigwedge x(\bigvee x\mathrm{F}(x) \leftrightarrow x = y)$
$\bigvee x\mathrm{F}(x)$

(g) $\sim \bigvee y \bigwedge x(x = z \leftrightarrow x = y)$
$\text{℩}x\, x = z = \text{℩}x \sim x = x$

(h) $\bigvee y \bigwedge x(\bigvee x\mathrm{F}(x) \vee \mathrm{G}(x) \leftrightarrow x = y)$
$\bigvee x\mathrm{F}(\text{℩}x[\bigvee x\mathrm{F}(x) \vee \mathrm{G}(x)]) \vee \mathrm{G}(\text{℩}x[\bigvee x\mathrm{F}(x) \vee \mathrm{G}(x)])$

**6. Theorems with unabbreviated proofs.** We may call that branch of logic which essentially involves the descriptive operator, as well as the identity sign, quantifiers, and sentential connectives, the *description calculus*. The directions for constructing an *unabbreviated derivation* within this calculus remain as before (see chapter IV, section 13, clauses (1) – (6)), with the understanding that PD and ID, as well as Id-1 and Id-2, are now to be included among the primitive inference rules, and the words 'term' and 'formula', throughout clauses (1) – (6) as well as in the formulation of the various inference rules, are to assume the broader senses of the present chapter.

The characterizations of a *complete* derivation, *derivability*, a *proof*, a *theorem*, an *argument*, a *symbolic argument*, a *valid symbolic argument*, an *English argument*, a *symbolization* of an English argument, and a *valid English argument* remain as before. (See pp. 95, 103, and 117.)

Theorems of the description calculus will receive numbers beginning with 401.

T401 1. ~~*Show*~~ $\bigwedge x(F(x) \leftrightarrow x = y) \rightarrow ℩xF(x) = y$

| | | |
|---|---|---|
| 2. | $\bigwedge x(F(x) \leftrightarrow x = y)$ | |
| 3. | $\bigvee y \bigwedge x(F(x) \leftrightarrow x = y)$ | 2, EG |
| 4. | $F(℩xF(x))$ | 3, PD |
| 5. | $F(℩xF(x)) \leftrightarrow ℩xF(x) = y$ | 2, UI |
| 6. | $F(℩xF(x)) \rightarrow ℩xF(x) = y$ | 5, BC |
| 7. | $℩xF(x) = y$ | 4, 6, MP |

T402 $\bigvee y \bigwedge x(F(x) \leftrightarrow x = y) \rightarrow F(℩xF(x))$

T403 $\sim \bigvee y \bigwedge x(F(x) \leftrightarrow x = y) \rightarrow ℩xF(x) = ℩w \sim w = w$

The next two theorems have an unusual form. They are conditionals with antecedents that are theorems of the identity calculus. We shall be interested only in their consequents, which obviously are also theorems and appear later as T406 and T407. At present, however, with only unabbreviated derivations at hand, proofs of the antecedents of these two theorems would run to exorbitant length; it is thus expedient to postpone the statement of the consequents until such time as we shall allow previously proved theorems to figure in a derivation.

T404 1. ~~*Show*~~ T328 $\rightarrow ℩xF(x) = ℩yF(y)$

2. T328

3. $\bigwedge b \bigwedge c[(\bigvee z \bigwedge x[F(x) \leftrightarrow x = z] \rightarrow F(℩xF(x))) \wedge$
$(\sim \bigvee z \bigwedge x[F(x) \leftrightarrow x = z] \rightarrow ℩xF(x) = c) \wedge$
$(\bigvee z \bigwedge y[F(y) \leftrightarrow y = z] \rightarrow F(b)) \wedge$
$(\sim \bigvee z \bigwedge y[F(y) \leftrightarrow y = z] \rightarrow b = c) \rightarrow$
$℩xF(x) = b]$ 2, UI

4. $\bigwedge c[(\bigvee z \bigwedge x[F(x) \leftrightarrow x = z] \rightarrow F(℩xF(x))) \wedge$
$(\sim \bigvee z \bigwedge x[F(x) \leftrightarrow x = z] \rightarrow ℩xF(x) = c) \wedge$
$(\bigvee z \bigwedge y[F(y) \leftrightarrow y = z] \rightarrow F(℩yF(y))) \wedge$
$(\sim \bigvee z \bigwedge y[F(y) \leftrightarrow y = z] \rightarrow ℩yF(y) = c) \rightarrow$
$℩xF(x) = ℩yF(y)]$ 3, UI

5. $(\bigvee z \bigwedge x[F(x) \leftrightarrow x = z] \rightarrow F(℩xF(x))) \wedge$
$(\sim \bigvee z \bigwedge x[F(x) \leftrightarrow x = z] \rightarrow$
$℩xF(x) = ℩w \sim w = w) \wedge$
$(\bigvee z \bigwedge y[F(y) \leftrightarrow y = z] \rightarrow F(℩yF(y))) \wedge$
$(\sim \bigvee z \bigwedge y[F(y) \leftrightarrow y = z] \rightarrow$
$℩yF(y) = ℩w \sim w = w) \rightarrow ℩xF(x) = ℩yF(y)$ 4, UI

6. ~~*Show*~~ $\bigvee z \bigwedge x[F(x) \leftrightarrow x = z] \rightarrow F(℩xF(x))$

7. $\bigvee z \bigwedge x[F(x) \leftrightarrow x = z]$

8. $F(℩xF(x))$ — 7, PD

9. ~~*Show*~~ $\sim \bigvee z \bigwedge x[F(x) \leftrightarrow x = z] \rightarrow ℩xF(x) = ℩w \sim w = w$

10. $\sim \bigvee z \bigwedge x[F(x) \leftrightarrow x = z]$

11. $℩xF(x) = ℩w \sim w = w$ — 10, ID

12. ~~*Show*~~ $\bigvee z \bigwedge y[F(y) \leftrightarrow y = z] \rightarrow F(℩yF(y))$

13. $\bigvee z \bigwedge y[F(y) \leftrightarrow y = z]$

14. $F(℩yF(y))$ — 13, PD

15. ~~*Show*~~ $\sim \bigvee z \bigwedge y[F(y) \leftrightarrow y = z] \rightarrow ℩yF(y) = ℩w \sim w = w$

16. $\sim \bigvee z \bigwedge y[F(y) \leftrightarrow y = z]$

17. $℩yF(y) = ℩w \sim w = w$ — 16, ID

18. $(\bigvee z \bigwedge x[F(x) \leftrightarrow x = z] \rightarrow F(℩xF(x))) \wedge (\sim \bigvee z \bigwedge x[F(x) \leftrightarrow x = z] \rightarrow ℩xF(x) = ℩w \sim w = w)$ — 6, 9, Adj

19. $(\bigvee z \bigwedge x[F(x) \leftrightarrow x = z] \rightarrow F(℩xF(x))) \wedge (\sim \bigvee z \bigwedge x[F(x) \leftrightarrow x = z] \rightarrow ℩xF(x) = ℩w \sim w = w) \wedge (\bigvee z \bigwedge y[F(y) \leftrightarrow y = z] \rightarrow F(℩yF(y)))$ — 18, 12, Adj

20. $(\bigvee z \bigwedge x[F(x) \leftrightarrow x = z] \rightarrow F(℩xF(x))) \wedge (\sim \bigvee z \bigwedge x[F(x) \leftrightarrow x = z] \rightarrow ℩xF(x) = ℩w \sim w = w) \wedge (\bigvee z \bigwedge y[F(y) \leftrightarrow y = z] \rightarrow F(℩yF(y))) \wedge (\sim \bigvee z \bigwedge y[F(y) \leftrightarrow y = z] \rightarrow ℩yF(y) = ℩w \sim w = w)$ — 19, 15, Adj

21. $℩xF(x) = ℩yF(y)$ — 5, 20, MP

T405 1. ~~*Show*~~ $T329 \rightarrow [\bigwedge x[F(x) \leftrightarrow G(x)] \rightarrow ℩xF(x) = ℩xG(x)]$

2. T329

3. ~~*Show*~~ $\bigwedge x[F(x) \leftrightarrow G(x)] \rightarrow ℩xF(x) = ℩xG(x)$

4. $\bigwedge x[F(x) \leftrightarrow G(x)]$

5. $\bigwedge b \bigwedge c[(\bigvee y \bigwedge x[F(x) \leftrightarrow x = y] \rightarrow F(℩xF(x))) \wedge (\sim \bigvee y \bigwedge x[F(x) \leftrightarrow x = y] \rightarrow ℩xF(x) = c) \wedge (\bigvee y \bigwedge x[G(x) \leftrightarrow x = y] \rightarrow G(b)) \wedge (\sim \bigvee y \bigwedge x[G(x) \leftrightarrow x = y] \rightarrow b = c) \wedge \bigwedge x[F(x) \leftrightarrow G(x)] \rightarrow ℩xF(x) = b]$ — 2, UI

| | | |
|---|---|---|
| 6. | $\bigwedge c[(\bigvee y \bigwedge x[F(x) \leftrightarrow x = y] \rightarrow F(\iota x F(x))) \wedge$ $(\sim \bigvee y \bigwedge x[F(x) \leftrightarrow x = y] \rightarrow \iota x F(x) = c) \wedge$ $(\bigvee y \bigwedge x[G(x) \leftrightarrow x = y] \rightarrow G(\iota x G(x))) \wedge$ $(\sim \bigvee y \bigwedge x[G(x) \leftrightarrow x = y] \rightarrow \iota x G(x) = c) \wedge$ $\bigwedge x[F(x) \leftrightarrow G(x)] \rightarrow$ $\iota x F(x) = \iota x G(x)]$ | 5, UI |
| 7. | $(\bigvee y \bigwedge x[F(x) \leftrightarrow x = y] \rightarrow F(\iota x F(x))) \wedge$ $(\sim \bigvee y \bigwedge x[F(x) \leftrightarrow x = y] \rightarrow$ $\iota x F(x) = \iota w \sim w = w) \wedge$ $(\bigvee y \bigwedge x[G(x) \leftrightarrow x = y] \rightarrow G(\iota x G(x))) \wedge$ $(\sim \bigvee y \bigwedge x[G(x) \leftrightarrow x = y] \rightarrow$ $\iota x G(x) = \iota w \sim w = w) \wedge$ $\bigwedge x[F(x) \leftrightarrow G(x)] \rightarrow$ $\iota x F(x) = \iota x G(x)$ | 6, UI |
| 8. | ~~Show~~ $\bigvee y \bigwedge x[F(x) \leftrightarrow x = y] \rightarrow F(\iota x F(x))$ | |
| 9. | $\bigvee y \bigwedge x[F(x) \leftrightarrow x = y]$ | |
| 10. | $F(\iota x F(x))$ | 9, PD |
| 11. | ~~Show~~ $\sim \bigvee y \bigwedge x[F(x) \leftrightarrow x = y] \rightarrow$ $\iota x F(x) = \iota w \sim w = w$ | |
| 12. | $\sim \bigvee y \bigwedge x[F(x) \leftrightarrow x = y]$ | |
| 13. | $\iota x F(x) = \iota w \sim w = w$ | 12, ID |
| 14. | ~~Show~~ $\bigvee y \bigwedge x[G(x) \leftrightarrow x = y] \rightarrow G(\iota x G(x))$ | |
| 15. | $\bigvee y \bigwedge x[G(x) \leftrightarrow x = y]$ | |
| 16. | $G(\iota x G(x))$ | 15, PD |
| 17. | ~~Show~~ $\sim \bigvee y \bigwedge x[G(x) \leftrightarrow x = y] \rightarrow$ $\iota x G(x) = \iota w \sim w = w$ | |
| 18. | $\sim \bigvee y \bigwedge x[G(x) \leftrightarrow x = y]$ | |
| 19. | $\iota x G(x) = \iota w \sim w = w$ | 18, ID |
| 20. | $(\bigvee y \bigwedge x[F(x) \leftrightarrow x = y] \rightarrow F(\iota x F(x))) \wedge$ $(\sim \bigvee y \bigwedge x[F(x) \leftrightarrow x = y] \rightarrow$ $\iota x F(x) = \iota w \sim w = w)$ | 8, 11, Adj |
| 21. | $(\bigvee y \bigwedge x[F(x) \leftrightarrow x = y] \rightarrow F(\iota x F(x))) \wedge$ $(\sim \bigvee y \bigwedge x[F(x) \leftrightarrow x = y] \rightarrow$ $\iota x F(x) = \iota w \sim w = w) \wedge$ $(\bigvee y \bigwedge x[G(x) \leftrightarrow x = y) \rightarrow G(\iota x G(x)))$ | 20, 14, Adj |
| 22. | $(\bigvee y \bigwedge x[F(x) \leftrightarrow x = y] \rightarrow F(\iota x F(x))) \wedge$ $(\sim \bigvee y \bigwedge x[F(x) \leftrightarrow x = y] \rightarrow$ $\iota x F(x) = \iota w \sim w = w) \wedge$ $\bigvee y \bigwedge x[G(x) \leftrightarrow x = y] \rightarrow G(\iota x G(x))) \wedge$ | |

| | | |
|---|---|---|
| | $(\sim \bigvee y \bigwedge x[G(x) \leftrightarrow x = y] \rightarrow$ | |
| | $\rceil x G(x) = \rceil w \sim w = w)$ | 21, 17, Adj |
| 23. | $(\bigvee y \bigwedge x[F(x) \leftrightarrow x = y] \rightarrow F(\rceil x F(x))) \wedge$ | |
| | $(\sim \bigvee y \bigwedge x[F(x) \leftrightarrow x = y] \rightarrow$ | |
| | $\rceil x F(x) = \rceil w \sim w = w) \wedge$ | |
| | $(\bigvee y \bigwedge x[G(x) \leftrightarrow x = y] \rightarrow G(\rceil x G(x))) \wedge$ | |
| | $(\sim \bigvee y \bigwedge x[G(x) \leftrightarrow x = y] \rightarrow$ | |
| | $\rceil x G(x) = \rceil w \sim w = w) \wedge$ | |
| | $\bigwedge x[F(x) \leftrightarrow G(x)]$ | 22, 4, Adj |
| 24. | $\rceil x F(x) = \rceil x G(x)$ | 7, 23, MP |

## EXERCISES

17. Prove T402 and T403.

**7. Alphabetic variance; proper substitution; abbreviated derivations.** In chapter III we introduced the relation of *equivalence* between *formulas*. Two symbolic *formulas* $\phi$ and $\psi$ are said to be *equivalent* just in case the biconditional

$$\phi \leftrightarrow \psi$$

is a theorem. This notion can be extended naturally so as to apply as well to terms. The symbolic *terms* $\zeta$ and $\eta$ are *equivalent* just in case the identity

$$\zeta = \eta$$

is a theorem.

T404, together with T328, will enable us to establish the equivalence of two descriptive terms differing only in their initial bound variables. This suggests that we extend to terms related in this way the notion of *alphabetic variance* introduced in chapter IV, section 7. Thus we call two symbolic *formulas immediate alphabetic variants* if (as before) they have the forms

$$\bigwedge \alpha \phi$$

and

$$\bigwedge \alpha' \phi' \quad ,$$

or else

$$\bigvee \alpha \phi$$

and

$$\bigvee \alpha' \phi' \quad ,$$

and we call two symbolic *terms immediate alphabetic variants* if they have the forms

$$\daleth\alpha\phi$$

and

$$\daleth\alpha'\phi' \quad ;$$

in each case $\alpha$ and $\alpha'$ are to be variables, $\phi$ and $\phi'$ are to be symbolic formulas, $\phi'$ is to come from $\phi$ by proper substitution of $\alpha'$ for $\alpha$, and $\phi$ is to come from $\phi'$ by proper substitution of $\alpha$ for $\alpha'$. Symbolic terms or formulas that differ only in parts that are immediate alphabetic variants of one another are also equivalent, and may be called *alphabetic variants.* Thus we replace the previous notion of alphabetic variance by the following: a symbolic term or formula $\psi$ is an *alphabetic variant* of a symbolic term or formula $\psi'$ just in case $\psi$ is like $\psi'$ except for having one or more occurrences of a term or formula $\phi$ where $\psi'$ has some immediate alphabetic variant of $\phi$.

For example,

$$\daleth x[\mathrm{F}(x) \wedge \mathrm{G}(x)]$$

is an immediate alphabetic variant of

$$\daleth y[\mathrm{F}(y) \wedge \mathrm{G}(y)] \quad ,$$

and therefore

$$z = \daleth x[\mathrm{F}(x) \wedge \mathrm{G}(x)]$$

is an alphabetic variant of

$$z = \daleth y[\mathrm{F}(y) \wedge \mathrm{G}(y)] \quad ;$$

further,

$$\bigvee x\mathrm{F}(xz)$$

is an immediate alphabetic variant of

$$\bigvee y\mathrm{F}(yz) \quad ,$$

and therefore

$$\daleth z\bigvee x\mathrm{F}(xz)$$

is an alphabetic variant of

$$\daleth z\bigvee y\mathrm{F}(yz) \quad .$$

To define the notion of an *instance* of a formula of the description calculus we can employ the notions of *proper substitution of a formula for a predicate letter* and *proper substitution on variables* as characterized in section 8 of chapter IV. (In the characterization of these notions we must now, of course, understand 'term', and 'formula', in the sense of the present chapter.) For *proper substitution of a term for an operation letter*, however, we need a slightly modified notion now that variables may be bound in terms. We say that a symbolic formula $\psi$ comes from a symbolic formula $\phi$ by *proper substitution of* a symbolic term $\eta$ *for* a $k$-place *operation letter* $\delta$ just in case there is no variable occurring in both $\phi$ and $\eta$, and $\psi$ can be obtained from $\phi$ in the following two steps:

I. Throughout $\phi$, replace $\delta$ by $\eta$ enclosed in braces.

II. In the expression resulting from I, successively consider each part of the form

$$\{\eta\}\,\zeta_1 \ldots \zeta_k \quad ,$$

where $\zeta_1, \ldots, \zeta_k$ are terms. Replace each such part by a certain term—in fact, by the term obtained from $\eta$ by replacing all *free* occurrences of '*a*' by $\zeta_1$, '*b*' by $\zeta_2$, etc., up to the *k*th variable, all of whose free occurrences are to be replaced by $\zeta_k$.

(The only alteration is that here we speak in step II only of *free* occurrences of variables; in chapter IV this qualification was unnecessary, for before the descriptive operator was introduced there was no way for a variable to be bound in a term.)

For example, from the formula

$$\mathsf{V}x\, x = \mathrm{A}(y)$$

the formula

$$\mathsf{V}x\, x = \text{\rotatebox{180}{$\iota$}}z[\mathsf{V}a\mathrm{F}(az) \vee \mathrm{G}(y)]$$

comes by proper substitution of the term

$$\text{\rotatebox{180}{$\iota$}}z[\mathsf{V}a\mathrm{F}(az) \vee \mathrm{G}(a)]$$

for the operation letter 'A'.

We say as before that a symbolic formula $\psi$ is an *instance* of a symbolic formula $\phi$ just in case $\psi$ is $\phi$ or obtainable from $\phi$ by one or more operations of proper substitution—whether on predicate letters, on operation letters, or on variables.

We once again remind the reader that he should mentally restore official notation when checking applications of the preceding notions.

Now that the notions of instance and alphabetic variance have been extended, we may reintroduce three of the abbreviatory clauses of chapter IV:

(7) *If $\phi$ is an instance of an alphabetic variant of a theorem which has already been proved, then $\phi$ may occur as a line.* (*Annotation as prescribed in chapter IV, section 9.*)

(8) *A symbolic formula may occur as a line if it is an alphabetic variant of an antecedent line.* (*Annotation as in chapter IV.*)

(9) *If* $\phi$, $\phi'$ *are symbolic formulas such that*

$$\phi \leftrightarrow \phi'$$

*is an instance of an alphabetic variant of a previously proved theorem, and* $\psi$, $\psi'$ *are symbolic formulas such that* $\psi'$ *is like* $\psi$ *except for having one or more occurrences of* $\phi'$ *where* $\psi$ *has occurrences of* $\phi$, *then*

$$\psi \leftrightarrow \psi'$$

*may occur as a line.* (*As before, we call*

$$\psi \leftrightarrow \psi'$$

*an* interchange of equivalents *on the basis of the previously proved theorem and use the annotation 'IE' with a parenthetical reference to this theorem.*)

We shall now introduce an abbreviation analogous to the derived rule known as Leibniz' law (p. 223). The reader will recall that in an application

$$\frac{\zeta = \zeta'}{\phi \leftrightarrow \phi'}$$

of LL, the occurrences of $\zeta$ and $\zeta'$ which are interchanged in passing from $\phi$ to $\phi'$ must all be free. By strengthening the premise, we can dispense with this limitation. For example, the inference from

$$\Lambda x\, \mathrm{A}(x) = \mathrm{B}(x)$$

to

$$\mathrm{V}x\, \mathrm{F}(\mathrm{A}(x)) \leftrightarrow \mathrm{V}x\, \mathrm{F}(\mathrm{B}(x)) \quad ,$$

though not justified by LL, is correct in view of T327. The premise of this inference is no longer an identity, but a universal generalization of an identity; the important feature is that all variables bound in the conclusion (in this case, only '$x$') are universally quantified in the premise.

More generally, the pattern of inference presently contemplated has the form

$$(1) \qquad \frac{\Lambda\alpha_1 \ldots \Lambda\alpha_n\, \zeta = \zeta'}{\phi \leftrightarrow \phi'} \;,$$

where $\zeta$ and $\zeta'$ are symbolic terms, $\phi$ and $\phi'$ are symbolic formulas, $\phi'$ is like $\phi$ except for having one or more occurrences of $\zeta'$ where $\phi$ has occurrences of $\zeta$, and $\alpha_1, \ldots, \alpha_n$ are all the variables bound in $\phi$. (If no variables are bound in $\phi$, the premise reduces of course to

$$\zeta = \zeta' \quad .)$$

We may consider a similar analogue to Euclid's law. Here the inference has the form

$$(2) \qquad \frac{\Lambda\alpha_1 \ldots \Lambda\alpha_n\, \zeta = \zeta'}{\eta = \eta'} ,$$

where $\zeta$, $\zeta'$, $\eta$, $\eta'$ are symbolic terms, $\eta'$ is like $\eta$ except for having one or more occurrences of $\zeta'$ where $\eta$ has occurrences of $\zeta$, and $\alpha_1, \ldots, \alpha_n$ are all the variables bound in $\eta$.

In the inferences (1) and (2) the premises contain terms. Comparable forms of inference are available whose premises instead contain formulas. They are

$$(3) \qquad \frac{\Lambda\alpha_1 \ldots \Lambda\alpha_n(\psi \leftrightarrow \psi')}{\phi \leftrightarrow \phi'} ,$$

and

$$(4) \qquad \frac{\Lambda\alpha_1 \ldots \Lambda\alpha_n(\psi \leftrightarrow \psi')}{\eta = \eta'} .$$

Here $\psi$, $\psi'$, $\phi$, $\phi'$ are to be symbolic formulas, and $\eta$, $\eta'$ are to be symbolic terms. We assume in addition that, in the first case, $\phi'$ is like $\phi$ except for having one or more occurrences of $\psi'$ where $\phi$ has occurrences of $\psi$ and $\alpha_1, \ldots, \alpha_n$ are all the variables bound in $\phi$, and, in the second case, $\eta'$ is like $\eta$ except for having one or more occurrences of $\psi'$ where $\eta$ has occurrences of $\psi$ and $\alpha_1, \ldots, \alpha_n$ are all the variables bound in $\eta$.

For example, the inference from

$$\Lambda x[\mathrm{F}(x) \leftrightarrow \mathrm{G}(x)]$$

to

$$\mathrm{V}x\mathrm{F}(x) \leftrightarrow \mathrm{V}x\mathrm{G}(x) ,$$

correct in view of T214, is comprehended under (3).

In each of the inference patterns (1) – (4), we shall say that the conclusion follows from the premise by *interchange*, and we incorporate these patterns into our system by the following abbreviatory clause:

(*10*) *A symbolic formula may occur as a line if it follows from an antecedent line by interchange.* (*Annotation:* 'Int' *and the number of the antecedent line.*)

As before, we add a clause permitting the compression of several steps into one:

(*11*) *A symbolic formula may occur as a line if it follows from antecedent lines by a succession of steps, and each intermediate step can be justified by one of clauses* (*2*), (*5a*), (*7*), (*8*), (*9*), *or* (*10*). (*Annotation as in chapter IV.*)

The special annotations 'SC', 'CD', 'QN', 'Sm', 'T', and 'LL' will be used as before. We shall also use 'EL', as in chapter VI, to stand for the pattern of inference corresponding to T316. Now that variables can have bound occurrences in terms, a slight reformulation is required:

Euclid's law (EL): $$\frac{\zeta = \zeta'}{\eta = \eta'}$$

Here $\zeta$, $\zeta'$, $\eta$, $\eta'$ are to be symbolic terms, and $\eta'$ is to be like $\eta$ except for having one or more *free* occurrences of $\zeta'$ where $\eta$ has *free* occurrences of $\zeta$.

There are close connections among Leibniz' law, Euclid's law, interchange of equivalents, and interchange. Indeed, the first three can rather easily be subsumed under interchange, given the principle of alphabetic variance (clause (8)).

Many inferences justified by LL, such as that from

(5) $$x = y$$

to

(6) $$[\bigvee x F(x) \vee G(x)] \leftrightarrow [\bigvee x F(x) \vee G(y)] \quad ,$$

are not immediately comprehended under Int; but this rule, assisted by familiar laws, leads from (5) to (6) by the following sequence of steps:

| | | |
|---|---|---|
| (i) | $x = y$ | |
| (ii) | $\bigwedge z\, x = y$ | i, T227, BC, MP |
| (iii) | $[\bigvee z F(z) \vee G(x)] \leftrightarrow [\bigvee z F(z) \vee G(y)]$ | ii, Int (form (1)) |
| (iv) | $[\bigvee x F(x) \vee G(x)] \leftrightarrow [\bigvee x F(x) \vee G(y)]$ | iii, AV |

The second form of Int bears a similar relationship to EL. The fact that a formula is a theorem just in case all its closures are theorems clarifies the relation between IE and the third form of Int.

Clauses (7) – (11), like their earlier counterparts, satisfy the requirements for abbreviatory clauses that are given on page 58. A proof of the theoretical dispensability of these clauses would be rather involved and is not attempted here; a few relevant remarks, however, will perhaps be appropriate.

To show clause (8) to be dispensable, we should have to show the existence of unabbreviated proofs, not only of T231 and T232, but also of the formula

$$\daleth x F(x) = \daleth y F(y)$$

and its instances. But this would be relatively easy, in view of the unabbreviated proof of T404 given in the previous section, and the fact that the antecedent of T404 is a theorem of the identity calculus (for whose theorems we already suppose unabbreviated proofs to exist).

Similar remarks apply to the relation between clause (10) and the formula

$$\bigwedge x[F(x) \leftrightarrow G(x)] \rightarrow \iota x F(x) = \iota x G(x) \quad ,$$

which has an unabbreviated proof in view of the unabbreviated proof of T405 given in the previous section. (Of course, several other theorems would also be involved in eliminating applications of clause (10), but they are all theorems of preceding chapters.)

The easiest way to show the eliminability of clause (9) is first to treat clause (10) and then show that clause (9) is reducible to it.

We mentioned earlier that without the rule ID we should be unable to justify the use of AV and IE for the description calculus. The reason is now relatively clear. We could not claim the dispensability of clauses (8) and (9) without having T404 and T405 at our disposal, and the proofs of these theorems essentially involve ID.

The four forms of Int are respectively illustrated by the following arguments and accompanying derivations.

(7) $A(xy) = A(yx) \quad \therefore \bigvee x \bigwedge y F[A(xy)] \leftrightarrow \bigvee x \bigwedge y F[A(yx)]$

1. $\cancel{Show}\ \bigvee x \bigwedge y F[A(xy)] \leftrightarrow \bigvee x \bigwedge y F[A(yx)]$
2. $\cancel{Show}\ \bigwedge x \bigwedge y\ A(xy) = A(yx)$
3. $A(xy) = A(yx)$ — Premise
4. $\bigvee x \bigwedge y F[A(xy)] \leftrightarrow \bigvee x \bigwedge y F[A(yx)]$ — 2, Int (form (1))

$A(xy) = A(yx) \quad \therefore \iota x\ A(xy) = y = \iota x\ A(yx) = y$

1. $\cancel{Show}\ \iota x\ A(xy) = y = \iota x\ A(yx) = y$
2. $\cancel{Show}\ \bigwedge x \bigwedge y\ A(xy) = A(yx)$
3. $A(xy) = A(yx)$ — Premise
4. $\iota x\ A(xy) = y = \iota x\ A(yx) = y$ — 2, Int (form (2))

$\sim F(xy) \quad . \quad \sim G(xy) \quad \therefore \bigvee y \bigwedge x F(xy) \leftrightarrow \bigvee y \bigwedge x G(xy)$

1. $\cancel{Show}\ \bigvee y \bigwedge x F(xy) \leftrightarrow \bigvee y \bigwedge x G(xy)$
2. $\cancel{Show}\ \bigwedge x \bigwedge y[F(xy) \leftrightarrow G(xy)]$
3. $F(xy) \leftrightarrow G(xy)$ — 1st premise, 2nd premise, Adj, T85, MP
4. $\bigvee y \bigwedge x F(xy) \leftrightarrow \bigvee y \bigwedge x G(xy)$ — 2, Int (form (3))

$\sim F(xy)$ . $\sim G(xy)$ $\therefore$ $\iota x \wedge y F(xy) = \iota x \wedge y G(xy)$

1. ~~*Show*~~ $\iota x \wedge y F(xy) = \iota x \wedge y G(xy)$
2. | ~~*Show*~~ $\wedge x \wedge y[F(xy) \leftrightarrow G(xy)]$
3. | | $F(xy) \leftrightarrow G(xy)$ — 1st premise, 2nd premise, Adj, T85, MP
4. | $\iota x \wedge y F(xy) = \iota x \wedge y G(xy)$ — 2, Int (form (4))

To appreciate the economy gained by Int the reader should, for example, construct a derivation corresponding to (7) employing LL rather than Int.

## EXERCISES

18. Which of the following pairs of expressions are alphabetic variants?

(a) $\vee x\, \iota x F(x) = x$
$\quad\;\; \vee x\, \iota y F(y) = x$

(b) $\vee x\, \iota x F(xy) = x$
$\quad\;\; \vee y\, \iota x F(xy) = y$

(c) $\iota x F(x) = \iota y G(y)$
$\quad\;\; \iota x F(x) = \iota x G(x)$

(d) $G(\iota x F(x)\, \iota x F(x)\, y)$
$\quad\;\; G(\iota x F(x)\, \iota y F(y)\, y)$

19. For each of the terms (i) – (v) below indicate the instance, if any, that can be obtained from the theorem

$$\wedge x\; y = A(x) \rightarrow \vee x\; y = A(x)$$

by proper substitution of that term for 'A'.

(i) $\iota z\; z = a$
(ii) $\iota a\; a = a$
(iii) $\iota z\; x = a$
(iv) $\iota z[\vee a F(az) \vee G(a)]$
(v) $\iota w\; a = a$

20. The second of each of the following pairs of formulas is an instance of the first; indicate in each case a sequence of substitutions by which the instance can be obtained.

(i) $\vee y\; A(x) = y$ ; $\vee y\; \iota x\; z = x = y$
(ii) $\vee y\; A(x) = y$ ; $\vee y\; \iota x\; x = x = y$
(iii) $\vee y\; A(x) = y$ ; $\vee y\; \iota z\; x = z = y$

21. Give short proofs of the following theorems.

(i) $\bigwedge x(F(x) \leftrightarrow P) \rightarrow (\bigwedge xF(x) \leftrightarrow P)$
(ii) $\iota x \bigwedge y[F(y) \rightarrow G(x)] = \iota y[\bigvee xF(x) \rightarrow G(y)]$
(iii) $\bigwedge x\, A(x) = B(x) \rightarrow \iota y\, A(y) = y = \iota z\, B(z) = z$

**8. Theorems with abbreviated proofs.** The consequents of T404 and T405 can now be proved without pain. It is only necessary to employ clause (7).

T406 $\iota xF(x) = \iota yF(y)$

T407 $\bigwedge x[F(x) \leftrightarrow G(x)] \rightarrow \iota xF(x) = \iota xG(x)$

T408 gives a necessary and sufficient condition for a descriptive term to be proper.

| | | | |
|---|---|---|---|
| T408 | 1. | ~~Show~~ $\bigwedge x[F(x) \leftrightarrow x = \iota xF(x)] \leftrightarrow \bigvee y \bigwedge x(F(x) \leftrightarrow x = y)$ | |
| | 2. | ~~Show~~ $\bigwedge x[F(x) \leftrightarrow x = \iota xF(x)] \rightarrow \bigvee y \bigwedge x(F(x) \leftrightarrow x = y)$ | |
| | 3. | $\bigwedge x[F(x) \leftrightarrow x = \iota xF(x)]$ | |
| | 4. | $\bigvee y \bigwedge x(F(x) \leftrightarrow x = y)$ | 3, EG |
| | 5. | ~~Show~~ $\bigvee y \bigwedge x(F(x) \leftrightarrow x = y) \rightarrow \bigwedge x[F(x) \leftrightarrow x = \iota xF(x)]$ | |
| | 6. | $\bigvee y \bigwedge x(F(x) \leftrightarrow x = y)$ | |
| | 7. | $\bigwedge x(F(x) \leftrightarrow x = z)$ | 6, EI |
| | 8. | $\iota xF(x) = z$ | 7, T401, MP |
| | 9. | $\bigwedge x[F(x) \leftrightarrow x = \iota xF(x)]$ | 8, 7, LL |
| | 10. | $\bigwedge x[F(x) \leftrightarrow x = \iota xF(x)] \leftrightarrow \bigvee y \bigwedge x(F(x) \leftrightarrow x = y)$ | 2, 5, CB |

T409 and T410 are useful corollaries of T408.

T409 $\bigvee y \bigwedge x(F(x) \leftrightarrow x = y) \wedge F(x) \rightarrow x = \iota xF(x)$

T410 $\bigvee y \bigwedge x(F(x) \leftrightarrow x = y) \wedge x = \iota xF(x) \rightarrow F(x)$

To obtain the converse of T407, we must assume that the descriptive terms involved are proper:

T411 $\bigvee y \bigwedge x(F(x) \leftrightarrow x = y) \wedge \bigvee y \bigwedge x(G(x) \leftrightarrow x = y) \rightarrow [\bigwedge x[F(x) \leftrightarrow G(x)] \leftrightarrow \iota xF(x) = \iota xG(x)]$

As we have seen (p. 243), an assertion that the object satisfying $\phi$ has the property expressed by $\phi$ will not always be true. T412, together with its instances, gives a necessary and sufficient condition for the truth of such a sentence.

T412

| | | |
|---|---|---|
| 1. | ~~*Show*~~ $F(\text{℩}xF(x)) \leftrightarrow \bigvee y\bigwedge x(F(x) \leftrightarrow x = y) \vee F(\text{℩}w \sim w = w)$ | |
| 2. | ~~*Show*~~ $F(\text{℩}xF(x)) \wedge \sim\bigvee y\bigwedge x(F(x) \leftrightarrow x = y) \rightarrow F(\text{℩}w \sim w = w)$ | |
| 3. | $F(\text{℩}xF(x)) \wedge \sim\bigvee y\bigwedge x(F(x) \leftrightarrow x = y)$ | |
| 4. | $F(\text{℩}w \sim w = w)$ | 3, S, ID, 3, S, LL |
| 5. | ~~*Show*~~ $\bigvee y\bigwedge x(F(x) \leftrightarrow x = y) \rightarrow F(\text{℩}xF(x))$ | |
| 6. | $\bigvee y\bigwedge x(F(x) \leftrightarrow x = y)$ | |
| 7. | $F(\text{℩}xF(x))$ | 6, PD |
| 8. | ~~*Show*~~ $\sim\bigvee y\bigwedge x(F(x) \leftrightarrow x = y) \wedge F(\text{℩}w \sim w = w) \rightarrow F(\text{℩}xF(x))$ | |
| 9. | $\sim\bigvee y\bigwedge x(F(x) \leftrightarrow x = y) \wedge F(\text{℩}w \sim w = w)$ | |
| 10. | $F(\text{℩}xF(x))$ | 9, S, ID, 9, S, LL |
| 11. | $F(\text{℩}xF(x)) \leftrightarrow \bigvee y\bigwedge x(F(x) \leftrightarrow x = y) \vee F(\text{℩}w \sim w = w)$ | 2, T27, BC, MP, IE(T45), 5, 8, Adj, T52, BC, MP, CB |

T413 is a corollary of T412; it gives alternative conditions under which the inference from

$$x = \text{℩}xF(x)$$

to

$$F(x)$$

is valid.

T413 $\qquad [\sim x = \text{℩}w \sim w = w \vee F(\text{℩}w \sim w = w)] \wedge x = \text{℩}xF(x) \rightarrow F(x)$

The converse of T401, like that of T407, requires a hypothesis of propriety.

T414 $\qquad \bigvee y\bigwedge x(F(x) \leftrightarrow x = y) \rightarrow [\bigwedge x(F(x) \leftrightarrow x = y) \leftrightarrow \text{℩}xF(x) = y]$

On the basis of ID we can show that any two improper definite descriptions designate the same thing.

T415 $\qquad \sim\bigvee y\bigwedge x(F(x) \leftrightarrow x = y) \wedge \sim\bigvee y\bigwedge x(G(x) \leftrightarrow x = y) \rightarrow \text{℩}xF(x) = \text{℩}xG(x)$

T416 gives a necessary and sufficient condition for the truth of any sentence containing a descriptive term; T324 and T325 are useful in its derivation. T417, which can be derived easily from an instance of

T416, along with an instance of T310, gives a necessary and sufficient condition for the truth of an identity sentence containing a descriptive term.

T416 $$G(\text{℩}x F(x)) \leftrightarrow \bigvee y[\bigwedge x(F(x) \leftrightarrow x = y) \wedge G(y)] \vee [\sim \bigvee y \bigwedge x(F(x) \leftrightarrow x = y) \wedge G(\text{℩}w \sim w = w)]$$

T417 $$z = \text{℩}x F(x) \leftrightarrow \bigwedge x(F(x) \leftrightarrow x = z) \vee [\sim \bigvee y \bigwedge x(F(x) \leftrightarrow x = y) \wedge z = \text{℩}w \sim w = w]$$

Each thing is the thing identical with it.

T418 $$\text{℩}x\, x = y = y$$

The self-identical thing and the non-self-identical things are the same.

T419 $$\text{℩}w\, w = w = \text{℩}w \sim w = w$$

T420 indicates the meaning which is to be assigned to *vacuous* descriptive terms.

T420 $$\text{℩}x F = \text{℩}w \sim w = w$$

Thus we conclude that

$$\text{℩}x \text{ Socrates is bald}$$

designates 0.

### EXERCISES

22. Prove T411, T413, and T414.
23. Prove T416 and T417.
24. Prove T418 – T420.
25. Show that the following two formulas are equivalent:

(a) $\sim(\bigvee y[\bigwedge x(F(x) \leftrightarrow x = y) \wedge G(y)] \vee [\sim \bigvee y \bigwedge x(F(x) \leftrightarrow x = y) \wedge G(\text{℩}w \sim w = w)])$

(b) $\bigvee y[\bigwedge x(F(x) \leftrightarrow x = y) \wedge \sim G(y)] \vee [\sim \bigvee y \bigwedge x(F(x) \leftrightarrow x = y) \wedge \sim G(\text{℩}w \sim w = w)]$

26. Show that the following two formulas are equivalent:

(a) $\bigwedge z(\bigvee y[\bigwedge x(F(x) \leftrightarrow x = y) \wedge G(yz)] \vee [\sim \bigvee y \bigwedge x(F(x) \leftrightarrow x = y) \wedge G(\text{℩}w \sim w = w\ z)])$

(b) $\bigvee y[\bigwedge x(F(x) \leftrightarrow x = y) \wedge \bigwedge z G(yz)] \vee [\sim \bigvee y \bigwedge x(F(x) \leftrightarrow x = y) \wedge \bigwedge z G(\text{℩}w \sim w = w\ z)]$

(Exercises 25 and 26 will be of interest to those readers who are familiar with Russell's theory of descriptions, as elaborated in Whitehead and Russell [1].)

**9. Arguments.** Consider again argument (1) of page 233. On the basis of the scheme of abbreviation

$F^1$ : *a* is a student
$G^1$ : *a* failed
$H^1$ : *a* is illogical
$A^0$ : Alfred ,

(1) has the symbolization

$$\bigvee y \bigwedge x[F(x) \wedge G(x) \leftrightarrow x = y] \quad . \qquad A = \daleth x[F(x) \wedge G(x)] \quad .$$
$$\bigwedge x[G(x) \rightarrow H(x)] \quad \therefore H(A) \quad ,$$

which, as the reader who solves exercise 28 will verify, is valid.

## EXERCISES

Show that the following arguments are valid. Exercise 27 is discussed for illustration.

27. Some problem on Examination I is harder than any other problem on that examination. Students who solve every problem on Examination II also solve every problem on Examination I. Alfred, a student, does not solve the hardest problem on Examination I. For each *x* and *y*, if *x* is harder than *y*, then *y* is not harder than *x*. ∴ There is a problem on Examination II that not every student solves.

The last premise of this argument is essential to its validity. It asserts explicitly the asymmetry of the relation *harder than*, implicit in the ordinary use of a comparative such as 'harder'. Many intuitively plausible arguments containing relation words are invalid; to make them valid one must add premises expressing obvious properties of the relations involved. For example, compare the symbolization of exercise 27 with exercise 38 (section 10), and the symbolization of exercise 32 below with exercise 39.

To derive the conclusion from the premises of No. 27 (or more precisely a symbolization of No. 27), derive first, using the fourth premise,

(1) At most one problem on Examination I is harder than any other problem on that examination.

The first premise and (1), together with T318 and PD, lead to the assertion

(2) The hardest problem on Examination I is a problem on Examination I;

and with (2) at one's disposal the derivation of the conclusion of the argument is straightforward.

28. Argument (1) of page 233.
29. Tully is the man who denounced Catiline. Cicero is a man.

Cicero preferred Pompey to Caesar. Everyone who preferred Pompey to Caesar denounced Catiline. There is exactly one man who denounced Catiline. ∴ Tully and Cicero are identical.

30. There is exactly one man whom Mary loves. There is exactly one man who loves Alice. The man whom Mary loves is the man whom Alice loves. Alfred is the man whom Alice loves. Mary loves the man who loves Alice. ∴ The man who loves Alice is Alfred.

31. 2 is the even prime. The positive square root of 4 is an even prime. There is exactly one even prime. ∴ 2 equals the positive square root of 4.

32. There is exactly one sophist who teaches without remuneration, and he is Socrates. Socrates argues better than any other sophist. Plato argues better than some sophist who teaches without remuneration. For each $x$ and $y$, if $x$ argues better than $y$, then $y$ does not argue better than $x$. ∴ Plato is not a sophist.

33. Any fish is faster than any smaller one. For each $x$ and $y$, if $x$ is faster than $y$, then $y$ is not faster than $x$. For each $x$ and $y$, if either $x$ is smaller than $y$ or $x$ is the same size as $y$, then $x$ is not larger than $y$. For each $x$ and $y$, if $x$ is neither smaller than $y$ nor the same size as $y$, then $y$ is smaller than $x$. ∴ If there is a largest fish, then the largest fish is the fastest fish.

**10. *Invalidity.** The notions of an *English translation* of a symbolic argument and a *closure axiom* are to be characterized as before (chapter IV, section 10), but the notions of a *relativization* of a sentence and an argument must be modified as follows.

If $\phi$ is a symbolic sentence and $\pi$ a 1-place predicate letter not occurring in $\phi$, then by the *relativization of* $\phi$ *to* $\pi$ we now understand that sentence which is obtained from $\phi$ by replacing each part of the form

$$\Lambda\alpha\psi \ ,$$
$$V\alpha\psi \ ,$$

or

$$\urcorner\alpha\psi \ ,$$

where $\alpha$ is a variable and $\psi$ a symbolic formula, by

$$\Lambda\alpha(\pi\alpha \rightarrow \psi) \ ,$$
$$V\alpha(\pi\alpha \wedge \psi) \ ,$$

or

$$\urcorner\alpha\,(\pi\alpha \wedge \psi)$$

respectively. For example, the relativization of the formula

$$\Lambda x F[\urcorner y G(xy)]$$

to the predicate '$H^1$' is

$$\Lambda x[H(x) \rightarrow F(\text{℩} y[H(y) \wedge G(xy)])] \quad .$$

Let $A$ be a symbolic argument whose premises and conclusion are sentences, let $\pi$ be a 1-place predicate occurring in none of these, and let $\delta_1, \ldots, \delta_n$ be all the operation letters occurring in $A$. Then a *relativization of A* to $\pi$ is an argument obtained from $A$ by replacing its premises and conclusion by their relativizations to $\pi$ and adjoining, as additional premises, the closure axioms for $\delta_1, \ldots, \delta_n$ with respect to $\pi$, together with the sentence

$$\pi \text{℩} x \sim x = x \quad .$$

Consider, for example, the argument

(1) $\Lambda x[F(x) \rightarrow G(x)] \quad \therefore G(\text{℩} x[F(x) \wedge H(A[x])]) \quad .$

A relativization of (1) to 'J[1]' is the argument

(2) $\Lambda x(J(x) \rightarrow [F(x) \rightarrow G(x)]) \quad . \quad \Lambda x[J(x) \rightarrow J(A[x])] \quad .$
$J(\text{℩} x \sim x = x) \quad \therefore G(\text{℩} x[J(x) \wedge F(x) \wedge H(A[x])]) \quad .$

If $A$, $B$ are symbolic arguments whose premises and conclusion are sentences, then $A$ is said to be a *relativization* of $B$ if there is a 1-place predicate letter $\pi$ not occurring in $B$ such that $A$ is a relativization of $B$ to $\pi$.

A false translation of a closure of a symbolic argument is still sufficient to establish invalidity; but as in the case of the identity calculus there are many invalid symbolic arguments to which this approach will not apply.

It happens, however, that a symbolic argument is invalid in the description calculus just in case a relativization of one of its closures has a false English translation. (This assertion would not hold in the absence of the rule ID.) In particular, it follows that a symbolic formula $\phi$ is not a theorem just in case a relativization of a closure of the argument having $\phi$ as conclusion and no premises has a false translation.

The invalidity of (1) is established by the following false translation of (2):

(3) Every integer greater than 2 is greater than 1. The successor of an integer is an integer. The object $x$ such that $x$ differs from $x$ is an integer. $\therefore$ The object $x$ such that $x$ is an integer greater than 2 whose successor is an integer is greater than 1.

To verify that (3) is a false English argument it is sufficient to recognize that both definite descriptions occurring in it are improper and to recall that improper definite descriptions designate the number 0.

## EXERCISES

Show of each of the following arguments that it is valid or that it is invalid.

34. $\bigvee y \bigwedge x[F(x) \leftrightarrow x = y]$ . $\iota x F(x) = \iota x G(x)$ . $\bigwedge x[G(x) \to H(x)]$ $\therefore H[\iota x F(x)]$

35. $\sim \iota x F(x) = \iota w \sim w = w$ . $\iota x F(x) = \iota x G(x)$ $\therefore \bigvee y[\bigwedge x(F(x) \leftrightarrow x = y) \wedge G(y)]$

36. $F(\iota w \sim w = w) \vee \sim A = \iota w \sim w = w$ . $A = \iota x F(x)$ $\therefore \bigwedge x(F(x) \leftrightarrow x = A)$

37. $\therefore A(x) = B(x) \to \iota x\, A(x) = y = \iota x\, B(x) = y$

38. $\bigvee x(F(x) \wedge \bigwedge y[F(y) \wedge \sim y = x \to G(xy)])$ . $\bigwedge x(H(x) \wedge \bigwedge y[J(y) \to K(xy)] \to \bigwedge y[F(y) \to K(xy)])$ . $H(A) \wedge \sim K[A\, \iota x(F(x) \wedge \bigwedge y[F(y) \wedge \sim y = x \to G(xy)])]$ $\therefore \bigvee x(J(x) \wedge \sim \bigwedge y[H(x) \to K(yx)])$ (Compare exercise 27, p. 259.)

39. $\bigvee y[\bigwedge x(F(x) \wedge G(x) \leftrightarrow x = y) \wedge y = A]$ . $\bigwedge x[F(x) \wedge \sim x = A \to H(Ax)]$ . $\bigvee x[F(x) \wedge G(x) \wedge H(Bx)]$ $\therefore \sim F(B)$ (Compare exercise 32, p. 260.)

40. $\bigvee x F(xA)$ . $G(B)$ . $\sim H[B\ \iota x F(xA)]$ $\therefore \bigvee x(F(xA) \wedge \bigvee z[G(z) \wedge \sim H(zx)])$

41. $H(A) \wedge A = \iota x \sim \bigvee y[H(y) \wedge G(yx)]$ . $\bigwedge x(\sim F(x) \wedge H(x) \to \bigvee y[G(yx) \wedge H(y)])$ $\therefore F(A)$

On the basis of the scheme of abbreviation

$A^0$ : God
$F^1$ : *a* exists
$G^2$ : *a* is greater than *b*
$H^1$ : *a* can be conceived

exercise 41 has the following translation:

> God can be conceived, and is that than which nothing greater can be conceived. If a nonexistent object can be conceived, then something greater can also be conceived. Therefore God exists.

(This English argument is, essentially, St. Anselm's ontological argument; compare page 4.)

Show that each of the following inference rules, if incorporated into our system of logic, would permit the validation of a symbolic argument that, by methods now available, can be shown invalid. (The restrictions on variables involved in PD and ID are thus seen to be necessary.) In all cases, $\alpha$, $\beta$, $\gamma$ are to be variables, and $\phi$, $\psi$ symbolic formulas. (Exercise 42 is solved for illustration.)

42. $$\frac{\bigvee\beta\bigwedge\alpha(\phi \leftrightarrow \alpha = \beta)}{\psi},$$

where $\psi$ comes from $\phi$ by proper substitution of

$$\rceil\alpha\phi$$

for $\alpha$ (but $\beta$ is permitted to be free in $\phi$).

Argument:

$$\bigwedge x[\mathrm{F}(x) \wedge \mathrm{A}(x) = \mathrm{B} \leftrightarrow x = \mathrm{B}] \quad \therefore \bigvee x[\mathrm{F}(x) \wedge \mathrm{A}(x) = \mathrm{C}]$$

Derivation:

1. ~~Show~~ $\bigvee x[\mathrm{F}(x) \wedge \mathrm{A}(x) = \mathrm{C}]$
2. $\bigvee y\bigwedge x[\mathrm{F}(x) \wedge \mathrm{A}(x) = y \leftrightarrow x = y]$ — Premise, EG
3. ~~Show~~ $\bigwedge y[\mathrm{F}(\rceil x[\mathrm{F}(x) \wedge \mathrm{A}(x) = y]) \wedge \mathrm{A}(\rceil x[\mathrm{F}(x) \wedge \mathrm{A}(x) = y]) = y]$
4. $\mathrm{F}(\rceil x[\mathrm{F}(x) \wedge \mathrm{A}(x) = y]) \wedge \mathrm{A}(\rceil x[\mathrm{F}(x) \wedge \mathrm{A}(x) = y]) = y$ — 2, the variant of PD above
5. $\mathrm{F}(\rceil x[\mathrm{F}(x) \wedge \mathrm{A}(x) = \mathrm{C}]) \wedge \mathrm{A}(\rceil x[\mathrm{F}(x) \wedge \mathrm{A}(x) = \mathrm{C}]) = \mathrm{C}$ — 3, UI
6. $\bigvee x[\mathrm{F}(x) \wedge \mathrm{A}(x) = \mathrm{C}]$ — 5, EG

Translation:

> Zero is the only integer whose square is zero. Therefore 3 is the square of some integer.

Clearly the translation is a false argument, and hence the symbolic argument is invalid.

43. $$\frac{\sim\bigvee\beta\bigwedge\alpha(\phi \leftrightarrow \alpha = \beta)}{\rceil\alpha\phi = \rceil\gamma \sim \gamma = \gamma}$$

(Here $\beta$ is permitted to be free in $\phi$.)

44. $$\frac{\bigvee\beta\bigwedge\alpha(\phi \leftrightarrow \alpha = \beta)}{\psi},$$

where $\beta$ is not free in $\phi$, and $\psi$ is like $\phi$ except for having free occurrences of

$$\rceil\alpha\phi$$

at some (but not all) places where $\phi$ has free occurrences of $\alpha$.

45. $$\frac{\bigvee\beta\bigwedge\alpha(\phi \leftrightarrow \alpha = \beta)}{\psi},$$

where $\beta$ is not free in $\phi$, and $\psi$ is like $\phi$ except for having occurrences of

$$\mathsf{7}\alpha\phi$$

(free or bound) wherever $\phi$ has free occurrences of $\alpha$.

46. Find a symbolic argument that, by methods now available, can be shown invalid but that could be validated if we were to incorporate into our present system of logic Euclid's law as formulated on page 224.

**11. *Historical remarks.** The earliest theories of descriptions are those of Frege (summarized in Carnap [2]) and Russell [1]. In both treatments, descriptive terms were regarded as defined expressions. Frege's theory has the disadvantage of requiring, in any language to which it is applied, the presence of at least one name that is not a descriptive term. Russell's theory has more serious drawbacks. In the first place, the rules of the quantifier calculus must be significantly curtailed when applied to formulas containing descriptive terms. In addition, a descriptive term must always be accompanied by a *scope* indicator, and formulas differing only in the scope of their descriptive terms will not always be equivalent. (See in this context exercises 25 and 26, p. 258.) In Rosser [1] the descriptive operator, '7', is treated as a primitive symbol; Rosser's system, however, is not *complete* in the sense that a symbolic argument is valid if none of its closures has a relativization with a false English translation. The present treatment is like Rosser's in that '7' is taken as primitive, but in other respects resembles Frege's theory. (T416 corresponds to Frege's contextual definition for descriptive terms.)

The completeness of the calculus of this chapter is proved in Montague and Kalish [1].

The principle of interchange, in its full generality, is justified in Quine [2].

**12. Appendix: summary of the system of logic developed in chapters I-VII.**

## INFERENCE RULES

(Here $\alpha$, $\beta$, $\gamma$ are to be variables, $\zeta$, $\zeta'$, $\eta$, $\eta'$, $\theta$ symbolic terms, and $\phi$, $\phi'$, $\psi$, $\chi$ symbolic formulas.)

PRIMITIVE SENTENTIAL RULES:

$$\begin{array}{l} \phi \rightarrow \psi \\ \phi \\ \hline \psi \end{array}$$

*Modus ponens* (MP)

$$\begin{array}{c} \phi \to \psi \\ \sim\psi \\ \hline \sim\phi \end{array}$$

*Modus tollens* (MT)

$$\begin{array}{c} \sim\sim\phi \\ \hline \phi \end{array} \qquad \begin{array}{c} \phi \\ \hline \sim\sim\phi \end{array}$$

Double negation (DN)

$$\begin{array}{c} \phi \\ \hline \phi \end{array}$$

Repetition (R)

$$\begin{array}{c} \phi \wedge \psi \\ \hline \phi \end{array} \qquad \begin{array}{c} \phi \wedge \psi \\ \hline \psi \end{array}$$

Simplification (S)

$$\begin{array}{c} \phi \\ \psi \\ \hline \phi \wedge \psi \end{array}$$

Adjunction (Adj)

$$\begin{array}{c} \phi \\ \hline \phi \vee \psi \end{array} \qquad \begin{array}{c} \psi \\ \hline \phi \vee \psi \end{array}$$

Addition (Add)

$$\begin{array}{c} \phi \vee \psi \\ \sim\phi \\ \hline \psi \end{array} \qquad \begin{array}{c} \phi \vee \psi \\ \sim\psi \\ \hline \phi \end{array}$$

*Modus tollendo ponens* (MTP)

$$\begin{array}{c} \phi \leftrightarrow \psi \\ \hline \phi \to \psi \end{array} \qquad \begin{array}{c} \phi \leftrightarrow \psi \\ \hline \psi \to \phi \end{array}$$

Biconditional-conditional (BC)

$$\begin{array}{c} \phi \to \psi \\ \psi \to \phi \\ \hline \phi \leftrightarrow \psi \end{array}$$

Conditional-biconditional (CB)

DERIVED SENTENTIAL RULES:

$$\begin{array}{c} \phi \to \psi \\ \sim\phi \to \psi \\ \hline \psi \end{array} \qquad \begin{array}{c} \phi \vee \psi \\ \phi \to \chi \\ \psi \to \chi \\ \hline \chi \end{array} \qquad \begin{array}{c} \phi \to \chi \\ \psi \to \chi \\ \hline \phi \vee \psi \to \chi \end{array}$$

Separation of cases (SC)

$$\begin{array}{c} \sim\phi \to \psi \\ \hline \phi \vee \psi \end{array}$$

Conditional-disjunction (CD)

PRIMITIVE QUANTIFICATIONAL RULES:

$$\begin{array}{c} \bigwedge\alpha\phi \\ \hline \psi \end{array},$$

Universal instantiation (UI)

$$\begin{array}{c} \psi \\ \hline \bigvee\alpha\phi \end{array},$$

Existential generalization (EG)

where $\psi$ comes from $\phi$ by proper substitution of a term for $\alpha$;

$$\frac{\bigvee\alpha\phi}{\psi}\ ,$$ Existential instantiation (EI)

where $\psi$ comes from $\phi$ by proper substitution of a variable for $\alpha$. (See page 238 for a definition of 'proper substitution'.)

DERIVED QUANTIFICATIONAL RULES:

$$\frac{\sim\bigwedge\alpha\phi}{\bigvee\alpha\sim\phi}\qquad\frac{\bigvee\alpha\sim\phi}{\sim\bigwedge\alpha\phi}$$

$$\frac{\sim\bigvee\alpha\phi}{\bigwedge\alpha\sim\phi}\qquad\frac{\bigwedge\alpha\sim\phi}{\sim\bigvee\alpha\phi}$$

Quantifier negation (QN)

PRIMITIVE RULES OF IDENTITY:

$$\frac{\psi}{\bigwedge\alpha(\alpha = \zeta \rightarrow \phi)}\ ,$$ Identity-1 (Id-1)

$$\frac{\bigwedge\alpha(\alpha = \zeta \rightarrow \phi)}{\psi}\ ,$$ Identity-2 (Id-2)

where $\psi$ comes from $\phi$ by proper substitution of $\zeta$ for $\alpha$, and $\alpha$ does not occur in $\zeta$.

DERIVED RULES OF IDENTITY:

$$\frac{\zeta = \eta}{\eta = \zeta}$$ Symmetry (Sm)

$$\frac{\begin{matrix}\zeta = \eta\\ \eta = \theta\end{matrix}}{\zeta = \theta}\qquad\frac{\begin{matrix}\zeta = \eta\\ \theta = \eta\end{matrix}}{\zeta = \theta}$$

$$\frac{\begin{matrix}\eta = \zeta\\ \eta = \theta\end{matrix}}{\zeta = \theta}\qquad\frac{\begin{matrix}\eta = \zeta\\ \theta = \eta\end{matrix}}{\zeta = \theta}$$

Transitivity (T)

$$\frac{\zeta = \zeta'}{\phi \leftrightarrow \phi'}\ ,$$ Leibniz' law (LL)

where $\phi'$ is like $\phi$ except for having one or more free occurrences of $\zeta'$ where $\phi$ has free occurrences of $\zeta$.

$$\frac{\zeta = \zeta'}{\eta = \eta'}\ ,$$ Euclid's law (EL)

where $\eta'$ is like $\eta$ except for having one or more free occurrences of $\zeta'$ where $\eta$ has free occurrences of $\zeta$.

PRIMITIVE RULES OF DESCRIPTION:

$$\frac{\bigvee\beta\bigwedge\alpha(\phi \leftrightarrow \alpha = \beta)}{\psi}\ , \qquad \text{Proper descriptions (PD)}$$

where $\psi$ comes from $\phi$ by proper substitution of

$$\daleth\alpha\phi$$

for $\alpha$, and $\beta$ is not free in $\phi$.

$$\frac{\sim\bigvee\beta\bigwedge\alpha(\phi \leftrightarrow \alpha = \beta)}{\daleth\alpha\phi = \daleth\gamma \sim\gamma = \gamma}\ , \qquad \text{Improper descriptions (ID)}$$

where $\beta$ is not free in $\phi$.

## INTERCHANGE

One formula follows from another by *interchange* (Int) if and only if the two formulas are respectively the conclusion and the premise of one of the following four patterns of inference (in all cases $\zeta$, $\zeta'$, $\eta$, $\eta'$ are to be symbolic terms, and $\phi$, $\phi'$, $\psi$, $\psi'$ symbolic formulas):

$$\frac{\bigwedge\alpha_1 \ldots \bigwedge\alpha_n\ \zeta = \zeta'}{\phi \leftrightarrow \phi'}\ ,$$

where $\phi'$ is like $\phi$ except for having one or more occurrences of $\zeta'$ where $\phi$ has occurrences of $\zeta$, and $\alpha_1, \ldots, \alpha_n$ are all the variables bound in $\phi$;

$$\frac{\bigwedge\alpha_1 \ldots \bigwedge\alpha_n\ \zeta = \zeta'}{\eta = \eta'}\ ,$$

where $\eta'$ is like $\eta$ except for having one or more occurrences of $\zeta'$ where $\eta$ has occurrences of $\zeta$, and $\alpha_1, \ldots, \alpha_n$ are all the variables bound in $\eta$;

$$\frac{\bigwedge\alpha_1 \ldots \bigwedge\alpha_n(\psi \leftrightarrow \psi')}{\phi \leftrightarrow \phi'}\ ,$$

where $\phi'$ is like $\phi$ except for having one or more occurrences of $\psi'$ where $\phi$ has occurrences of $\psi$, and $\alpha_1, \ldots, \alpha_n$ are all the variables bound in $\phi$;

$$\frac{\bigwedge\alpha_1 \ldots \alpha_n(\psi \leftrightarrow \psi')}{\eta = \eta'}\ ,$$

where $\eta'$ is like $\eta$ except for having one or more occurrences of $\psi'$ where $\eta$ has occurrences of $\psi$, and $\alpha_1, \ldots, \alpha_n$ are all the variables bound in $\eta$.

## DIRECTIONS FOR CONSTRUCTING A DERIVATION FROM A CLASS *K* OF SYMBOLIC FORMULAS

(1) If $\phi$ is any symbolic formula, then

$$Show\ \phi$$

may occur as a line. (Annotation: 'Assertion'.)

(2) Any member of *K* may occur as a line. (Annotation: 'Premise'.)

(3) If $\phi$, $\psi$ are symbolic formulas such that

$$Show\ (\phi \rightarrow \psi)$$

occurs as a line, then $\phi$ may occur as the next line. (Annotation: 'Assumption'.)

(4) If $\phi$ is a symbolic formula such that

$$Show\ \phi$$

occurs as a line, then

$$\sim \phi$$

may occur as the next line; if $\phi$ is a symbolic formula such that

$$Show \sim \phi$$

occurs as a line, then $\phi$ may occur as the next line. (Annotation: 'Assumption'.)

(5a) A symbolic formula may occur as a line if it follows from antecedent lines (see p. 21) by a primitive inference rule other than EI.

(5b) A symbolic formula may occur as a line if it follows from an antecedent line by the inference rule EI, provided that the variable of instantiation (see p. 100) does not occur in any preceding line. (The annotation for (5a) and (5b) should refer to the inference rule employed and the numbers of the antecedent lines involved.)

(6) When the following arrangement of lines has appeared:

$$\begin{array}{l} Show\ \phi \\ \chi_1 \\ \cdot \\ \cdot \\ \cdot \\ \chi_m \quad , \end{array}$$

where none of $\chi_1$ through $\chi_m$ contains uncancelled '*Show*' and either

(i) $\phi$ occurs unboxed among $\chi_1$ through $\chi_m$;

(ii) $\phi$ is of the form

$$(\psi_1 \rightarrow \psi_2)$$

and $\psi_2$ occurs unboxed among $\chi_1$ through $\chi_m$;

(iii) for some formula $\chi$, both $\chi$ and its negation occur unboxed among $\chi_1$ through $\chi_m$; or

(iv) $\phi$ is of the form

$$\wedge\alpha_1 \ldots \wedge\alpha_k \psi \quad ,$$

$\psi$ occurs unboxed among the lines $\chi_1$ through $\chi_m$, and the variables $\alpha_1$ through $\alpha_k$ are not free in lines antecedent to the displayed occurrence of

$$Show\ \phi \quad ,$$

then one may simultaneously cancel the displayed occurrence of '*Show*' and box all subsequent lines.

The remaining clauses are abbreviatory (in the sense of page 58):

(7) If $\phi$ is an instance of an alphabetic variant of a theorem that has already been proved, then $\phi$ may occur as a line. (Annotation: the number of the theorem in question, sometimes together with a diagrammatic indication of the sequence of substitutions involved.) (For the notion of *instance* see chapter VII, pp. 249 – 50.)

(8) A symbolic formula may occur as a line if it is an alphabetic variant of an antecedent line. (Annotation: 'AV' and the number of the antecedent line.) (For the notion of *alphabetic variance* see chapter VII, pp. 248 – 49.)

(9) If $\phi$, $\phi'$ are symbolic formulas such that

$$\phi \leftrightarrow \phi'$$

is an instance of an alphabetic variant of a previously proved theorem, and $\psi$, $\psi'$ are symbolic formulas such that $\psi'$ is like $\psi$ except for having one or more occurrences of $\phi'$ where $\psi$ has occurrences of $\phi$, then

$$\psi \leftrightarrow \psi'$$

may occur as a line. (Annotation: 'IE' ('interchange of equivalents') together with a parenthetical reference to the theorem involved.)

(10) A symbolic formula may occur as a line if it follows from an antecedent line by interchange. (Annotation: 'Int' and the number of the antecedent line.)

(11) A symbolic formula may occur as a line if it follows from antecedent lines by a succession of steps, and each intermediate step can be justified by one of clauses (2), (5a), (7), (8), (9), or (10). (The annotation should determine the omitted succession of steps by indicating, in order of application, the antecedent lines, the premises, the inference rules, and the previously proved theorems employed.) (The use of derived inference rules is comprehended under this clause.)

## DERIVABILITY

A derivation is *complete* just in case every line either is boxed or contains cancelled '*Show*'. A symbolic formula $\phi$ is *derivable* from a class $K$ of symbolic formulas just in case one can construct a complete derivation from $K$ in which

~~*Show*~~ $\phi$

occurs as an unboxed line.

### 13. Appendix: list of theorems of chapter VII.

| | |
|---|---|
| T401 | $\bigwedge x(F(x) \leftrightarrow x = y) \rightarrow \iota x F(x) = y$ |
| T402 | $\bigvee y \bigwedge x(F(x) \leftrightarrow x = y) \rightarrow F(\iota x F(x))$ |
| T403 | $\sim \bigvee y \bigwedge x(F(x) \leftrightarrow x = y) \rightarrow \iota x F(x) = \iota w \sim w = w$ |
| T404 | $T328 \rightarrow \iota x F(x) = \iota y F(y)$ |
| T405 | $T329 \rightarrow [\bigwedge x[F(x) \leftrightarrow G(x)] \rightarrow \iota x F(x) = \iota x G(x)]$ |
| T406 | $\iota x F(x) = \iota y F(y)$ |
| T407 | $\bigwedge x[F(x) \leftrightarrow G(x)] \rightarrow \iota x F(x) = \iota x G(x)$ |
| T408 | $\bigwedge x[F(x) \leftrightarrow x = \iota x F(x)] \leftrightarrow \bigvee y \bigwedge x(F(x) \leftrightarrow x = y)$ |
| T409 | $\bigvee y \bigwedge x(F(x) \leftrightarrow x = y) \wedge F(x) \rightarrow x = \iota x F(x)$ |
| T410 | $\bigvee y \bigwedge x(F(x) \leftrightarrow x = y) \wedge x = \iota x F(x) \rightarrow F(x)$ |
| T411 | $\bigvee y \bigwedge x(F(x) \leftrightarrow x = y) \wedge \bigvee y \bigwedge x(G(x) \leftrightarrow x = y) \rightarrow$ <br> $[\bigwedge x[F(x) \leftrightarrow G(x)] \leftrightarrow \iota x F(x) = \iota x G(x)]$ |
| T412 | $F(\iota x F(x)) \leftrightarrow \bigvee y \bigwedge x(F(x) \leftrightarrow x = y) \vee F(\iota w \sim w = w)$ |
| T413 | $[\sim x = \iota w \sim w = w \vee F(\iota w \sim w = w)] \wedge$ <br> $x = \iota x F(x) \rightarrow F(x)$ |
| T414 | $\bigvee y \bigwedge x(F(x) \leftrightarrow x = y) \rightarrow [\bigwedge x(F(x) \leftrightarrow x = y) \leftrightarrow \iota x F(x) = y]$ |
| T415 | $\sim \bigvee y \bigwedge x(F(x) \leftrightarrow x = y) \wedge \sim \bigvee y \bigwedge x(G(x) \leftrightarrow x = y) \rightarrow$ <br> $\iota x F(x) = \iota x G(x)$ |
| T416 | $G(\iota x F(x)) \leftrightarrow \bigvee y[\bigwedge x(F(x) \leftrightarrow x = y) \wedge G(y)] \vee$ <br> $[\sim \bigvee y \bigwedge x(F(x) \leftrightarrow x = y) \wedge G(\iota w \sim w = w)]$ |
| T417 | $z = \iota x F(x) \leftrightarrow \bigwedge x(F(x) \leftrightarrow x = z) \vee$ <br> $[\sim \bigvee y \bigwedge x(F(x) \leftrightarrow x = y) \wedge z = \iota w \sim w = w]$ |
| T418 | $\iota x\, x = y = y$ |
| T419 | $\iota w\, w = w = \iota w \sim w = w$ |
| T420 | $\iota x F = \iota w \sim w = w$ |

# Chapter VIII
# Definitions; formal theories

**1. The vocabulary of formal languages.** In previous chapters we have considered ordinary English supplemented by various symbols. To avoid excessive complication, we must at this point renounce English and limit ourselves to purely symbolic languages. It is in connection with such languages that the notion of a *definition* can most conveniently be treated.

In symbolic languages, as in earlier developments, we distinguish two kinds of meaningful expressions, *terms* and *formulas*. Reverting to our informal characterization of these expressions, we may say that a term is an expression that becomes a name once its free variables are replaced by names, and a formula is an expression that becomes a sentence once its free variables are replaced by names. Before giving a precise characterization of terms and formulas, we must consider the basic symbols from which symbolic languages will be constructed. One category of such symbols is that of *variables*, which are as before lower-case Latin letters with or without numerical subscripts. It is clear that by the allowance for subscripts there are infinitely many variables. We retain the standard order established earlier, in fact, the order

$$a, b, \ldots, z, a_0, b_0, \ldots, z_0, a_1, \ldots \quad .$$

We may thus speak of the *first variable* (which is '*a*'), the *second variable* (which is '*b*'), and so on.

Variables are the simplest terms. Formulas, together with more complicated terms, are formed with the aid of *constants*, which fall into two classes, *formula-makers* and *term-makers*, according to the kind of expression that they generate. Each constant may be used in combination with a certain number of variables and previously generated terms and formulas to construct a new term or formula. To make this procedure precise, we shall associate with every constant a fixed *degree*, which will be a quadruple $\langle i, m, n, p\rangle$ of nonnegative integers, in which $i$ is either 0 or 1. Here $i$ is 0 or 1 according as the constant in question is a term-maker or a formula-maker, and $m$, $n$, and $p$ are respectively the number of variables, the number of terms, and the number of formulas that the constant demands.

Our *terms* and *formulas* can then be exhaustively characterized as follows.

(*1*) *Every variable is a term.*

(*2*) *If* $\delta$ *is a constant of degree* $\langle 0, m, n, p \rangle$, $\alpha_1, \ldots, \alpha_m$ *are distinct variables,* $\zeta_1, \ldots, \zeta_n$ *are terms, and* $\phi_1, \ldots, \phi_p$ *are formulas, then the expression*

$$\delta\alpha_1 \ldots \alpha_m \, \zeta_1 \ldots \zeta_n \, \phi_1 \ldots \phi_p$$

*is a term.*

(*3*) *If* $\delta$ *is a constant of degree* $\langle 1, m, n, p \rangle$, $\alpha_1, \ldots, \alpha_m$ *are distinct variables,* $\zeta_1, \ldots, \zeta_n$ *are terms, and* $\phi_1, \ldots, \phi_p$ *are formulas, then the expression*

$$\delta\alpha_1 \ldots \alpha_m \, \zeta_1 \ldots \zeta_n \, \phi_1 \ldots \phi_p$$

*is a formula.*

(Each of $m$, $n$, $p$ may take on the value zero, in which case one of the strings $\alpha_1 \ldots \alpha_m$, $\zeta_1 \ldots \zeta_n$, or $\phi_1 \ldots \phi_p$ will disappear. Thus, for example, if $m = p = 0$, the expression

$$\delta\alpha_1 \ldots \alpha_m \, \zeta_1 \ldots \zeta_n \, \phi_1 \ldots \phi_p$$

will become simply

$$\delta\zeta_1 \ldots \zeta_n \quad .)$$

The preceding characterization employs the notion of a *constant of degree* $\langle i, m, n, p \rangle$ and, before becoming completely intelligible, would have to be supplemented by a characterization of this notion. The most appropriate way of providing the required characterization is by giving a list of symbols that are to be regarded as constants, together with the specification of a degree for each. Such a list, sufficient for the purposes of this and the following chapter, is to be found in an appendix to chapter IX. The list has an arbitrary character, stemming from the accidental features of chapter IX and the later sections of the present chapter. If the developments there had been more extensive, the list would have been longer. Indeed, for some purposes it would be convenient to have an infinite list of constants. (Such a list could be given, here as in the case of variables, by a general characterization of the structure of the symbols comprised in it.)

At this point, however, we shall not give a complete list of the constants that will be used in the formal languages of later sections; for many constants will be introduced by definition, and it is in connection with their definitions that their meaning can most conveniently be elucidated. We wish to include among our constants all the special symbols introduced in the preceding chapters: '$\sim$', '$\rightarrow$', '$\wedge$', '$\vee$', '$\leftrightarrow$', '$\bigwedge$', '$\bigvee$', '$=$', '$\daleth$'. These symbols are called *logical constants* and are all formula-makers except for '⅂'. The identity sign '=' is of degree ⟨1, 0, 2, 0⟩; that is, it is a formula-maker that demands two terms to produce a formula. Thus, for example, '= $x\,y$' is a formula. (The reader will note that the order of symbols, both here and in the case of the sentential connectives other than '∼', differs from that with which he has become familiar.) The negation sign '∼' is of degree ⟨1, 0, 0, 1⟩, and '→', '∧', '∨', and '↔' are of degree ⟨1, 0, 0, 2⟩; that is, the negation sign is a formula-maker requiring one formula to produce a new formula, and the other sentential connectives are formula-makers demanding two formulas to produce a new formula. Thus, for example, '∼ = $x\,y$' is a formula (the *negation* of ' = $x\,y$') and '∧ = $x\,y$ = $y\,z$' is a formula (the *conjunction* of ' = $x\,y$' and '= $y\,z$'). The quantifiers are symbols of degree ⟨1, 1, 0, 1⟩; that is, they are formula-makers that demand one variable and one formula to produce a new formula. Thus, for example, '⋀$x$ = $x\,x$' is a formula. The descriptive operator '⅂' is a symbol of degree ⟨0, 1, 0, 1⟩; that is, it is a term-maker that demands one variable and one formula to produce a new term. Thus, for example, '⅂$x$ = $x\,x$' is a term. Constants other than '∼', '→', '∧', '∨', '↔', '⋀', '⋁', '=', and '⅂' are called *nonlogical*.

We wish also to include among our constants the *operation letters* and *predicate letters* of previous chapters, that is, the symbols

$$A^0, \ldots, E^0, A^1, \ldots, E^1, \ldots, F^0, \ldots, Z^0, F^1, \ldots, Z^1, \ldots ,$$

together with their subscripted variants. Each $n$-place operation letter is of degree ⟨0, 0, $n$, 0⟩, and each $n$-place predicate letter of degree ⟨1, 0, $n$, 0⟩.

As additional examples of nonlogical constants we list the following:

| | |
|---|---|
| + | (degree: ⟨0, 0, 2, 0⟩) |
| $\sqrt[3]{}$ | (degree: ⟨0, 0, 1, 0⟩) |
| ϵ | (degree: ⟨1, 0, 2, 0⟩) |
| I | (degree: ⟨1, 0, 1, 0⟩) |
| $\overset{1}{\forall}$ | (degree: ⟨1, 1, 0, 1⟩) |
| E | (degree: ⟨0, 1, 0, 1⟩) |
| lim | (degree: ⟨0, 1, 1, 0⟩) |

'+' and '$\sqrt[3]{}$' are to be understood in their familiar mathematical senses, as signs for addition and cube root, and

$$\epsilon\, x\, y \quad ,$$
$$\mathrm{I}\, x \quad ,$$
$$\overset{1}{\forall} x\, F^1 x \quad ,$$
$$\mathrm{E} x\, F^1 x \quad ,$$
$$\lim\ n\ A^1 n$$

are to be read respectively

$x$ is a member of the set $y$ ,
$x$ is an integer ,
there is exactly one $x$ such that $F^1x$ ,
the set of objects $x$ such that $F^1x$ ,
the limit of $A^1n$ as $n$ approaches infinity .

It will be convenient to provide a terminology for certain kinds of constants. Those constants which must be followed immediately by one or more variables (that is, which have degree $\langle i, m, n, p \rangle$ with $m > 0$) will be called *operators;* these variables will turn out, in the light of section 2, to be *bound*. Among the logical constants, '⋀', '⋁', and '⅂' are the only operators; '$\overset{\downarrow}{V}$', 'E', and 'lim' are also operators. The identity sign is not an operator; for even though it *may* be followed immediately by variables, as in '$= x\ y$', these variables will be free. Further, it is not required that the identity sign be followed immediately by variables, as the formula '$=$ ⅂$x = x\,x\,y$' indicates. Operators will sometimes for emphasis be called *variable-binding operators.*

A constant of degree $\langle 1, 0, n, 0 \rangle$ is called an *n-place predicate;* that is, an $n$-place predicate is a formula-maker requiring exactly $n$ terms to produce a formula. The identity symbol, then, is a 2-place predicate; predicate letters are predicates; '$\epsilon$' and 'I' are predicates. A constant of degree $\langle 0, 0, n, 0 \rangle$ is called an *n-place operation symbol;* that is, an $n$-place operation symbol is a term-maker requiring exactly $n$ terms to produce a new term. Operation letters are operation symbols, and so are '$+$' and '$\sqrt[3]{\ }$'. Finally, we shall call a 0-place operation symbol an *individual constant,* and a constant of degree $\langle 1, 0, 0, p \rangle$ with $p > 0$ a *sentential connective.*

**2. Bondage and freedom; proper substitution; alphabetic variance.** Variables may now be bound not only by quantifiers and '⅂' but by arbitrary operators. For example, if $\delta$ is an operator of degree $\langle 0, 1, 0, 1 \rangle$, then each occurrence of the variable $\alpha$ in the term

$$\delta\,\alpha = \alpha\,\beta$$

is bound. Thus we must give a more general characterization of bondage and freedom than the earlier treatment provides.

An *occurrence* of a *variable* $\alpha$ is said to be *bound in* a term or formula $\phi$ just in case it stands within an occurrence in $\phi$ of some expression

$$\delta\alpha_1 \ldots \alpha_m\, \zeta_1 \ldots \zeta_n\, \phi_1 \ldots \phi_p \quad ,$$

where $\delta$ is a constant of degree $\langle 0, m, n, p \rangle$ or $\langle 1, m, n, p \rangle$, $\alpha_1, \ldots, \alpha_m$ are distinct variables, $\zeta_1, \ldots, \zeta_n$ are terms, $\phi_1, \ldots, \phi_p$ are formulas, and $\alpha$ is one of $\alpha_1, \ldots, \alpha_m$. An *occurrence* of a *variable* is *free in* $\phi$ just in case it stands within $\phi$ but is not bound in $\phi$. A *variable* is *bound* or *free in* $\phi$ according as it has a bound or free occurrence in $\phi$. Thus, in

$$\overset{\downarrow}{V}x\ F^2x\,y \quad ,$$

both occurrences of '$x$' are bound, the only occurrence of '$y$' is free, the variable '$x$' is bound but not free, and the variable '$y$' is free but not bound.

We now consider freedom and bondage of arbitrary terms. If $\phi$ is a formula or term, then an *occurrence* of a *term* $\zeta$ is *bound in* $\phi$ just in case it stands within an occurrence in $\phi$ of some expression

$$\delta\alpha_1 \ldots \alpha_m \zeta_1 \ldots \zeta_n \phi_1 \ldots \phi_p \quad ,$$

where $\delta$ is a constant of degree $\langle 0, m, n, p \rangle$ or $\langle 1, m, n, p \rangle$, $\alpha_1, \ldots, \alpha_m$ are distinct variables, $\zeta_1, \ldots, \zeta_n$ are terms, $\phi_1, \ldots, \phi_p$ are formulas, and at least one of the variables $\alpha_1, \ldots, \alpha_m$ is free in $\zeta$. An *occurrence* of a *term* is said to be *free in* $\phi$ if it stands within $\phi$ but is not bound in $\phi$. A *term* is *bound* or *free in* $\phi$ according as it has a bound or free occurrence in $\phi$. For example, in

(1) $$\Lambda x \; F^1 \, EyG^2xy \quad ,$$

the occurrence of '$EyG^2xy$' is bound, but in

(2) $$\Lambda x \; F^1 \, EyG^2zy \quad ,$$

the occurrence of '$EyG^2zy$' is free. Observe that the characterization of bondage and freedom for the special case of variables and their occurrences is subsumed under the more general characterization just given. Thus in (1) and (2) both '$x$' and '$y$' are bound, and '$z$' is free in (2), no matter whether we apply the definitions pertaining only to variables or the more general definitions pertaining to arbitrary terms.

A *sentence* is as before a formula in which no variable is free, and a *name* is a term in which no variable is free.

*Proper substitution* will be useful in this chapter, and in chapter IX will acquire additional importance in connection with definitions. The following characterization of this notion is essentially identical with the earlier treatment.

If $\alpha$ is a variable, $\zeta$ a term, and $\phi$, $\psi$ formulas, then $\psi$ is said to *come from* $\phi$ *by proper substitution of* $\zeta$ *for* $\alpha$ just in case $\psi$ is like $\phi$ except for containing free occurrences of $\zeta$ wherever $\phi$ contains free occurrences of $\alpha$. For example, if $\phi$ is the formula

$$\Lambda x \in x \, y \quad ,$$

then

$$\Lambda x \in x \; EzG^2wz$$

comes from $\phi$ by proper substitution of '$EzG^2wz$' for '$y$'.

If $\delta$ is an $n$-place operation symbol or predicate, $\chi$ is correspondingly a term or a formula, and $\phi$, $\psi$ are formulas, then $\psi$ is said to *come from* $\phi$ by *proper substitution of* $\chi$ *for* $\delta$ just in case $\psi$ can be obtained from $\phi$ by

(I) replacing, throughout $\phi$, the constant $\delta$ by

$$\{\chi\} \quad ,$$

and,

(II) in the expression resulting from (I), successively replacing each part of the form

$$\{\chi\}\, \zeta_1 \ldots \zeta_n \quad ,$$

where $\zeta_1, \ldots, \zeta_n$ are terms, by the expression obtained from $\chi$ by replacing all free occurrences of '*a*' by $\zeta_1$, '*b*' by $\zeta_2$, etc., up to the *n*th variable, whose free occurrences are to be replaced by $\zeta_n$;

we require in addition that $\chi$ and $\phi$ have no variables in common. For example, consider the formula

$$\text{(3)} \qquad \Lambda x \; G^2 xy \quad .$$

The formula

$$\Lambda x = Ez {=} yz \; x$$

comes from (3) by proper substitution of

$$\text{(4)} \qquad = Ez {=} bz \; a$$

for the 2-place predicate '$G^2$'. In performing the substitution we obtain in the first step

$$\Lambda x \, \{ = Ez {=} bz \; a \} \, xy \quad ,$$

and in the second step

$$\Lambda x = Ez {=} yz \;\; x \quad ;$$

we observe, moreover, that (3) and (4) have no variables in common.

If *K* is a class consisting of variables, operation symbols, and predicates, then a formula $\psi$ is said to *come from* a formula $\phi$ *by iterated proper substitution on members of K* just in case $\psi$ can be obtained from $\phi$ by zero or more applications of proper substitution, each of which is either the substitution of (1) a term for a variable belonging to *K*, (2) a term for an operation symbol belonging to *K*, or (3) a formula for a predicate belonging to *K*. A formula $\psi$ is an *instance* of a formula $\phi$ if and only if $\psi$ comes from $\phi$ by iterated proper substitution on members of some class consisting of variables, operation symbols, and predicates.

The notion of alphabetic variance must also be extended. Let $\delta$ be an operator of degree $\langle 0, m, n, p \rangle$ or $\langle 1, m, n, p \rangle$, let $i \leq m$, and let $\phi$, $\psi$ be expressions of the respective forms

$$\delta \alpha_1 \ldots \alpha_m \, \phi_1 \ldots \phi_n \, \phi_{n+1} \ldots \phi_{n+p} \quad ,$$
$$\delta \beta_1 \ldots \beta_m \, \psi_1 \ldots \psi_n \, \psi_{n+1} \ldots \psi_{n+p} \quad ,$$

where $\alpha_1, \ldots, \alpha_m$ are distinct variables, $\beta_1, \ldots, \beta_m$ are also distinct variables which, except for $\beta_i$, are identical with $\alpha_1, \ldots, \alpha_m$ respectively, $\phi_1, \ldots, \phi_n, \psi_1, \ldots, \psi_n$ are terms, and $\phi_{n+1}, \ldots, \phi_{n+p}, \psi_{n+1}, \ldots, \psi_{n+p}$ are formulas. (Thus $\phi$, $\psi$ are both either terms or formulas.) We say that $\phi$, $\psi$ are *immediate alphabetic variants* if, for each $j \leqslant n+p$, $\psi_j$ comes from $\phi_j$ by proper substitution of $\beta_i$ for $\alpha_i$ and, conversely, $\phi_j$ comes from $\psi_j$ by proper substitution of $\alpha_i$ for $\beta_i$.

This definition is the natural extension to arbitrary operators of the notion of immediate alphabetic variance considered in chapters IV, VI, and VII.

More generally, a term or formula $\psi$ is said to be an *alphabetic variant* of a term or formula $\phi$ if there is a finite sequence of expressions beginning with $\phi$ and ending with $\psi$, and such that each expression of the sequence (except the first) is obtained from its predecessor by replacing an occurrence of a term or formula by an immediate alphabetic variant of that term or formula.

For example,

$$= \mathrm{E}x\mathrm{F}^1x\urcorner y\Lambda x\mathrm{G}^2xy$$

is an alphabetic variant of

$$= \mathrm{E}y\mathrm{F}^1y\urcorner x\Lambda z\mathrm{G}^2zx$$

by virtue of the sequence

$$\begin{aligned} &= \mathrm{E}y\mathrm{F}^1y\urcorner x\Lambda z\mathrm{G}^2zx \quad , \\ &= \mathrm{E}x\mathrm{F}^1x\urcorner x\Lambda z\mathrm{G}^2zx \quad , \\ &= \mathrm{E}x\mathrm{F}^1x\urcorner y\Lambda z\mathrm{G}^2zy \quad , \\ &= \mathrm{E}x\mathrm{F}^1x\urcorner y\Lambda x\mathrm{G}^2xy \quad . \end{aligned}$$

**3. Informal notational conventions.** The official characterization in section 1 of *term* and *formula* introduces a notation that, though highly convenient for the formulation of the general definitions of the last two sections, leads to virtual unreadability when we consider specific formulas of any length. This remark applies even to the formula (4) of the previous section, which according to a more customary notational style would assume the following perspicuous form:

$$\mathrm{E}z[b = z] = a \quad .$$

A more compelling example can be provided by considering the associative law for the addition of integers. According to the style of section 1 this law has the following awkward appearance:

$$\Lambda x \Lambda y \Lambda z \rightarrow \wedge \wedge \mathrm{I}x\mathrm{I}y\mathrm{I}z = + + x\,y\,z + x + y\,z \quad ,$$

rather than the more customary:

$$\Lambda x \Lambda y \Lambda z\, [\mathrm{I}x \wedge \mathrm{I}y \wedge \mathrm{I}z \rightarrow (x + y) + z = x + (y + z)] \quad .$$

To simplify the appearance of formulas and to bring them into closer accord with customary usage, we shall adopt some informal notational conventions.

Although official notation requires that a constant always stand *before* the variables, terms, and formulas to which it applies, we shall often depart from this order. For example, sentential connectives of degree $\langle 1, 0, 0, 2 \rangle$ will be written between their attendant formulas, as in chapters I – VII, and operation symbols and predicates of some mathematical currency will be transposed to the position that mathematical usage prescribes. Parentheses and brackets, unnecessary in official notation, must now be introduced to avoid ambiguity and, along with other symbols of punctuation, will sometimes also be used purely for perspicuity. The number of parentheses required to avoid ambiguity will be somewhat reduced by adoption of the conventions given in earlier chapters for their omission. For instance, '$\rightarrow$' and '$\leftrightarrow$' will be regarded as marking a greater break than '$\wedge$' and '$\vee$', and in an iterated conjunction or disjunction without parentheses the components are to be understood as associated to the left.

As in earlier chapters, operation letters and predicate letters will be relieved of their superscripts provided no ambiguity ensues.

4. **Derivability.** We must now specify the circumstances under which a formula will be considered *derivable* from a class of formulas. The characterization will follow the lines of the earlier development. Indeed, clauses (1) – (11) of chapter VII (see pp. 268 – 69) are taken as the directions for constructing a *derivation from a class K of formulas*. Of course, we must now understand 'symbolic term' and 'symbolic formula' in the sense of 'term' and 'formula' given in the present chapter. The *primitive inference rules*, to which reference is made in clause (5a), are to be just those of chapter VII (see pp. 264 – 67), and *interchange* (clause (10)) is to be defined as before (see p. 267); throughout the characterizations, however, 'symbolic formula' and 'symbolic term' are to be replaced by 'formula' and 'term' respectively. *Alphabetic variance* (clauses (7), (8), and (9)) is to be understood in the sense of the present chapter.

It should be pointed out that clauses (8) and (10) can no longer be regarded as theoretically dispensable (in the sense of page 58). The passage from

$$\overset{1}{\vee} x \, Fx$$

to

$$\overset{1}{\vee} y \, Fy$$

and the passage from

$$\wedge x (Fx \leftrightarrow Gx)$$

to

$$\text{E}x\text{F}x = \text{E}x\text{G}x$$

are justified by AV and Int respectively; but neither of these inferences can be reduced to applications of clauses (1) – (6). Thus we must regard the directions for constructing an unabbreviated derivation as consisting of clauses (1) – (6), (8), and (10). The remaining clauses will then be theoretically dispensable. (Clause (9), though not reducible to (1) – (6), can be eliminated once (10) is available.) In practice, however, we shall draw no distinction between an abbreviated and an unabbreviated derivation.

As before, a derivation is said to be *complete* if every line either is boxed or contains cancelled '*Show*', and a formula $\phi$ is *derivable* from a class $K$ of formulas if one can construct a complete derivation from $K$ in which

~~*Show*~~ $\phi$

occurs as an unboxed line.

**5. Formal theories; the theory of commutative ordered fields.** A *theory* consists of two things: (1) a class L of *constants*, which may be any class of nonlogical constants, and (2) a class of *axioms*, which may be any class of formulas containing no nonlogical constants beyond those in L.

As an example, we consider the *theory of commutative ordered fields*, which has the following constants:

| | |
|---|---|
| $+$ | (2-place operation symbol) |
| $-$ | (1-place operation symbol) |
| 0 | (individual constant) |
| $\cdot$ | (2-place operation symbol) |
| $^{-1}$ | (1-place operation symbol) |
| 1 | (individual constant) |
| $\leqslant$ | (2-place predicate) |

(In characterizing a theory, it is unnecessary to specify any interpretation. To facilitate comprehension, however, a few remarks concerning what may be regarded as the standard interpretation of this theory will be useful. The domain of discourse, or the class of objects to which the variables of the theory are construed as referring, is to consist either of the rational numbers (that is, numbers expressible as the quotient of two integers) or of the real numbers (that is, the numbers ordinarily considered in elementary algebra; in particular, beyond the rationals, such numbers as $\pi$ and $\sqrt{2}$ are included, but complex numbers, built up with the aid of $\sqrt{-1}$, are not); and the constants listed above are to have their usual mathematical meaning. The symbol '$-$', which is often used in two ways, as both a 1-place operation symbol (as in '$-2$') and a 2-place operation symbol (as in '$3 - 2$'), is here restricted to the first usage; thus we read '$-x$' as 'the negative of $x$'. The symbol '$^{-1}$' occurs as a superscript, as in '$x^{-1}$', which is read 'the reciprocal of $x$', and designates the number $\frac{1}{x}$. The term '$0^{-1}$', to which no meaning is ordinarily assigned, may be interpreted as designating any fixed number; the axioms below place no restriction on the choice of that number.)

The axioms of the theory of commutative ordered fields are the following formulas, *A*1 – *A*15:

| | |
|---|---|
| *A*1 | $x + (y + z) = (x + y) + z$ |
| *A*2 | $x + y = y + x$ |
| *A*3 | $x + 0 = x$ |
| *A*4 | $x + -x = 0$ |
| *A*5 | $x \cdot (y \cdot z) = (x \cdot y) \cdot z$ |
| *A*6 | $x \cdot y = y \cdot x$ |
| *A*7 | $x \cdot 1 = x$ |
| *A*8 | $\sim x = 0 \rightarrow x \cdot x^{-1} = 1$ |
| *A*9 | $x \cdot (y + z) = (x \cdot y) + (x \cdot z)$ |
| *A*10 | $\sim 0 = 1$ |
| *A*11 | $0 \leqslant x \vee 0 \leqslant -x$ |
| *A*12 | $\sim x = 0 \rightarrow \sim 0 \leqslant x \vee \sim 0 \leqslant -x$ |
| *A*13 | $0 \leqslant x \wedge 0 \leqslant y \rightarrow 0 \leqslant x + y$ |
| *A*14 | $0 \leqslant x \wedge 0 \leqslant y \rightarrow 0 \leqslant x \cdot y$ |
| *A*15 | $x \leqslant y \leftrightarrow 0 \leqslant y + -x$ |

Now if $T$ is an arbitrary theory, then a *term, formula*, or *sentence of T* is respectively a term, formula, or sentence that contains no nonlogical constants beyond the constants of $T$, and a *theorem of T* is a formula of $T$ that is derivable from the axioms of $T$. (In clause (7) of the directions for constructing a derivation, which permits the use of previously proved theorems, a *theorem* is, as always, to be understood as a *theorem of logic*—that is, a formula derivable from the empty class of premises—and not as a theorem of a theory. A limited use of previously proved theorems of a theory will be introduced below, after *T*1.)

For example, let us return to the theory of commutative ordered fields and derive a few theorems. Theorems 1 and 2 are *cancellation laws for addition*, and 3 and 4 *cancellation laws for multiplication*. (Theorems of the present theory and its extensions will be numbered *T*1, *T*2, and so on; we italicize 'T' here to prevent confusion with the numbering system of the theorems of chapters I – VII.)

*T*1

| | | |
|---|---|---|
| 1. | ~~Show~~ $x + z = y + z \rightarrow x = y$ | |
| 2. | ~~Show~~ $\wedge x \wedge y \wedge z\ x + (y + z) = (x + y) + z$ | |
| 3. | $x + (y + z) = (x + y) + z$ | *A*1 |
| 4. | ~~Show~~ $\wedge x\ x + -x + 0$ | |
| 5. | $x + -x = 0$ | *A*4 |
| 6. | ~~Show~~ $\wedge x\ x + 0 = x$ | |
| 7. | $x + 0 = x$ | *A*3 |
| 8. | ~~Show~~ $x + z = y + z \rightarrow x = y$ | |
| 9. | $x + z = y + z$ | |
| 10. | $(x + z) + -z = (y + z) + -z$ | 9, EL |
| 11. | $x + (z + -z) = y + (z + -z)$ | 2, UI, UI, UI, 2, UI, UI, UI, 10, T, T |
| 12. | $x + 0 = y + 0$ | 4, UI, 11, LL |
| 13. | $x = y$ | 6, UI, 6, UI, 12, T, T |

The proof of *T*1 is made unduly long by the need for subsidiary derivations of closures of axioms. For example, line 2 is required to obtain

(1) $$(x + z) + -z = x + (z + -z) \quad ,$$

which is in turn required to infer line 11 from line 10. Now (1) comes from *A*1 by iterated substitution on variables but is not itself an axiom. We shall in future dispense with such subsidiary derivations as those beginning with lines 2, 4, 6, and, when showing a formula to be a theorem of a theory *T*, permit as a line of a derivation any formula that comes by iterated substitution on variables from an alphabetic variant of an axiom of *T* or a previously proved theorem of *T*. For annotation we shall simply refer to the axiom or theorem involved.

Using this informal abbreviation, we may simplify as follows the derivation of *T*1:

| | | |
|---|---|---|
| 1. | ~~Show~~ $x + z = y + z \rightarrow x = y$ | |
| 2. | $x + z = y + z$ | |
| 3. | $(x + z) + -z = (y + z) + -z$ | 2, EL |
| 4. | $x + (z + -z) = y + (z + -z)$ | *A*1, *A*1, 3, T, T |
| 5. | $x + 0 = y + 0$ | *A*4, 4, LL |
| 6. | $x = y$ | *A*3, *A*3, 5, T, T |

*T*2 $z + x = z + y \to x = y$

*T*3 $\sim z = 0 \land x \cdot z = y \cdot z \to x = y$

*T*4 $\sim z = 0 \land z \cdot x = z \cdot y \to x = y$

*T*5

| | | |
|---|---|---|
| 1. | ~~*Show*~~ $-0 = 0$ | |
| 2. | $0 + -0 = 0$ | *A*4 |
| 3. | $0 + 0 = 0$ | *A*3 |
| 4. | $-0 = 0$ | 2, 3, T, *T*2, MP |

*T*6

| | | |
|---|---|---|
| 1. | ~~*Show*~~ $x \cdot 0 = 0$ | |
| 2. | $(x \cdot 1) + (x \cdot 0) = x \cdot (1 + 0)$ | *A*9, Sym |
| 3. | $(x \cdot 1) + (x \cdot 0) = x \cdot 1$ | *A*3, 2, LL |
| 4. | $(x \cdot 1) + 0 = x \cdot 1$ | *A*3 |
| 5. | $x \cdot 0 = 0$ | 3, 4, T, *T*2, MP |

*T*7 $x + -y = 0 \leftrightarrow x = y$

*T*8 $x = -y \leftrightarrow y = -x$

*T*9 $x = 0 \leftrightarrow -x = 0$

The next group of theorems, *T*10 – *T*14, are laws of *signs*.

*T*10 $--x = x$

*T*11

| | | |
|---|---|---|
| 1. | ~~*Show*~~ $-(x + y) = -x + -y$ | |
| 2. | $(x + y) + (-x + -y) =$ $(x + -x) + (y + -y)$ | *A*1, *A*2, LL |
| 3. | $(x + y) + (-x + -y) = 0$ | *A*4, 2, LL, *A*3, T |
| 4. | $(x + y) + -(x + y) = 0$ | *A*4 |
| 5. | $-(x + y) = (-x + -y)$ | 4, 3, T, *T*2, MP |

(To obtain line 2 above, several applications of *A*1, *A*2, and LL are required. But here and henceforth we shall for the most part omit repetitions in annotations.)

*T*12 $(-x) \cdot y = -(x \cdot y)$

*T*13 $x \cdot (-y) = -(x \cdot y)$

*T*14

| | | |
|---|---|---|
| 1. | ~~*Show*~~ $(-x) \cdot (-y) = x \cdot y$ | |
| 2. | $(-x) \cdot (-y) = -(x \cdot -(y))$ | *T*12 |
| 3. | $= --(x \cdot y)$ | 2, *T*13, LL |
| 4. | $= x \cdot y$ | 3, *T*10, *T* |

(In the derivation above, the left side of line 2 is imagined to be repeated in the blanks of lines 3 and 4.)

*T*15

| | | |
|---|---|---|
| 1. | ~~Show~~ $\sim x = 0 \rightarrow (-x)^{-1} = -(x^{-1})$ | |
| 2. | $\sim x = 0$ | |
| 3. | $x \cdot (x^{-1}) = 1$ | 2, *A*8, MP |
| 4. | $(-x) \cdot -(x^{-1}) = 1$ | 3, *T*14, T |
| 5. | $\sim -x = 0$ | 2, *T*9, BC, MT |
| 6. | $(-x) \cdot (-x)^{-1} = 1$ | 5, *A*8, MP |
| 7. | $(-x)^{-1} = -(x^{-1})$ | 4, 6, T, 5, Adj, *T*4, MP |

We state next two basic principles of multiplication.

*T*16 $\quad x \cdot y = 0 \leftrightarrow x = 0 \vee y = 0$

*T*17 $\quad \sim x \cdot y = 0 \rightarrow (x \cdot y)^{-1} = x^{-1} \cdot y^{-1}$

We turn now to theorems on order. *T*18 – *T*21 assert that $\leqslant$ is respectively *reflexive*, *antisymmetric*, *transitive*, and *connected*, and hence is a *simple ordering*.

*T*18

| | | |
|---|---|---|
| 1. | ~~Show~~ $x \leqslant x$ | |
| 2. | ~~Show~~ $0 \leqslant 0$ | |
| 3. | $\sim 0 \leqslant 0$ | |
| 4. | $0 \leqslant -0$ | 3, *A*11 |
| 5. | $0 \leqslant 0$ | 4, *T*5, LL |
| 6. | $0 \leqslant x + -x$ | 2, *A*4, LL |
| 7. | $x \leqslant x$ | 6, *A*15 |

In the annotation for lines 4 and 7 above, reference to theorems and inference rules of the sentential calculus has been omitted; this practice will be adopted henceforth.

*T*19

| | | |
|---|---|---|
| 1. | ~~Show~~ $x \leqslant y \wedge y \leqslant x \rightarrow x = y$ | |
| 2. | $x \leqslant y \wedge y \leqslant x$ | |
| 3. | $0 \leqslant y + -x \wedge 0 \leqslant x + -y$ | 2, *A*15 |
| 4. | $0 \leqslant -(x + -y) \wedge 0 \leqslant x + -y$ | 3, *T*10, *T*11, *A*2, LL |
| 5. | $x + -y = 0$ | 4, *A*12 |
| 6. | $x = y$ | 5, *T*7 |

*T*20

| | | |
|---|---|---|
| 1. | ~~Show~~ $x \leqslant y \wedge y \leqslant z \rightarrow x \leqslant z$ | |
| 2. | $x \leqslant y \wedge y \leqslant z$ | |
| 3. | $0 \leqslant y + -x \wedge 0 \leqslant z + -y$ | 2, *A*15 |
| 4. | $0 \leqslant (y + -x) + (z + -y)$ | 3, *A*13 |
| 5. | $0 \leqslant z + -x$ | 4, *A*1, *A*2, *A*3, *A*4, LL |
| 6. | $x \leqslant z$ | 5, *A*15 |

*T*21

1. ~~*Show*~~ $x \leqslant y \vee y \leqslant x$
2. $0 \leqslant y + -x \vee 0 \leqslant -(y + -x)$ — *A*11
3. $0 \leqslant y + -x \vee 0 \leqslant x + -y$ — 2, T10, *T*11, *A*2, LL
4. $x \leqslant y \vee y \leqslant x$ — 3, *A*15

*T*22 asserts that squares are always nonnegative; hence (*T*23) 1 is nonnegative.

*T*22

1. ~~*Show*~~ $0 \leqslant x \cdot x$
2. ~~*Show*~~ $0 \leqslant x \to 0 \leqslant x \cdot x$
3. $0 \leqslant x$
4. $0 \leqslant x \cdot x$ — 3, *A*14
5. ~~*Show*~~ $0 \leqslant -x \to 0 \leqslant x \cdot x$
6. $0 \leqslant -x$
7. $0 \leqslant (-x) \cdot (-x)$ — 6, *A*14
8. $0 \leqslant x \cdot x$ — 7, *T*14, LL
9. $0 \leqslant x \cdot x$ — 2, 5, *A*11

*T*23 $\quad 0 \leqslant 1$

The remaining theorems of this section are familiar laws of *inequality.*

*T*24

1. ~~*Show*~~ $x \leqslant y \to x + z \leqslant y + z$
2. $x \leqslant y$
3. $0 \leqslant y + -x$ — 2, *A*15
4. $(y + z) + -(x + z)$
   $= (y + z) + (-x + -z)$ — *T*11, EL
5. $= y + -x$ — 4, *A*1, *A*2, *A*3, *A*4, LL
6. $0 \leqslant (y + z) + -(x + z)$ — 3, 5, LL
7. $x + z \leqslant y + z$ — 6, *A*15

*T*25 $\quad x + z \leqslant y + z \to x \leqslant y$

*T*26 $\quad x \leqslant y \wedge z \leqslant w \to x + z \leqslant y + w$

*T*27 $\quad \sim y + 1 \leqslant y$

*T*28 $\quad \sim y \leqslant y + -1$

*T*29 $\quad x \leqslant y \to -y \leqslant -x$

*T*30 $\quad -y \leqslant -x \to x \leqslant y$

*T*31 $\quad x \leqslant 0 \leftrightarrow 0 \leqslant -x$

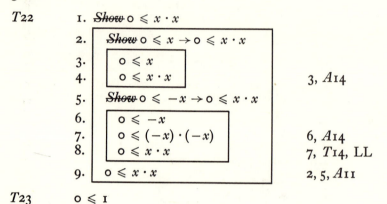

*T*32 $0 \leqslant x \wedge y \leqslant 0 \rightarrow x \cdot y \leqslant 0$

*T*33

1. ~~*Show*~~ $x \leqslant y \wedge 0 \leqslant z \rightarrow x \cdot z \leqslant y \cdot z$
2. $x \leqslant y \wedge 0 \leqslant z$
3. $0 \leqslant y + -x \wedge 0 \leqslant z$ — 2, *A*15
4. $0 \leqslant (y + -x) \cdot z$ — 3, *A*14
5. $0 \leqslant (y \cdot z) + -(x \cdot z)$ — 4, *A*6, *A*9, *T*13, *A*6, LL
6. $x \cdot z \leqslant y \cdot z$ — 5, *A*15

*T*34

1. ~~*Show*~~ $x \leqslant y \wedge z \leqslant 0 \rightarrow y \cdot z \leqslant x \cdot z$
2. $x \leqslant y \wedge z \leqslant 0$
3. $0 \leqslant -z$ — 2, *T*31
4. $x \cdot (-z) \leqslant y \cdot (-z)$ — 2, 3, *T*33
5. $-(x \cdot z) \leqslant -(y \cdot z)$ — 4, *T*13, LL
6. $y \cdot z \leqslant x \cdot z$ — 5, *T*30

*T*35

1. ~~*Show*~~ $0 \leqslant z \wedge \sim z = 0 \rightarrow 0 \leqslant z^{-1}$
2. $0 \leqslant z \wedge \sim z = 0$
3. ~~*Show*~~ $0 \leqslant z^{-1}$
4. $\sim 0 \leqslant z^{-1}$
5. $0 \leqslant -(z^{-1})$ — 4, *A*11
6. $0 \leqslant z \cdot -(z^{-1})$ — 2, 5, *A*14
7. $0 \leqslant -(z \cdot z^{-1})$ — 6, *T*13, LL
8. $0 \leqslant -1$ — 7, 2, *A*8, LL
9. $\sim 0 \leqslant -1$ — *T*23, *A*10, *A*12

*T*36

1. ~~*Show*~~ $x \cdot z \leqslant y \cdot z \wedge 0 \leqslant z \wedge \sim z = 0 \rightarrow x \leqslant y$
2. $x \cdot z \leqslant y \cdot z \wedge 0 \leqslant z \wedge \sim z = 0$
3. $(x \cdot z) \cdot z^{-1} \leqslant (y \cdot z) \cdot z^{-1}$ — 2, *T*35, *T*33
4. $x \leqslant y$ — 3, *A*5, 2, *A*8, *A*7, LL

*T*37 $y \cdot z \leqslant x \cdot z \wedge z \leqslant 0 \wedge \sim z = 0 \rightarrow x \leqslant y$

*T*38 $x = y \rightarrow x \leqslant y$

**6. Extensions of theories; the theory of real numbers.** If $T$ is a theory and $\phi$ an arbitrary formula (whether of $T$ or not), then by an *instance of* $\phi$ *within* $T$ is understood any formula of $T$ that comes from $\phi$ by iterated proper substitution on some class of symbols that contains no constant of $T$.

If $T$ and $U$ are theories, then $U$ is said to be an *extension of* $T$ if all constants of $T$ are constants of $U$ and all axioms of $T$ are axioms of $U$. For example, the theory of commutative ordered fields has as an extension the *theory of real numbers*, which can be characterized as follows: the *constants of the theory of real numbers* are those of the theory of commutative ordered fields together with the 1-place predicate 'I', which is read 'is an integer'. The *axioms of the theory of real numbers* are $A1 - A15$ above, together with

$$A16 \qquad \mathrm{I}(0) \quad ,$$

$$A17 \qquad \mathrm{I}(x) \rightarrow \mathrm{I}(x + 1) \wedge \mathrm{I}(x + -1) \quad ,$$

$$A18 \qquad \mathrm{I}(x) \wedge \mathrm{I}(y) \wedge x \leqslant y \wedge y \leqslant x + 1 \rightarrow y = x \vee y = x + 1 \quad ,$$

and all instances within the present theory of the formula

$$AS19 \qquad \bigvee x\mathrm{F}(x) \wedge \bigvee y \bigwedge x(\mathrm{F}(x) \rightarrow x \leqslant y) \rightarrow \bigvee z[\bigwedge x(\mathrm{F}(x) \rightarrow x \leqslant z) \wedge \bigwedge y(\bigwedge x[\mathrm{F}(x) \rightarrow x \leqslant y] \rightarrow z \leqslant y)] \quad .$$

Axioms 16 through 18 assert that 0 is an integer, that the successor and the predecessor of an integer are again integers, and that between an integer and its successor there is no other integer. Note that the formula called $AS19$ is not a formula of the theory of real numbers because it contains the constant 'F' and hence is not itself an axiom of this theory. It enables us, however, to describe easily an infinity of axioms of the theory, that is, all its instances within the theory. Thus, in view of $AS19$, the following formula is an axiom of the theory of real numbers:

$$(1) \quad \bigvee x\, x \cdot x \leqslant 1 + 1 \wedge \bigvee y \bigwedge x(x \cdot x \leqslant 1 + 1 \rightarrow x \leqslant y) \rightarrow \bigvee z[\bigwedge x(x \cdot x \leqslant 1 + 1 \rightarrow x \leqslant z) \wedge \bigwedge y(\bigwedge x[x \cdot x \leqslant 1 + 1 \rightarrow x \leqslant y] \rightarrow z \leqslant y)]$$

We may refer to $AS19$ itself as an *axiom schema;* this accounts for the designation '$AS19$'.

$AS19$ is often called the *Continuity Schema* and has the following intuitive content: every nonempty set of numbers that is bounded above has a least upper bound. For example, the instance (1) asserts that on the hypothesis that some number is such that its square is at most 2 and there is a number greater than all numbers fulfilling this condition, there is a least upper bound of the set of numbers whose square is at most 2. (Such a least upper bound indeed exists; it is the square root of 2.)

It should be observed that all of $A1 - A18$ are true in the domain of rational numbers as well as in the domain of real numbers. It is the Continuity Schema that distinguishes between these two domains; it has, for example, the consequence that every positive number is a square ($T129$). On the other hand, the positive rational number 2 has no square root among the rationals ($T128$). Accordingly, we must henceforth regard our variables as referring to the real numbers.

Theorems of the theory of real numbers:

*T*39

1. ~~*Show*~~ $\mathrm{I}(y) \wedge 0 \leqslant y \wedge \sim y = 0 \rightarrow 1 \leqslant y$
2. $\mathrm{I}(y) \wedge 0 \leqslant y \wedge \sim y = 0$
3. ~~*Show*~~ $1 \leqslant y$
4. $\sim 1 \leqslant y$
5. $y \leqslant 0 + 1$ — 4, *T*21, *A*3, *A*2, LL
6. $y = 0 + 1$ — *A*18, *A*16, 2, 5
7. $1 \leqslant y$ — *T*18, 6, *A*2, *A*3, LL

Since the theory of real numbers is an extension of the theory of commutative ordered fields, all theorems of the latter are theorems of the former; this accounts for the use of *T*21 and *T*18 in the derivation above.

*T*40 $\mathrm{I}(y) \wedge y \leqslant 0 \wedge \sim y = 0 \rightarrow y \leqslant -1$

It will frequently be convenient to consider theorem schemata as well as axiom schemata. By a *theorem schema* of a theory *T* we understand a formula containing, besides constants of *T*, some additional predicates or operation symbols, and such that all of its instances within *T* are theorems of *T*. For example, *TS*41 – *TS*45 below are theorem schemata of the theory of real numbers.

The derivation of theorem schemata of a theory *T* is simplified by considering a theory *U* that is an extension of *T* and that satisfies the following two conditions: (i) every constant of *U* that is not a constant of *T* is either a predicate or an operation symbol; (ii) every instance within *T* of an axiom of *U* is a theorem of *T*. Then any theorem of *U* that is not a formula of *T* will be a theorem schema of *T*. (This remark requires a small proof, which will not, however, be given here.)

In connection with the theory of real numbers we consider a theory *U* whose constants are those of the theory of real numbers together with the 1-place predicate letter 'F', and whose axioms are *A*1 – *A*18 and all instances within *U* of the formula *AS*19. Thus *AS*19 is itself an axiom of *U*, and so is the formula

$$\bigvee x \mathrm{F}(-x) \wedge \bigvee y \bigwedge x(\mathrm{F}(-x) \rightarrow x \leqslant y) \rightarrow \bigvee z[\bigwedge x(\mathrm{F}(-x) \rightarrow x \leqslant z) \wedge \bigwedge y(\bigwedge x[\mathrm{F}(-x) \rightarrow x \leqslant y] \rightarrow z \leqslant y)]$$

Clearly, *U* is an extension of the theory of real numbers that satisfies conditions (i) and (ii) above.

*TS*41 provides a simple illustration of these points. It is a theorem of *U*, as the proof below indicates, and hence, by the observations above, a theorem schema of the theory of real numbers.

*TS*41

1. ~~*Show*~~ $\bigvee x \mathrm{F}(x) \rightarrow \bigvee x \mathrm{F}(-x)$

2. $\bigvee x F(x)$
3. $Fy$ — 2, EI
4. $F(--y)$ — 3, *T*10, LL
5. $\bigvee x F(-x)$ — 4, EG

*TS*42 is the dual of *AS*19; it asserts, roughly speaking, that every nonempty set of numbers that is bounded below has a greatest lower bound.

*TS*42 1. ~~Show~~ $\bigvee x F(x) \wedge \bigvee y \bigwedge x(F(x) \rightarrow y \leqslant x) \rightarrow \bigvee z[\bigwedge x(F(x) \rightarrow z \leqslant x) \wedge \bigwedge y(\bigwedge x[F(x) \rightarrow y \leqslant x] \rightarrow y \leqslant z)]$

2. $\bigvee x F(x) \wedge \bigvee y \bigwedge x(F(x) \rightarrow y \leqslant x)$
3. $\bigwedge x(F(x) \rightarrow u \leqslant x)$ — 2, EI
4. ~~Show~~ $\bigwedge x(F(-x) \rightarrow x \leqslant -u)$
5. $F(-x) \rightarrow x \leqslant -u$ — 3, UI, *T*10, LL, *T*30
6. $\bigvee y \bigwedge x(F(-x) \rightarrow x \leqslant y)$ — 4, EG
7. $\bigvee z[\bigwedge x(F(-x) \rightarrow x \leqslant z) \wedge \bigwedge y(\bigwedge x[F(-x) \rightarrow x \leqslant y] \rightarrow z \leqslant y)]$ — 2, *TS*41, 6, *AS*19
8. $\bigwedge x(F(-x) \rightarrow x \leqslant w) \wedge \bigwedge y(\bigwedge x[F(-x) \rightarrow x \leqslant y] \rightarrow w \leqslant y)$ — 7, EI
9. ~~Show~~ $\bigwedge x(F(x) \rightarrow -w \leqslant x)$
10. $F(x) \rightarrow -w \leqslant x$ — 8, UI, *T*10, LL, *T*30
11. ~~Show~~ $\bigwedge y(\bigwedge x[F(x) \rightarrow y \leqslant x] \rightarrow y \leqslant -w)$
12. $\bigwedge x[F(x) \rightarrow y \leqslant x]$
13. ~~Show~~ $\bigwedge x(F(-x) \rightarrow x \leqslant -y)$
14. $F(-x) \rightarrow x \leqslant -y$ — 12, UI, *T*10, LL, *T*30
15. $y \leqslant -w$ — 8, UI, 13, *T*10, LL, *T*30
16. $\bigvee z[\bigwedge x(F(x) \rightarrow z \leqslant x) \wedge \bigwedge y(\bigwedge x[F(x) \rightarrow y \leqslant x] \rightarrow y \leqslant z)]$ — 9, 11, EG

In the preceding derivation a new informal abbreviation was introduced. The subsidiary derivation of line 11 has the following form:

~~*Show*~~ $\wedge\alpha(\phi \to \psi)$

$\phi$
.
.
.
$\psi$ ,

rather than the explicit form:

~~*Show*~~ $\wedge\alpha(\phi \to \psi)$

~~*Show*~~ $\phi \to \psi$

$\phi$
.
.
.
$\psi$

This sort of abbreviation corresponds to mathematical practice and will be used frequently in the sequel.

The following theorem schema, whose proof involves a new abbreviation that will be discussed below, makes approximately the following assertion: a greatest lower bound of a class of integers belongs to that class. *TS*44 makes the dual assertion concerning least upper bounds. These assertions, it should be observed, do not hold for arbitrary classes of real numbers. Consider, for example, the class of real numbers greater than 0. It is easily seen that 0 is the greatest lower bound of this class but not a member of it.

*TS*43

1. ~~*Show*~~ $\wedge x[F(x) \to I(x)] \wedge \wedge x(F(x) \to y \leqslant x) \wedge \wedge z(\wedge x[F(x) \to z \leqslant x] \to z \leqslant y) \to F(y)$
2. $\wedge x[F(x) \to I(x)] \wedge \wedge x(F(x) \to y \leqslant x) \wedge \wedge z(\wedge x[F(x) \to z \leqslant x] \to z \leqslant y)$
3. ~~*Show*~~ $F(y)$
4. $\sim F(y)$
5. ~~*Show*~~ $\wedge x(F(x) \to y + 1 \leqslant x)$
6. $F(x) \wedge \sim y + 1 \leqslant x$
7. ~~*Show*~~ $\wedge z(F(z) \to x \leqslant z)$
8. $F(z) \wedge \sim x \leqslant z$
9. $y \leqslant z$ — 2, UI, 8
10. $y + 1 \leqslant z + 1$ — 9, *T*24
11. $x \leqslant y + 1$ — 6, *T*21
12. $x \leqslant z + 1$ — 10, 11, *T*20

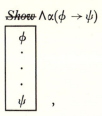

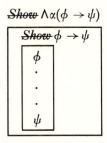

| | | |
|---|---|---|
| 13. | $z \leqslant x$ | 8, $T21$ |
| 14. | $I(x) \wedge I(z)$ | 6, 2, UI, 8, 2, UI |
| 15. | $x = z \vee x = z + 1$ | 14, 13, 12, $A18$ |
| 16. | $\sim x = z$ | 8, $T38$ |
| 17. | $\sim x = z + 1$ | 6, 10, T308 of ch. VI |
| 18. | $x = z + 1$ | 15, 16 |
| 19. | $x \leqslant y$ | 7, 2, AV, UI |
| 20. | $y \leqslant x$ | 6, 2, UI |
| 21. | $x = y$ | 19, 20, $T19$ |
| 22. | $\sim F(x)$ | 4, 21, LL |
| 23. | $F(x)$ | 6 |
| 24. | $y + 1 \leqslant y$ | 5, 2, UI |
| 25. | $\sim y + 1 \leqslant y$ | $T27$ |

In the preceding derivation we permit the following variant of indirect derivation:

~~Show~~ $\wedge\alpha\,(\phi \rightarrow \psi)$

$\phi \wedge \sim \psi$
.
.
.
$\chi$
$\sim \chi$

which, executed in full detail, would read:

~~Show~~ $\wedge\alpha(\phi \rightarrow \psi)$

~~Show~~ $\phi \rightarrow \psi$

$\phi$
~~Show~~ $\psi$

$\sim \psi$
.
.
.
$\chi$
$\sim \chi$

(An analogous variant of indirect derivation, in which the formula derived is not

$$\wedge\alpha(\phi \rightarrow \psi)$$

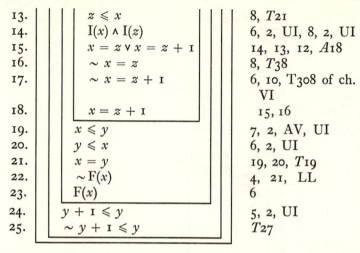

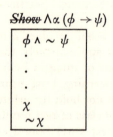

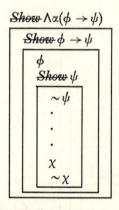

but simply

$$\phi \rightarrow \psi$$

is also occasionally useful and is in fact employed in the derivations of *T*53 and *T*55 below.) The new abbreviation is employed in the subsidiary derivations of lines 5 and 7 above.

*TS*44 $\quad \bigwedge x[F(x) \rightarrow I(x)] \wedge \bigwedge x(F(x) \rightarrow x \leqslant y) \wedge \bigwedge z(\bigwedge x[F(x) \rightarrow x \leqslant z] \rightarrow y \leqslant z) \rightarrow F(y)$

The next two theorem schemata are the natural principles of *mathematical induction* over all integers. *TS*45 is a simple consequence of the Continuity Schema, together with *TS*43 and *TS*44. The stronger principle *TS*49 can be obtained in a simple way from *TS*45; we leave the derivation to the reader. *T*46 – *T*48 are applications of the induction principle *TS*45; the proofs of *T*47 and *T*48 are left to the reader. Henceforth we shall for the most part omit annotative reference to principles of logic, that is, principles of chapters I – VII.

*TS*45

1. ~~Show~~ $F(0) \wedge \bigwedge x[I(x) \wedge F(x) \rightarrow F(x+1) \wedge F(x + -1)] \rightarrow \bigwedge x[I(x) \rightarrow F(x)]$
2. $F(0) \wedge \bigwedge x[I(x) \wedge F(x) \rightarrow F(x+1) \wedge F(x + -1)] \wedge \sim \bigwedge x[I(x) \rightarrow F(x)]$
3. $I(u) \wedge \sim F(u)$ — 2, QN, EI
4. ~~Show~~ $\sim 0 \leqslant u$
5. $0 \leqslant u$
6. ~~Show~~ $\bigwedge x(0 \leqslant x \wedge I(x) \wedge \sim F(x) \rightarrow 0 \leqslant x)$
7. $0 \leqslant x \wedge I(x) \wedge \sim F(x) \rightarrow 0 \leqslant x$
8. $\bigvee x[0 \leqslant x \wedge I(x) \wedge \sim F(x)] \wedge \bigvee y \bigwedge x[0 \leqslant x \wedge I(x) \wedge \sim F(x) \rightarrow y \leqslant x]$ — 3, 5, EG, 6, EG
9. $\bigwedge x(0 \leqslant x \wedge I(x) \wedge \sim F(x) \rightarrow w \leqslant x) \wedge \bigwedge y(\bigwedge x[0 \leqslant x \wedge I(x) \wedge \sim F(x) \rightarrow y \leqslant x] \rightarrow y \leqslant w)$ — 8, *TS*42, EI
10. ~~Show~~ $\bigwedge x[0 \leqslant x \wedge I(x) \wedge \sim F(x) \rightarrow I(x)]$
11. $0 \leqslant x \wedge I(x) \wedge \sim F(x) \rightarrow I(x)$
12. $0 \leqslant w \wedge I(w) \wedge \sim F(w)$ — 10, 9, *TS*43

| | | |
|---|---|---|
| 13. | $\sim 0 = w$ | 2, 12, T308 of ch. VI |
| 14. | $I(w + -1)$ | 12, *A*17 |
| 15. | $0 \leqslant w + -1$ | 12, 13, *T*39, *T*24, *A*4, LL |
| 16. | ~~Show~~ $F(w + -1)$ | |
| 17. | $\sim F(w + -1)$ | |
| 18. | $w \leqslant w + -1$ | 9, UI, 15, 14, 17 |
| 19. | $\sim w \leqslant w + -1$ | *T*28 |
| 20. | $F((w + -1) + 1)$ | 2, UI, 14, 16 |
| 21. | $F(w)$ | 20, *A*1, *A*2, *A*4, *A*3, LL |
| 22. | $\sim F(w)$ | 12 |
| 23. | ~~Show~~ $\sim u \leqslant 0$ | |
| | Derivation similar to that of line 4, using *AS*19 and *TS*44 in place of *TS*42 and *TS*43. | |
| 24. | $0 \leqslant u$ | 23, *T*21 |

*T*46

| | | |
|---|---|---|
| 1. | ~~Show~~ $I(x) \rightarrow I(-x)$ | |
| 2. | $I(-0)$ | *A*16, *T*5 |
| 3. | ~~Show~~ $\bigwedge x[I(x) \wedge I(-x) \rightarrow I(-(x + 1)) \wedge I(-(x + -1))]$ | |
| 4. | $I(x) \wedge I(-x)$ | |
| 5. | $-(x + 1) = -x + -1$ | *T*11 |
| 6. | $I(-(x + 1))$ | 4, *A*17, 5 |
| 7. | $-(x + -1) = -x + 1$ | *T*11, *T*10 |
| 8. | $I(-(x + -1))$ | 4, *A*17, 7 |
| 9. | $I(x) \rightarrow I(-x)$ | 2, 3, *TS*45 |

*T*47 $I(x) \wedge I(y) \rightarrow I(x + y)$

*T*48 $I(x) \wedge I(y) \rightarrow I(x \cdot y)$

*TS*49 $\bigvee x[I(x) \wedge F(x)] \wedge \bigwedge x[I(x) \wedge F(x) \rightarrow F(x + 1) \wedge F(x + -1)] \rightarrow \bigwedge x[I(x) \rightarrow F(x)]$

The inductive proofs of the next two theorems, which are included only for use in proving *T*52, are left to the reader. According to *T*52, an integer is always either odd or even but never both. (Here and henceforth parentheses within an iterated sum or product will be dropped.)

*T*50 $I(x) \rightarrow \bigvee y(I(y) \wedge x = y + y) \vee \bigvee y(I(y) \wedge x = y + y + 1)$

*T*51 $\quad \mathrm{I}(x) \rightarrow \sim\bigvee y(\mathrm{I}(y) \wedge x + x = y + y + 1)$

*T*52 $\quad \mathrm{I}(x) \rightarrow [\bigvee y(\mathrm{I}(y) \wedge x = y + y) \leftrightarrow \sim\bigvee y(\mathrm{I}(y) \wedge x = y + y + 1)]$

We conclude this section on the Continuity Schema with a proof of the *Principle of Archimedes*, which asserts the following: if $y$ is any number, and $x$ any number other than 0, then $y$ can be exceeded by an integral multiple of $x$. This assertion is *T*54; it is convenient to prove first the special case in which $x$ is assumed to be positive.

*T*53

| | | |
|---|---|---|
| 1. | ~~Show~~ $0 \leqslant x \wedge \sim x = 0 \rightarrow \bigvee n(\mathrm{I}(n) \wedge y \leqslant n \cdot x)$ | |
| 2. | $0 \leqslant x \wedge \sim x = 0 \wedge \sim\bigvee n(\mathrm{I}(n) \wedge y \leqslant n \cdot x)$ | |
| 3. | $\mathrm{I}(0) \wedge 0 = 0 \cdot x$ | *A*16, *T*6, *A*6 |
| 4. | $\bigvee z \bigvee n(\mathrm{I}(n) \wedge z = n \cdot x)$ | 3 |
| 5. | ~~Show~~ $\bigwedge z[\bigvee n(\mathrm{I}(n) \wedge z = n \cdot x) \rightarrow z \leqslant y]$ | |
| 6. | $\bigvee n(\mathrm{I}(n) \wedge z = n \cdot x)$ | |
| 7. | $\mathrm{I}(n_0) \wedge z = n_0 \cdot x$ | 6, EI |
| 8. | $z \leqslant y$ | 2, 7, *T*21 |
| 9. | $\bigwedge z[\bigvee n(\mathrm{I}(n) \wedge z = n \cdot x) \rightarrow z \leqslant v_0] \wedge \bigwedge v(\bigwedge z[\bigvee n(\mathrm{I}(n) \wedge z = n \cdot x) \rightarrow z \leqslant v] \rightarrow v_0 \leqslant v)$ | 4, 5, *AS*19, EI |
| 10. | ~~Show~~ $\sim\bigwedge z[\bigvee n(\mathrm{I}(n) \wedge z = n \cdot x) \rightarrow z \leqslant v_0 + -x]$ | |
| 11. | $\bigwedge z[\bigvee n(\mathrm{I}(n) \wedge z = n \cdot x) \rightarrow z \leqslant v_0 + -x]$ | |
| 12. | $v_0 \leqslant v_0 + -x$ | 9 (2nd conjunct), 11 |
| 13. | $v_0 + x \leqslant v_0$ | 12, *T*24, *A*2, *A*3, *A*4 |
| 14. | $v_0 \leqslant v_0 + x$ | 2, *T*24, *A*1, *A*2, *A*3 |
| 15. | $v_0 = v_0 + x$ | 13, 14, *T*19 |
| 16. | $x = 0$ | 15, *A*3, *T*2 |
| 17. | $\sim x = 0$ | 2 |
| 18. | $\bigvee n(\mathrm{I}(n) \wedge z_0 = n \cdot x) \wedge \sim z_0 \leqslant v_0 + -x$ | 10, T242 of ch. III, EI |
| 19. | $\mathrm{I}(n_1) \wedge z_0 = n_1 \cdot x$ | 18, EI |
| 20. | $\bigvee n(\mathrm{I}(n) \wedge (n_1 + 1) \cdot x = n \cdot x)$ | 19, *A*17 |
| 21. | $(n_1 + 1) \cdot x \leqslant v_0$ | 9 (1st conjunct), 20 |
| 22. | $z_0 + x \leqslant v_0$ | 21, *A*6, *A*9, 19, *A*7 |

| | | |
|---|---|---|
| 23. | $z_0 \leqslant v_0 + -x$ | 22, *T*24, *A*1, *A*4, *A*3 |
| 24. | $\sim z_0 \leqslant v_0 + -x$ | 18 |

*T*54

| | | |
|---|---|---|
| 1. | ~~*Show*~~ $\sim x = 0 \rightarrow \bigvee n(\mathrm{I}(n) \wedge y \leqslant n \cdot x)$ | |
| 2. | $\sim x = 0$ | |
| 3. | $0 \leqslant x \rightarrow \bigvee n(\mathrm{I}(n) \wedge y \leqslant n \cdot x)$ | 2, *T*53 |
| 4. | ~~*Show*~~ $x \leqslant 0 \rightarrow \bigvee n(\mathrm{I}(n) \wedge y \leqslant n \cdot x)$ | |
| 5. | $x \leqslant 0$ | |
| 6. | $0 \leqslant -x$ | 5, *T*31 |
| 7. | $\sim -x = 0$ | 2, *T*9 |
| 8. | $\mathrm{I}(n_0) \wedge y \leqslant n_0 \cdot (-x)$ | 6, 7, *T*53, EI |
| 9. | $y \leqslant (-n_0) \cdot x$ | 8, *T*13, *T*12 |
| 10. | $\mathrm{I}(-n_0)$ | 8, *T*46 |
| 11. | $\bigvee n(\mathrm{I}(n) \wedge y \leqslant n \cdot x)$ | 10, 9 |
| 12. | $\bigvee n(\mathrm{I}(n) \wedge y \leqslant n \cdot x)$ | 3, 4, *T*21 |

**7. Definitions.** Of considerable importance, both from the intuitive point of view and for later purposes, is the notion of a *definition* of one constant in terms of others. Loosely speaking, a definition is a formula, more specifically, a biconditional or an identity, which elucidates the meaning of the constant it defines. We shall exclude from consideration, both here and in the next chapter, definitions of logical constants. These symbols will be regarded as completely understood, and accordingly will be used freely in the formulation of definitions. In this chapter we shall further restrict ourselves to definitions of predicates and operation symbols. Definitions of variable-binding operators present special difficulties whose consideration we prefer to postpone till the next chapter.

Thus in the following $\delta$ is assumed to be a nonlogical constant and either a predicate or an operation symbol; that is, $\delta$ is to be either an operation symbol or a predicate other than '='.

If $\delta$ is an $n$-place predicate (other than '=') and $L$ any class of nonlogical constants not containing $\delta$, then a *possible definition of* $\delta$ *in terms of* $L$ is a formula of the form

$$\delta\alpha_1 \ldots \alpha_n \leftrightarrow \phi \quad ,$$

where $\alpha_1, \ldots, \alpha_n$ are distinct variables, and $\phi$ is a formula all of whose nonlogical constants are members of $L$ and which contains no free variables beyond $\alpha_1, \ldots, \alpha_n$.

For example, let $L$ be the class consisting of the symbols '+', '·', and '0'. Then a possible definition of the 2-place predicate '<' in terms of $L$ is the biconditional

$$x < y \leftrightarrow \bigvee z(\sim z = 0 \wedge x + (z \cdot z) = y) \quad .$$

Definitions of *operation symbols* will assume the form of identities rather than biconditionals. Thus, if $\delta$ is an $n$-place operation symbol and $L$ any class of nonlogical constants not containing $\delta$, then a *possible definition of* $\delta$ *in terms of* $L$ is a formula of the form

$$\delta\alpha_1 \ldots \alpha_n = \zeta \quad ,$$

where $\alpha_1, \ldots, \alpha_n$ are distinct variables, and $\zeta$ is a term all of whose nonlogical constants are members of $L$ and which contains no free variables beyond $\alpha_1, \ldots, \alpha_n$.

For example, let $L$ be the class whose only member is the symbol '$\cdot$'. Then a possible definition of the 1-place operation symbol '$\sqrt[3]{}$' in terms of $L$ is the formula

$$\sqrt[3]{x} = \text{℩} y[(y \cdot y) \cdot y = x] \quad .$$

Generally speaking, the right side of a possible definition is regarded as elucidating the meaning of the constant occurring on the left side; hence the requirement that the latter constant not occur on the right side of the definition.

It is often useful to introduce new constants into a theory $T$ by definition. This consists in passing to an extension of $T$ obtained from $T$ by adding possible definitions of the new constants in terms of the constants of $T$. More exactly, we say that a theory $U$ is a *definitional extension* of a theory $T$ just in case (i) $U$ is an extension of $T$, (ii) every constant of $U$ that is not a constant of $T$ is either a predicate or an operation symbol, (iii) for each constant $\delta$ of $U$ that is not a constant of $T$, there occurs among the axioms of $U$ exactly one possible definition of $\delta$ in terms of the constants of $T$, and (iv) every axiom of $U$ that is not an axiom of $T$ is a possible definition, in terms of the constants of $T$, of some nonlogical constant that is not a constant of $T$.

The distinctive properties of definitions stem from the following three facts, which concern definitional extensions and do not hold for arbitrary extensions.

(1) If $\phi$ is a formula of a definitional extension $U$ of a theory $T$, then there is a formula $\psi$ of $T$ such that

$$\phi \leftrightarrow \psi$$

is a theorem of $U$. Thus the power of expression of a theory is not essentially increased by the addition of defined symbols.

(2) If $\phi$ is a theorem of a definitional extension of a theory $T$ and at the same time a formula of $T$, then $\phi$ is already a theorem of $T$. Thus the deductive power of a theory is not essentially enhanced by the addition of definitions.

(3) The assertion (2) has as an immediate consequence the fact that the addition of definitions can never introduce a contradiction into a

theory. Accordingly, let us call a theory $T$ *consistent* if there is no sentence $\phi$ such that both $\phi$ and its negation are theorems of $T$. Then a definitional extension of a consistent theory is always consistent.

We should also call attention to the fact that (4) if $V$ is a definitional extension of $U$ and $U$ is a definitional extension of $T$, then $V$ is equivalent to a definitional extension of $T$. (Two theories are said to be *equivalent* if they have the same constants and the same theorems.)

The theory of real numbers has as a definitional extension the theory $T_1$, characterized as follows. The *constants of* $T_1$ are those of the theory of real numbers, together with:

| | |
|---|---|
| $\neq$ | (2-place predicate) |
| $<$ | (2-place predicate) |
| 2 | (individual constant) |
| $^2$ | (1-place operation symbol) |
| $\mid\;\mid$ | (1-place operation symbol) |
| — | (2-place operation symbol) |
| $-$ | (2-place operation symbol) . |

The *axioms of* $T_1$ are those of the theory of real numbers, together with:

*D*1 $\quad x \neq y \leftrightarrow \sim x = y$

*D*2 $\quad x < y \leftrightarrow x \leqslant y \wedge \sim x = y$

*D*3 $\quad 2 = 1 + 1$

*D*4 $\quad x^2 = x \cdot x$

*D*5 $\quad |x| = \text{℩}z[(0 \leqslant x \wedge z = x) \vee (\sim 0 \leqslant x \wedge z = -x)]$

*D*6 $\quad x - y = x + -y$

*D*7 $\quad \dfrac{x}{y} = x \cdot y^{-1}$

The intended reading of the new constants should be clear from the possible definitions *D*1 – *D*7. For example, the operation symbol introduced by *D*5 is read 'the absolute value of *x*'. *D*7 has as a consequence

$$\frac{x}{0} = x \cdot 0^{-1} .$$

Thus division by zero, like the expression '$0^{-1}$', is regarded as meaningful. Our axioms, however, do not determine the value of '$\frac{x}{0}$' because of the antecedent of *A*8. In deference to mathematical custom, we depart further than usual from our official notation and use '$-$' both as a 1-place and as a 2-place operation symbol. *D*6 defines binary '$-$' in terms of singulary '$-$' and addition. The context will always determine which of the two senses is intended.

We list a number of theorems of $T_1$, proving a few and leaving the others (which can be easily obtained from preceding theorems and definitions) to the reader.

*T*55 1. ~~Show~~ $\mathrm{I}(n) \wedge \mathrm{I}(m) \wedge n < m \rightarrow n + 1 \leqslant m$

| | | |
|---|---|---|
| 2. | $\mathrm{I}(n) \wedge \mathrm{I}(m) \wedge n < m \wedge \sim n + 1 \leqslant m$ | |
| 3. | $n \leqslant m$ | 2, *D*2 |
| 4. | $m \leqslant n + 1$ | 2, *T*21 |
| 5. | $m = n \vee m = n + 1$ | 3, 4, *A*18 |
| 6. | $\sim m = n$ | 2, *D*2 |
| 7. | $n + 1 \leqslant m$ | 5, 6, *T*38 |
| 8. | $\sim n + 1 \leqslant m$ | 2 |

*T*56 $x - x = 0$

*T*57 $-(x - y) = y - x$

*T*58 $x - y = (z - y) - (z - x)$

*T*59 $(x - y) + (z - w) = (x + z) - (y + w)$

*T*60 $(x + y)^2 = x^2 + 2xy + y^2$

In *T*60, as in later statements, we adopt the mathematical practices of omitting the multiplication sign and of regarding addition and subtraction as marking greater breaks than multiplication.

*T*61 $1^2 = 1$

*T*62 $(x - y)^2 = x^2 - 2xy + y^2$

*T*63 $(x \cdot y)^2 = x^2 \cdot y^2$

*T*64 $y \neq 0 \rightarrow \left(\frac{x}{y}\right)^2 = \frac{x^2}{y^2}$

*T*65 $\frac{x}{2} + \frac{x}{2} = x$

*T*66 $\frac{0}{x} = 0$

*T*67 $\frac{x+y}{z} = \frac{x}{z} + \frac{y}{z}$

*T*68 $xz \neq 0 \rightarrow \frac{xy}{xz} = \frac{y}{z}$

| | |
|---|---|
| *T*69 | $y \neq 0 \rightarrow -\left(\frac{x}{y}\right) = \frac{-x}{y} \wedge -\left(\frac{x}{y}\right) = \frac{x}{-y}$ |
| *T*70 | $0 \leqslant x \wedge x \leqslant 1 \rightarrow x^2 \leqslant x$ |
| *T*71 | $0 \leqslant x^2$ |
| *T*72 | $x \neq 0 \rightarrow 0 < x^2$ |
| *T*73 | $x \leqslant x + y^2$ |
| *T*74 | $0 < 1$ |
| *T*75 | $1 < 2$ |
| *T*76 | $\sim x < x$ |
| *T*77 | $x < y \wedge y < z \rightarrow x < z$ |
| *T*78 | $x \leqslant y \wedge y < z \rightarrow x < z$ |
| *T*79 | $x < y \wedge y \leqslant z \rightarrow x < z$ |
| *T*80 | $x \leqslant y \leftrightarrow \sim y < x$ |
| *T*81 | $x \leqslant y \leftrightarrow x < y \vee x = y$ |
| *T*82 | $x < y \leftrightarrow x + z < y + z$ |
| *T*83 | $x < y \wedge z < w \rightarrow x + z < y + w$ |
| *T*84 | $0 < x \rightarrow x < 2x$ |
| *T*85 | $x < y \leftrightarrow -y < -x$ |
| *T*86 | $x < 0 \leftrightarrow 0 < -x$ |
| *T*87 | $0 < x \rightarrow 0 < x^{-1}$ |
| *T*88 | $x < 0 \rightarrow x^{-1} < 0$ |
| *T*89 | $0 < z \rightarrow [x < y \leftrightarrow xz < yz]$ |
| *T*90 | $z < 0 \rightarrow [x < y \leftrightarrow yz < xz]$ |
| *T*91 | $0 < x \wedge 0 < y \rightarrow [x \leqslant y \leftrightarrow y^{-1} \leqslant x^{-1}]$ |
| *T*92 | $x < 0 \wedge y < 0 \rightarrow [x \leqslant y \leftrightarrow y^{-1} \leqslant x^{-1}]$ |
| *T*93 | $0 < x \wedge 0 < y \rightarrow [x < y \leftrightarrow y^{-1} < x^{-1}]$ |
| *T*94 | $x < 0 \wedge y < 0 \rightarrow [x < y \leftrightarrow y^{-1} < x^{-1}]$ |
| *T*95 | $0 < x \wedge x \leqslant y \wedge 0 \leqslant z \rightarrow \frac{z}{y} \leqslant \frac{z}{x}$ |

*T*96 $0 < x \wedge x < y \wedge 0 < z \rightarrow \frac{z}{y} < \frac{z}{x}$

*T*97 $x \leqslant y \wedge 0 < z \rightarrow \frac{x}{z} \leqslant \frac{y}{z}$

*T*98 $x < y \wedge 0 < z \rightarrow \frac{x}{z} < \frac{y}{z}$

*T*99 $0 < x \wedge x < y \rightarrow 0 < \frac{x}{y} \wedge \frac{x}{y} < 1$

*T*100

| | | |
|---|---|---|
| 1. | ~~Show~~ $0 \leqslant y \wedge x^2 \leqslant y^2 \rightarrow x \leqslant y$ | |
| 2. | $0 \leqslant y \wedge x^2 \leqslant y^2$ | |
| 3. | ~~Show~~ $0 < x \wedge 0 < y \rightarrow x \leqslant y$ | |
| 4. | $0 < x \wedge 0 < y \wedge \sim x \leqslant y$ | |
| 5. | $y < x$ | 4, *T*80 |
| 6. | $y \cdot y < y \cdot x$ | 4, 5, *T*89 |
| 7. | $y \cdot x < x \cdot x$ | 4, 5, *T*89 |
| 8. | $y^2 < x^2$ | 6, 7, *T*77, *D*4 |
| 9. | $\sim y^2 < x^2$ | 2, *T*80 |
| 10. | ~~Show~~ $x \leqslant 0 \rightarrow x \leqslant y$ | |
| 11. | $x \leqslant 0$ | |
| 12. | $x \leqslant y$ | 11, 2, *T*20 |
| 13. | ~~Show~~ $y = 0 \rightarrow x < y$ | |
| 14. | $y = 0$ | |
| 15. | $x^2 \leqslant 0$ | 2, 14, *D*4, *T*6 |
| 16. | $x^2 = 0$ | 15, *T*71, *T*19 |
| 17. | $x = 0$ | 16, *D*4, *T*16 |
| 18. | $x \leqslant y$ | 17, 14, *T*18 |
| 19. | $(0 < x \wedge 0 < y) \vee x \leqslant 0 \vee y = 0$ | 2, *T*81, *T*80 |
| 20. | $x \leqslant y$ | 3, 10, 13, 19 |

*T*101 $0 \leqslant x \wedge 0 < y \rightarrow 0 \leqslant \frac{x}{y}$

*T*102 $0 < x \wedge 0 < y \rightarrow 0 < \frac{x}{y}$

*T*103 $\bigvee t \bigwedge z[(0 \leqslant x \wedge z = x) \vee (\sim 0 \leqslant x \wedge z = -x) \leftrightarrow z = t]$

By the last theorem, the descriptive phrase in *D*5 is proper; using this fact, it is easy to obtain the following two theorems.

*T*104 $\quad 0 \leqslant x \rightarrow |x| = x$

*T*105 $\quad x < 0 \rightarrow |x| = -x$

Having obtained *T*104 and *T*105, it will no longer be necessary to refer to *D*5, which has a rather cumbersome form.

*T*106

| | | |
|---|---|---|
| 1. | ~~Show~~ $0 \leqslant |x|$ | |
| 2. | ~~Show~~ $0 \leqslant x \rightarrow 0 \leqslant |x|$ | |
| 3. | $0 \leqslant x$ | |
| 4. | $|x| = x$ | 3, *T*104 |
| 5. | $0 \leqslant |x|$ | 3, 4 |
| 6. | ~~Show~~ $x < 0 \rightarrow 0 \leqslant |x|$ | |
| 7. | $x < 0$ | |
| 8. | $|x| = -x$ | 7, *T*105 |
| 9. | $0 < -x$ | 7, *T*86 |
| 10. | $0 \leqslant |x|$ | 8, 9, *D*2 |
| 11. | $0 \leqslant |x|$ | 2, 6, *T*80 |

*T*107 $\quad |x| = 0 \leftrightarrow x = 0$

*T*108 $\quad x \leqslant |x|$

*T*109 $\quad -x \leqslant |x|$

*T*110 $\quad |-x| = |x|$

*T*111 $\quad |x| \leqslant y \leftrightarrow x \leqslant y \wedge -x \leqslant y$

*T*112

| | | |
|---|---|---|
| 1. | ~~Show~~ $|x + y| \leqslant |x| + |y|$ | |
| 2. | $x \leqslant |x|$ | *T*108 |
| 3. | $y \leqslant |y|$ | *T*108 |
| 4. | $x + y \leqslant |x| + |y|$ | 2, 3, *T*26 |
| 5. | $-x \leqslant |x|$ | *T*109 |
| 6. | $-y \leqslant |y|$ | *T*109 |
| 7. | $-x + -y \leqslant |x| + |y|$ | 5, 6, *T*26 |
| 8. | $-(x + y) \leqslant |x| + |y|$ | 7, *T*11 |
| 9. | $|x + y| \leqslant |x| + |y|$ | 4, 8, *T*111 |

*T*113, which is an immediate consequence of *T*112, is the familiar *triangular inequality*.

*T*113 $\quad |x - y| \leqslant |x - z| + |z - y|$

*T*114 $\quad |x - y| \leqslant |x| + |y|$

*T*115 $\quad |x - y| = |y - x|$

*T*116 $|x \cdot y| = |x| \cdot |y|$

*T*117 $y \neq 0 \rightarrow \left|\dfrac{x}{y}\right| = \dfrac{|x|}{|y|}$

*T*118

| | | |
|---|---|---|
| 1. | ~~Show~~ $\lvert x - y\rvert < z \leftrightarrow x - z < y \wedge y < x + z$ | |
| 2. | ~~Show~~ $x - z < y \wedge y < x + z \rightarrow \lvert x - y\rvert < z$ | |
| 3. | $x - z < y \wedge y < x + z$ | |
| 4. | $-(y - x) < z$ | 3, *T*82, *D*6, *A*2, *A*1, *A*4, *A*3, *T*85, *T*10 |
| 5. | $y - x < z$ | 3, *T*82, *D*6, *A*2, *A*1, *A*4, *A*3 |
| 6. | $\lvert y - x\rvert = y - x \vee \lvert y - x\rvert = -(y - x)$ | *T*104, *T*105, *T*80 |
| 7. | $\lvert x - y\rvert < z$ | 4, 5, 6 |
| 8. | ~~Show~~ $\lvert x - y\rvert < z \rightarrow x - z < y \wedge y < x + z$ | |
| 9. | $\lvert x - y\rvert < z$ | |
| 10. | $x - y < z$ | 9, *T*108, *T*78 |
| 11. | $x - z < y$ | 10, *T*82, *D*6, *A*2, *A*1, *A*4, *A*3 |
| 12. | $y - x < z$ | 9, *T*115, *T*108, *T*78 |
| 13. | $y < x + z$ | 12, *T*82, *D*6, *A*2, *A*1, *A*4, *A*3 |
| 14. | $x - z < y \wedge y < x + z$ | 11, 13 |

We pass now to a definitional extension $T_2$ of $T_1$, obtained by adjoining the 1-place predicates 'N' and 'R' to the constants of $T_1$, and the following possible definitions to the axioms of $T_1$:

*D*8 $\mathrm{N}(x) \leftrightarrow \mathrm{I}(x) \wedge 0 \leqslant x$

*D*9 $\mathrm{R}(x) \leftrightarrow \bigvee y \bigvee z[\mathrm{I}(y) \wedge \mathrm{I}(z) \wedge 0 < z \wedge x = \dfrac{y}{z}]$

*D*8 introduces the notion of a *natural number*, *D*9 that of a *rational number*. By remark (4) on page 296, $T_2$ is equivalent to a definitional extension of the theory of real numbers.

The following theorem schema (of $T_2$) is the *minimum principle* for natural numbers: every nonempty set of natural numbers has a least member.

*TS*119

| | | |
|---|---|---|
| 1. | ~~Show~~ $\bigvee x F(x) \wedge \bigwedge x[F(x) \rightarrow N(x)] \rightarrow \bigvee x(F(x) \wedge \bigwedge y[F(y) \rightarrow x \leqslant y])$ | |
| 2. | $\bigvee x F(x) \wedge \bigwedge x[F(x) \rightarrow N(x)]$ | |
| 3. | ~~Show~~ $\bigwedge x[F(x) \rightarrow 0 \leqslant x]$ | |
| 4. | $F(x)$ | |
| 5. | $0 \leqslant x$ | 2, 4, *D*8 |
| 6. | $\bigwedge x[F(x) \rightarrow z_0 \leqslant x] \wedge \bigwedge y(\bigwedge x[F(x) \rightarrow y \leqslant x] \rightarrow y \leqslant z_0)$ | 2, 3, *TS*42, EI |
| 7. | ~~Show~~ $\bigwedge x[F(x) \rightarrow I(x)]$ | |
| 8. | $F(x)$ | |
| 9. | $I(x)$ | 2, 8, *D*8 |
| 10. | $F(z_0)$ | 7, 6, *TS*43 |
| 11. | $\bigvee x(F(x) \wedge \bigwedge y[F(y) \rightarrow x \leqslant y])$ | 10, 6 (1st conjunct), AV, EG |

We now state two induction principles for natural numbers, both of which are immediate consequences of the minimum principle. *TS*120, whose proof is left to the reader, is the principle of *strong induction;* it asserts that every natural number has the property F, on the assumption that an arbitrary natural number has F if every smaller natural number has F. *TS*121 is the more familiar principle of *weak induction;* if 0 has the property F, and whenever a natural number $x$ has F, so does $x + 1$, then every natural number has F.

*TS*120 $\quad \bigwedge x(N(x) \wedge \bigwedge y[N(y) \wedge y < x \rightarrow F(y)] \rightarrow F(x)) \rightarrow \bigwedge x[N(x) \rightarrow F(x)]$

*TS*121

| | | |
|---|---|---|
| 1. | ~~Show~~ $F(0) \wedge \bigwedge x[N(x) \wedge F(x) \rightarrow F(x+1)] \rightarrow \bigwedge x[N(x) \rightarrow F(x)]$ | |
| 2. | $F(0) \wedge \bigwedge x[N(x) \wedge F(x) \rightarrow F(x+1)] \wedge \sim \bigwedge x[N(x) \rightarrow F(x)]$ | |
| 3. | $\bigvee x[N(x) \wedge \sim F(x)]$ | 2 (3rd conjunct.) |
| 4. | $\bigwedge x[N(x) \wedge \sim F(x) \rightarrow N(x)]$ | |
| 5. | $N(x_0) \wedge \sim F(x_0) \wedge \bigwedge x(N(x) \wedge \sim F(x) \rightarrow x_0 \leqslant x)$ | 3, 4, *TS*119, EI |
| 6. | $x_0 \neq 0$ | 2, 5 |
| 7. | $N(x_0 - 1)$ | 5, *D*8, *A*17, *D*6, 6, *T*39 |
| 8. | $F(x_0 - 1)$ | 5, 7, *T*28, *D*6 |
| 9. | $F(x_0)$ | 2, 7, 8 |
| 10. | $\sim F(x_0)$ | 5 |

In the annotations of lines 7 and 9 above, we have omitted reference to certain arithmetical principles that are by now completely familiar;

this practice will be pursued in the sequel. Also, obvious theorems of logic, such as line 4 above, will sometimes be used even if not listed in chapters I–VII.

The theory of natural numbers was axiomatized for the first time in Peano [1]. Peano's axioms consist of *TS*121, together with *T*122 – *T*125 below. *T*126 expresses the fact that the set of natural numbers is closed under addition and multiplication.

*T*122 $\quad \mathrm{N}(0)$

*T*123 $\quad \mathrm{N}(x) \rightarrow \mathrm{N}(x + 1)$

*T*124 $\quad \mathrm{N}(x) \wedge \mathrm{N}(y) \wedge x + 1 = y + 1 \rightarrow x = y$

*T*125 $\quad \mathrm{N}(x) \rightarrow x + 1 \neq 0$

*T*126 $\quad \mathrm{N}(x) \wedge \mathrm{N}(y) \rightarrow \mathrm{N}(x + y) \wedge \mathrm{N}(x \cdot y)$

The next theorem, which is a lemma for *T*128, asserts that if $x^2$ is an even integer, then so is $x$. *T*128, whose proof is due to Pythagoras and appears in Euclid [1], states that 2 has no rational square root.

*T*127

| | | |
|---|---|---|
| 1. | ~~*Show*~~ $\mathrm{I}(x) \wedge \bigvee y(\mathrm{I}(y) \wedge x^2 = 2y) \rightarrow \bigvee y(\mathrm{I}(y) \wedge x = 2y)$ | |
| 2. | $\mathrm{I}(x) \wedge \bigvee y(\mathrm{I}(y) \wedge x^2 = 2y) \wedge \sim\bigvee y(\mathrm{I}(y) \wedge x = 2y)$ | |
| 3. | $\mathrm{I}(y_0) \wedge x = 2y_0 + 1$ | 2, *T*52 |
| 4. | $x^2 = (2y_0)^2 + 2 \cdot 2 \cdot y_0 + 1$ | 3, *T*60, *T*61 |
| 5. | $= 2 \cdot 2 \cdot y_0^2 + 2 \cdot 2 \cdot y_0 + 1$ | 4, *T*63 |
| 6. | $= 2 \cdot (2 \cdot y_0^2 + 2 \cdot y_0) + 1$ | 5, *A*9 |
| 7. | $\mathrm{I}(2 \cdot y_0^2 + 2 \cdot y_0)$ | 3, *T*48, *A*16, *A*17, *T*47 |
| 8. | $\bigvee y(\mathrm{I}(y) \wedge x^2 = 2 \cdot y + 1)$ | 7, 6, EG |
| 9. | $\sim\bigvee y(\mathrm{I}(y) \wedge x^2 = 2 \cdot y + 1)$ | 2, *T*48, *T*52 |

*T*128

| | | |
|---|---|---|
| 1. | ~~*Show*~~ $\mathrm{R}(x) \rightarrow x^2 \neq 2$ | |
| 2. | $\mathrm{R}(x) \wedge x^2 = 2$ | |
| 3. | $\bigvee z \bigvee y[\mathrm{I}(y) \wedge \mathrm{I}(z) \wedge 0 < z \wedge x = \frac{y}{z}]$ | 2, *D*9 |
| 4. | ~~*Show*~~ $\bigwedge z(\bigvee y[\mathrm{I}(y) \wedge \mathrm{I}(z) \wedge 0 < z \wedge x = \frac{y}{z}] \rightarrow \mathrm{N}(z))$ | |
| | Elementary | |
| 5. | $\bigvee y[\mathrm{I}(y) \wedge \mathrm{I}(z_0) \wedge 0 < z_0 \wedge x = \frac{y}{z_0}] \wedge$ | |
| | $\bigwedge z(\bigvee y[\mathrm{I}(y) \wedge \mathrm{I}(z) \wedge 0 < z \wedge$ | |

| | | |
|---|---|---|
| | $x = \frac{y}{z}] \rightarrow z_0 \leqslant z)$ | 3, 4, *TS*119, EI |
| 6. | $\mathrm{I}(y_0) \wedge \mathrm{I}(z_0) \wedge 0 < z_0 \wedge x = \frac{y_0}{z_0}$ | 5 (1st conjunct), EI |
| 7. | $\frac{{y_0}^2}{{z_0}^2} = 2$ | 6, 2, *T*64 |
| 8. | ${y_0}^2 = 2 \cdot {z_0}^2$ | 7, 6, *T*16 |
| 9. | $\bigvee v(\mathrm{I}(v) \wedge {y_0}^2 = 2 \cdot v)$ | 6, *T*48, 8, EG |
| 10. | $\mathrm{I}(w_0) \wedge y_0 = 2 \cdot w_0$ | 6, 9, *T*127, EI |
| 11. | $2 \cdot 2 \cdot {w_0}^2 = 2 \cdot {z_0}^2$ | 8, 10, *T*63 |
| 12. | $2 \cdot {w_0}^2 = {z_0}^2$ | 11 |
| 13. | $\bigvee n(\mathrm{I}(n) \wedge {z_0}^2 = 2 \cdot n)$ | 10, *T*48, 12, EG |
| 14. | $\mathrm{I}(m_0) \wedge z_0 = 2 \cdot m_0$ | 6, 13, *T*127, EI |
| 15. | $x = \frac{2 \cdot w_0}{2 \cdot m_0}$ | 6, 10, 14 |
| 16. | $x = \frac{w_0}{m_0}$ | 14, 6, 15, *T*68 |
| 17. | ~~*Show*~~ $0 < m_0$ | |
| 18. | $\sim 0 < m_0$ | |
| 19. | $2 \cdot m_0 \leqslant 0$ | 18, *T*80, *T*33 |
| 20. | $z_0 \leqslant 0$ | 14, 19 |
| 21. | $\sim z_0 \leqslant 0$ | 6, *T*80 |
| 22. | $z_0 \leqslant m_0$ | 10, 14, 17, 16, EG, 5 (2nd conjunct) |
| 23. | $m_0 < m_0 + m_0$ | 17 |
| 24. | $m_0 < z_0$ | 23, 14 |
| 25. | $\sim z_0 \leqslant m_0$ | 24, *T*80 |

While many positive numbers resemble 2 in having no square root among the rationals, it is a consequence of the Continuity Schema that every positive number does have a square root:

*T*129

| | | |
|---|---|---|
| 1. | ~~*Show*~~ $0 < x \rightarrow \bigvee y\, x = y^2$ | |
| 2. | $0 < x$ | |
| 3. | $0^2 < x$ | 2 |
| 4. | $\bigvee y\, y^2 \leqslant x$ | 3 |
| 5. | ~~*Show*~~ $\bigwedge y(y^2 \leqslant x \rightarrow y \leqslant 1 \vee y \leqslant x)$ | |
| 6. | $y^2 \leqslant x \wedge \sim(y \leqslant 1 \vee y \leqslant x)$ | |
| 7. | $1 \leqslant y$ | 6 |
| 8. | $y \cdot 1 \leqslant y^2$ | 7 |
| 9. | $y \leqslant x$ | 8, 6 |
| 10. | $\sim y \leqslant x$ | 6 |

| | | |
|---|---|---|
| 11. | ~~Show~~ $1 \leqslant x \rightarrow \bigvee z \bigwedge y(y^2 \leqslant x \rightarrow y \leqslant z)$ | |
| 12. | $1 \leqslant x$ | |
| 13. | ~~Show~~ $\bigwedge y(y^2 \leqslant x \rightarrow y \leqslant x)$ | |
| | Elementary, from 5 and 12 | |
| 14. | $\bigvee z \bigwedge y(y^2 \leqslant x \rightarrow y \leqslant z)$ | 13 |
| 15. | ~~Show~~ $x \leqslant 1 \rightarrow \bigvee z \bigwedge y(y^2 \leqslant x \rightarrow y \leqslant z)$ | |
| | Similar to proof of 11 | |
| 16. | $\bigvee z \bigwedge y(y^2 \leqslant x \rightarrow y \leqslant z)$ | 11, 15, *T*21 |
| 17. | $\bigwedge y(y^2 \leqslant x \rightarrow y \leqslant z_0) \wedge \bigwedge u[\bigwedge y(y^2 \leqslant x \rightarrow y \leqslant u) \rightarrow z_0 \leqslant u]$ | 4, 16, *AS*19, EI |
| 18. | ~~Show~~ $1 \leqslant x \rightarrow 0 < z_0$ | |
| 19. | $1 \leqslant x$ | |
| 20. | $1^2 \leqslant x$ | 19 |
| 21. | $1 \leqslant z_0$ | 17(1st conjunct), 20 |
| 22. | $0 < z_0$ | 21 |
| 23. | ~~Show~~ $x \leqslant 1 \rightarrow 0 < z_0$ | |
| 24. | $x \leqslant 1$ | |
| 25. | $x^2 \leqslant x$ | 24, 2 |
| 26. | $x \leqslant z_0$ | 17(1st conjunct), 25 |
| 27. | $0 < z_0$ | 2, 26 |
| 28. | $0 < z_0$ | 18, 23, *T*21 |
| 29. | ~~Show~~ $x \leqslant z_0^2$ | |
| 30. | $\sim x \leqslant z_0^2$ | |
| 31. | $z_0^2 < x$ | 30 |
| 32. | $\bigvee h\, h = \frac{1}{2} \cdot \left(1 - \frac{z_0^2}{x}\right)$ | T321 of ch. VI |
| 33. | $h_0 = \frac{1}{2} \cdot \left(1 - \frac{z_0^2}{x}\right)$ | 32, EI |
| 34. | ~~Show~~ $0 < h_0 \wedge h_0 < 1$ | |
| 35. | $0 < \frac{z_0^2}{x} \wedge \frac{z_0^2}{x} < 1$ | 28, *T*72, 31, *T*99 |
| 36. | $-1 < -\frac{z_0^2}{x} \wedge -\frac{z_0^2}{x} < 0$ | 35, *T*85 |

| | | |
|---|---|---|
| 37. | $0 < 1 - \frac{z_0^2}{x} \wedge 1 - \frac{z_0^2}{x} < 1$ | 36 |
| 38. | $0 < h_0 \wedge h_0 < \frac{1}{2}$ | 37, 33 |
| 39. | $0 < h_0 \wedge h_0 < 1$ | 38, *T*74, *T*75, *T*96 |
| 40. | ~~*Show*~~ $\left(\frac{z_0}{1 - h_0}\right)^2 \leqslant x$ | |
| 41. | $\frac{z_0^2}{x} = 1 - \left(1 - \frac{z_0^2}{x}\right)$ | |
| 42. | $= 1 - 2h_0$ | 41, 33 |
| 43. | $1 - 2h_0 \leqslant 1 - 2h_0 + h_0^2$ | *T*73 |
| 44. | $1 - 2h_0 \leqslant (1 - h_0)^2$ | 43, *T*62 |
| 45. | $\frac{z_0^2}{x} \leqslant (1 - h_0)^2$ | 42, 44 |
| 46. | $z_0^2 \leqslant x \cdot (1 - h_0)^2$ | 45, 2 |
| 47. | $0 < 1 - h_0$ | 34 |
| 48. | $0 < (1 - h_0)^2$ | 47, *T*72 |
| 49. | $\frac{z_0^2}{(1 - h_0)^2} \leqslant x$ | 46, 48, *T*97 |
| 50. | $\left(\frac{z_0}{1 - h_0}\right)^2 \leqslant x$ | 49, 47, *T*64 |
| 51. | $\frac{z_0}{1 - h_0} \leqslant z_0$ | 17(1st conjunct), 40 |
| 52. | $-h_0 < 0$ | 34 |
| 53. | $1 - h_0 < 1$ | 52 |
| 54. | $0 < 1 - h_0$ | 34 |
| 55. | $z_0 < \frac{z_0}{1 - h_0}$ | 54, 53, 28, *T*96 |
| 56. | $\sim z_0 < \frac{z_0}{1 - h_0}$ | 51 |
| 57. | ~~*Show*~~ $z_0^2 \leqslant x$ | |
| 58. | $\sim z_0^2 \leqslant x$ | |
| 59. | $x < z_0^2$ | 58 |
| 60. | $h_1 = \frac{1}{2} \cdot \left(1 - \frac{x}{z_0^2}\right)$ | T321 of ch. VI, EI |

| | | |
|---|---|---|
| 61. | ~~Show~~ $0 < h_1 \wedge h_1 < 1$ | |
| 62. | $0 < \frac{x}{z_0^2} \wedge \frac{x}{z_0^2} < 1$ | 2, 59, *T*99 |
| 63. | $-1 < -\frac{x}{z_0^2} \wedge -\frac{x}{z_0^2} < 0$ | 62, *T*85 |
| 64. | $0 < 1 - \frac{x}{z_0^2} \wedge 1 - \frac{x}{z_0^2} < 1$ | 63 |
| 65. | $0 < h_1 \wedge h_1 < \frac{1}{2}$ | 64, 60 |
| 66. | $0 < h_1 \wedge h_1 < 1$ | 65 |
| 67. | ~~Show~~ $\bigwedge y(y^2 \leqslant x \rightarrow y \leqslant z_0 \cdot (1 - h_1))$ | |
| 68. | $y^2 \leqslant x$ | |
| 69. | $x = z_0^2 \cdot \frac{x}{z_0^2}$ | 28, *T*72 |
| 70. | $= z_0^2 \cdot \left(1 - \left(1 - \frac{x}{z_0^2}\right)\right)$ | 69 |
| 71. | $= z_0^2 \cdot (1 - 2h_1)$ | 70, 60 |
| 72. | $1 - 2h_1 \leqslant (1 - h_1)^2$ | *T*73, *T*62 |
| 73. | $x \leqslant z_0^2 \cdot (1 - h_1)^2$ | 72, *T*71, 71 |
| 74. | $x \leqslant (z_0 \cdot (1 - h_1))^2$ | 73, *T*63 |
| 75. | $y^2 \leqslant (z_0 \cdot (1 - h_1))^2$ | 68, 74 |
| 76. | $0 < 1 - h_1$ | 61 |
| 77. | $0 \leqslant z_0 \cdot (1 - h_1)$ | 28, 76, *A*14 |
| 78. | $y \leqslant z_0 \cdot (1 - h_1)$ | 75, 77, *T*100 |
| 79. | $z_0 \leqslant z_0 \cdot (1 - h_1)$ | 17(2nd conjunct), 67 |
| 80. | $-h_1 < 0$ | 61 |
| 81. | $1 - h_1 < 1$ | 80 |
| 82. | $z_0 \cdot (1 - h_1) < z_0$ | 81, 28 |
| 83. | $\sim z_0 \cdot (1 - h_1) < z_0$ | 79 |
| 84. | $x = z_0^2$ | 29, 57, *T*19 |
| 85. | $\bigvee y\, x = y^2$ | 84 |

It is a simple consequence of *T*22 and *T*129 that a number is non-negative just in case it is a square:

*T*130 $\qquad 0 \leqslant x \leftrightarrow \bigvee y\, x = y^2$

The following theorem, which follows immediately from *T*130, shows how '$\leqslant$' might have been defined in terms of '+' and '·'.

*T*131 $\qquad x \leqslant y \leftrightarrow \bigvee z\, x + z^2 = y$

For historical remarks pertaining to this chapter, see section 4 of chapter IX.

# Chapter IX
# Variable-binding operators

**1. Definitions reconsidered.** We turn now to the treatment, deferred from chapter VIII, of definitions of variable-binding operators. Roughly speaking, in order to define a nonlogical constant $\delta$ of degree $\langle i, m, n, p \rangle$, we must provide for each term or formula

$$\delta\alpha_1 \ldots \alpha_m \, \zeta_1 \ldots \zeta_n \, \phi_1 \ldots \phi_p$$

(where $\alpha_1, \ldots, \alpha_m$ are distinct variables, $\zeta_1, \ldots, \zeta_n$ are terms, and $\phi_1, \ldots, \phi_p$ are formulas) a synonymous term or formula not containing $\delta$. To do this precisely, we shall make use of operation and predicate letters and *definitional schemata.*

For simplicity, consider for a moment the special case of constants of degree $\langle 1, 1, 0, 1 \rangle$. Let $\delta$ be a nonlogical constant of this kind, and $L$ a class of nonlogical constants not containing $\delta$. Then a *possible definitional schema* for $\delta$ *in terms of* $L$ will be a formula of the form

$$\delta\alpha\pi\alpha \leftrightarrow \phi \quad ,$$

where $\alpha$ is a variable, $\pi$ is a 1-place predicate letter not in $L$, and $\phi$ is a formula containing no free variables and no nonlogical constants beyond $\pi$ and those in $L$.

More generally, consider an arbitrary nonlogical constant $\delta$, and again let $L$ be a class of nonlogical constants not containing $\delta$.

If $\delta$ is a nonlogical constant of degree $\langle 0, m, n, p \rangle$, then a *possible definitional schema for* $\delta$ *in terms of* $L$ will take the form

$$\delta\alpha_1 \ldots \alpha_m \, \zeta_1\alpha_1 \ldots \alpha_m \ldots \zeta_n\alpha_1 \ldots \alpha_m \, \pi_1\alpha_1 \ldots \alpha_m \ldots \pi_p\alpha_1 \ldots \alpha_m = \eta \quad .$$

Here $\alpha_1, \ldots, \alpha_m$ are to be distinct variables, $\zeta_1, \ldots, \zeta_n$ are to be distinct $m$-place operation letters, $\pi_1, \ldots, \pi_p$ are to be distinct $m$-place predicate letters, and $\eta$ is to be a term containing no free variables, and no nonlogical constants beyond $\zeta_1, \ldots, \zeta_n, \pi_1, \ldots, \pi_p$, and those in $L$; we assume in addition that $\zeta_1, \ldots, \zeta_n, \pi_1, \ldots, \pi_p$ are not in $L$, and are distinct from $\delta$.

If, on the other hand, $\delta$ is a nonlogical constant of degree $\langle 1, m, n, p \rangle$, then a *possible definitional schema for* $\delta$ *in terms of* $L$ will be a formula

$$\delta\alpha_1 \ldots \alpha_m\, \zeta_1\alpha_1 \ldots \alpha_m \;\ldots\; \zeta_n\alpha_1 \ldots \alpha_m\, \pi_1\alpha_1 \ldots \alpha_m \;\ldots\; \pi_p\alpha_1 \ldots \alpha_m \leftrightarrow \phi \quad .$$

As before, $\alpha_1, \ldots, \alpha_m$ are to be distinct variables, $\zeta_1, \ldots, \zeta_n$ are to be distinct *m*-place operation letters, and $\pi_1, \ldots, \pi_p$ are to be distinct *m*-place predicate letters. In addition, $\phi$ is to be a formula containing no free variables, and no nonlogical constants beyond $\zeta_1, \ldots, \zeta_n, \pi_1, \ldots, \pi_p$, and those in *L*; again $\zeta_1, \ldots, \zeta_n, \pi_1, \ldots, \pi_p$ are to be distinct from $\delta$ and the members of *L*.

For example, a possible definitional schema for the operator 'lim' (mentioned on page 273) in terms of

$$0, <, -, |\ |$$

would be the following:

$$\lim n\, A(n) = \text{⅂}x \wedge z(0 < z \rightarrow \vee k \wedge n[N(n) \wedge k < n \rightarrow |A(n) - x| < z]) \quad .$$

(This schema in fact reflects customary mathematical usage and will be included in theories developed below.) As another example, consider the operator '$\overset{1}{\vee}$' (also mentioned on page 273). The following formula is a possible definitional schema for this constant in terms of the empty set of nonlogical constants:

$$\overset{1}{\vee} x F(x) \leftrightarrow \vee y \wedge x[F(x) \leftrightarrow x = y] \quad .$$

Observe that the notions just introduced comprehend the case in which $\delta$ is a predicate or operation symbol, which arises when $m = 0$. Thus we can dispense with the *possible definitions* of chapter VIII in favor of a uniform system of possible definitional schemata.

In order to accommodate the definitional introduction of variable-binding operators, we must modify the characterization of a *definitional extension* of a theory; we now use the notion of a possible definitional schema.

Indeed, a theory *U* is now said to be a *definitional extension* of a theory *T* if (i) *U* is an extension of *T*, and, for some *D*, (ii) *D* is a class of possible definitional schemata, in terms of the constants of *T*, for constants of *U* that are not constants of *T*, (iii) for each constant $\delta$ of *U* that is not a constant of *T*, there is in *D* some possible definitional schema for $\delta$ in terms of the constants of *T*, (iv) each constant of *U* that is not a constant of *T* occurs in at most one member of *D*, and (v) the axioms of *U* consist of the axioms of *T* together with all instances within *U* of members of *D*.

As an example, we may consider a theory $T_3$ that closely resembles the theory $T_2$ of chapter VIII (p. 301). The *constants of* $T_3$ are those of $T_2$, and the *axioms of* $T_3$ consist of those of the theory of real numbers, together with all instances within $T_3$ of the following formulas ('A' and 'B' are here to be 0-place operation letters):

*DS1* $\quad A \neq B \leftrightarrow \sim A = B$

*DS2* $\quad A < B \leftrightarrow A \leqslant B \wedge \sim A = B$

*DS*3 $2 = 1 + 1$

*DS*4 $A^2 = A \cdot A$

*DS*5 $|A| = \daleth z[(0 \leqslant A \wedge z = A) \vee (\sim 0 \leqslant A \wedge z = -A)]$

*DS*6 $A - B = A + -B$

*DS*7 $\frac{A}{B} = A \cdot B^{-1}$

*DS*8 $N(A) \leftrightarrow I(A) \wedge 0 \leqslant A$

*DS*9 $R(A) \leftrightarrow \bigvee y \bigvee z[I(y) \wedge I(z) \wedge 0 \leqslant z \wedge \sim 0 = z \wedge A = y \cdot z^{-1}]$ .

It is clear that the formulas *DS*1 – *DS*9 are possible definitional schemata in terms of the constants of the theory of real numbers, and hence that T3 is a definitional extension of that theory. Further, T3 is easily seen to be equivalent to T2. Accordingly, in developing T3 and its extensions, we may employ T1 – T131 of chapter VIII.

(The construction of T3 illustrates a general principle: if *U* is a definitional extension of a theory *T* in the sense of chapter VIII, then there is a definitional extension of *T* in the present sense that is equivalent to *U*.)

Definitional extensions continue to enjoy the properties (1) – (4) mentioned on pages 295 – 96.

**2. The theory of convergence.** We are now in a position to extend T3 by introducing definitions of variable-binding operators; it is essential for this purpose to employ possible definitional schemata and not merely possible definitions. Accordingly, we construct as follows a definitional extension T4 of T3. The constants of T4 are those of T3 together with the formula-maker '$\Leftrightarrow$', which is to have degree $\langle 1, 1, 2, 0 \rangle$, and the axioms of T4 consist of those of T3 together with all instances within T4 of the formula:

*DS*10 $A(n) \underset{n}{\Leftrightarrow} B(n) \leftrightarrow$

$$\bigwedge z(0 < z \rightarrow \bigvee k \bigwedge n[N(n) \wedge k < n \rightarrow |A(n) - B(n)| < z])$$

(This symbolism departs from our official notational style in the direction of mathematical practice, which often introduces theoretically superfluous parentheses and which, generally speaking, requires the variables accompanying an operator to be written as subscripts. Strict adherence to the style of chapter VIII, section 1, would convert the left side of the biconditional *DS*10 into '$\Leftrightarrow$ *n* A*n* B*n*'.)

The formula

$$A(n) \underset{n}{\Leftrightarrow} B(n)$$

is read 'A(*n*) and B(*n*) converge (to one another, as *n* approaches infinity)'.

Intuitively, the terms 'A($n$)' and 'B($n$)' are regarded as representing two infinite sequences of numbers,

$$A(0), A(1), \ldots, A(n), \ldots$$

and

$$B(0), B(1), \ldots, B(n), \ldots \quad ;$$

according to *DS*10, these sequences are regarded as converging if, for any positive number $z$, there is a point beyond which the difference between any two corresponding terms of the two sequences is less than $z$.

The next two theorems give simple examples of convergence and non-convergence.

*T*132

1. ~~Show~~ $\frac{n^2 + 1}{n} \underset{n}{\Leftrightarrow} n$

2. ~~Show~~ $\bigwedge z\Big(0 < z \rightarrow \bigvee k \bigwedge n\Big[N(n) \wedge k < n \rightarrow \Big|\frac{n^2 + 1}{n} - n\Big| < z\Big]\Big)$

3. $0 < z$

4. ~~Show~~ $\bigwedge n\Big[N(n) \wedge \frac{1}{z} < n \rightarrow \Big|\frac{n^2 + 1}{n} - n\Big| < z\Big]$

5. $N(n) \wedge \frac{1}{z} < n$

6. $0 < n$ — 3, 5

7. $\frac{n^2 + 1}{n} - n = \frac{n^2}{n} + \frac{1}{n} - n$ — *T*67

8. $= n + \frac{1}{n} - n$ — 7, 6

9. $= \frac{1}{n}$ — 8

10. $0 < \frac{1}{n}$ — 6

11. $\Big|\frac{1}{n}\Big| = \frac{1}{n}$ — 10

12. $\left|\frac{n^2+1}{n} - n\right| = \frac{1}{n}$ 9, 11

13. $1 < n \cdot z$ 3, 5

14. $\frac{1}{n} < z$ 6, 13

15. $\left|\frac{n^2+1}{n} - n\right| < z$ 12, 14

16. $\bigvee k \bigwedge n\left[N(n) \wedge k < n \to \left|\frac{n^2+1}{n} - n\right| < z\right]$ 4, EG

17. $\frac{n^2+1}{n} \underset{n}{\Leftrightarrow} n$ 2, *DS*10

*T*133 1. ~~*Show*~~ $\sim n + 1 \underset{n}{\Leftrightarrow} n$

2. $n + 1 \underset{n}{\Leftrightarrow} n$

3. $\bigvee k \bigwedge n[N(n) \wedge k < n \to |n + 1 - n| < 1]$ *DS*10, *T*74

4. $\bigwedge n[N(n) \wedge k_0 < n \to |n + 1 - n| < 1]$ 3, EI

5. ~~*Show*~~ $0 \leqslant k_0 \to \bigvee n[N(n) \wedge k_0 < n]$

6. $0 \leqslant k_0$

7. $I(n_0) \wedge k_0 + 1 \leqslant n_0 \cdot 1$ *T*54, *A*10, EI

8. $k_0 < n_0$ 7

9. $N(n_0)$ 6, 7, 8

10. $\bigvee n[N(n) \wedge k_0 < n]$ 9, 8, EG

11. ~~*Show*~~ $k_0 < 0 \to \bigvee n[N(n) \wedge k_0 < n]$

12. $k_0 < 0$

13. $N(0)$ *T*122

14. $\bigvee n[N(n) \wedge k_0 < n]$ 12, 13, EG

15. $\bigvee n[N(n) \wedge k_0 < n]$ 5, 11

16. $N(n_1) \wedge k_0 < n_1$ 15, EI

17. $|n_1 + 1 - n_1| < 1$ 4, 16

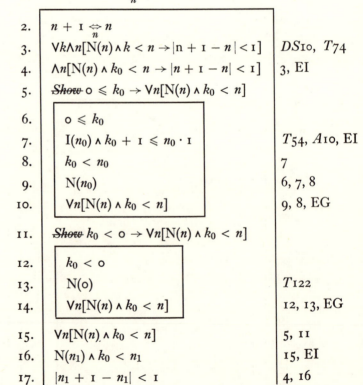

18. $n_1 + 1 - n_1 = 1$
19. $|1| < 1$ — 17, 18
20. $1 < 1$ — 19
21. $\sim 1 < 1$ — *T*76

Nonlogical operators, like the logical operators '$\bigwedge$', '$\bigvee$', and '$\daleth$', satisfy general principles of alphabetic variance and interchange. Particular cases of these principles are given in the next two theorem schemata. (By a *theorem schema* of T4 is understood, as before (see p. 287 of chapter VIII), a formula containing some additional predicates or operation symbols beyond the constants of T4 and such that all of its instances within T4 are theorems of T4. In showing particular formulas to be theorem schemata of T4, we shall consider an extension T4′ of T4 that contains the additional 1-place operation symbols 'C', '$C_1$', 'D', '$D_1$', and 'E', together with the 1-place predicate 'F' (which figures in *AS*19), and whose additional axioms are the instances within this enlarged vocabulary of *AS*19 and *DS*1 – *DS*10. Then each theorem of T4′ will be a theorem schema of T4.)

*TS*134

1. ~~Show~~ $C(n) \underset{n}{\Leftrightarrow} D(n) \leftrightarrow C(m) \underset{m}{\Leftrightarrow} D(m)$
2. $C(n) \underset{n}{\Leftrightarrow} D(n) \leftrightarrow C(n) \underset{n}{\Leftrightarrow} D(n)$
3. $C(n) \underset{n}{\Leftrightarrow} D(n) \leftrightarrow C(m) \underset{m}{\Leftrightarrow} D(m)$ — 2, AV

*TS*135

1. ~~Show~~ $\bigwedge n\, C(n) = C_1(n) \wedge \bigwedge n\, D(n) = D_1(n) \rightarrow [C(n) \underset{n}{\Leftrightarrow} D(n) \leftrightarrow C_1(n) \underset{n}{\Leftrightarrow} D_1(n)]$
2. $\bigwedge n\, C(n) = C_1(n) \wedge \bigwedge n\, D(n) = D_1(n)$
3. $C(n) \underset{n}{\Leftrightarrow} D(n) \leftrightarrow C_1(n) \underset{n}{\Leftrightarrow} D(n)$ — 2, Int
4. $C(n) \underset{n}{\Leftrightarrow} D(n) \leftrightarrow C_1(n) \underset{n}{\Leftrightarrow} D_1(n)$ — 2, Int, 3

According to the next three theorem schemata, the relation of convergence is an equivalence relation, that is, reflexive (*TS*136), symmetric (*TS*137), and transitive (*TS*138). The proofs of *TS*136 and *TS*137 are very simple, and are left to the reader.

*TS*136 $C(n) \underset{n}{\Leftrightarrow} C(n)$

*TS*137 $C(n) \underset{n}{\Leftrightarrow} D(n) \rightarrow D(n) \underset{n}{\Leftrightarrow} C(n)$

*TS*138 1. ~~*Show*~~ $C(n) \underset{n}{\Leftrightarrow} D(n) \wedge D(n) \underset{n}{\Leftrightarrow} E(n) \rightarrow C(n) \underset{n}{\Leftrightarrow} E(n)$

2. $C(n) \underset{n}{\Leftrightarrow} D(n) \wedge D(n) \underset{n}{\Leftrightarrow} E(n)$

3. ~~*Show*~~ $\bigwedge z(0 < z \rightarrow \bigvee k \bigwedge n[N(n) \wedge k < n \rightarrow |C(n) - E(n)| < z])$

4. $0 < z$

5. $0 < \frac{z}{2}$ — 4

6. $\bigwedge n\left[N(n) \wedge k_0 < n \rightarrow |C(n) - D(n)| < \frac{z}{2}\right]$ — 5, 2, *DS*10, EI

7. $\bigwedge n\left[N(n) \wedge k_1 < n \rightarrow |D(n) - E(n)| < \frac{z}{2}\right]$ — 5, 2, *DS*10, EI

8. ~~*Show*~~ $k_0 \leqslant k_1 \rightarrow \bigvee k \bigwedge n[N(n) \wedge k < n \rightarrow |C(n) - E(n)| < z]$

9. $k_0 \leqslant k_1$

10. ~~*Show*~~ $\bigwedge n[N(n) \wedge k_1 < n \rightarrow |C(n) - E(n)| < z]$

11. $N(n) \wedge k_1 < n$

12. $|C(n) - D(n)| < \frac{z}{2}$ — 6, 11, 9

13. $|D(n) - E(n)| < \frac{z}{2}$ — 7, 11

14. $|C(n) - D(n)| + |D(n) - E(n)| < z$ — 12, 13

15. $|C(n) - E(n)| < z$ — *T*113, 14

16. $\bigvee k \bigwedge n[N(n) \wedge k < n \rightarrow |C(n) - E(n)| < z]$ — 10, EG

17. ~~*Show*~~ $k_1 \leqslant k_0 \rightarrow \bigvee k \bigwedge n[N(n) \wedge k < n \rightarrow |C(n) - E(n)| < z]$

Similar to derivation of line 8

18. $\bigvee k \bigwedge n[N(n) \wedge k < n \rightarrow |C(n) - E(n)| < z]$ — 8, 17

19. $C(n) \underset{n}{\Leftrightarrow} E(n)$ — 3, *DS*10

The relation of convergence is not only an equivalence relation but also, according to the next theorem schema, a congruence relation under the operation of addition. The proof of *TS*139, which is similar to that of *TS*138, we leave to the reader.

*TS*139 $\quad \mathrm{C}(n) \underset{n}{\Leftrightarrow} \mathrm{C}_1(n) \wedge \mathrm{D}(n) \underset{n}{\Leftrightarrow} \mathrm{D}_1(n) \rightarrow$

$$\mathrm{C}(n) + \mathrm{D}(n) \underset{n}{\Leftrightarrow} \mathrm{C}_1(n) + \mathrm{D}_1(n)$$

If two sequences are ultimately identical (that is, differ only in an initial segment), then they converge:

*TS*140 $\quad \bigvee k \bigwedge n[\mathrm{N}(n) \wedge k < n \rightarrow \mathrm{C}(n) = \mathrm{D}(n)] \rightarrow \mathrm{C}(n) \underset{n}{\Leftrightarrow} \mathrm{D}(n)$

This is a simple consequence of *DS*10.

The next theorem schema is a lemma for *TS*142; it asserts that if $\mathrm{E}(n)$ is a strictly increasing sequence of natural numbers, then, for every natural number $n$, $n \leqslant \mathrm{E}(n)$.

*TS*141 1. ~~Show~~ $\bigwedge n \bigwedge m[\mathrm{N}(n) \wedge \mathrm{N}(m) \wedge n < m \rightarrow \mathrm{E}(n) < \mathrm{E}(m)] \wedge$
$\bigwedge n[\mathrm{N}(n) \rightarrow \mathrm{N}(\mathrm{E}(n))] \rightarrow \bigwedge n[\mathrm{N}(n) \rightarrow n \leqslant \mathrm{E}(n)]$

2. $\bigwedge n \bigwedge m[\mathrm{N}(n) \wedge \mathrm{N}(m) \wedge n < m \rightarrow \mathrm{E}(n) < \mathrm{E}(m)] \wedge \bigwedge n[\mathrm{N}(n) \rightarrow \mathrm{N}(\mathrm{E}(n))]$
3. $0 \leqslant \mathrm{E}(0)$ — 2, *DS*8
4. ~~Show~~ $\bigwedge n[\mathrm{N}(n) \wedge n \leqslant \mathrm{E}(n) \rightarrow n + 1 \leqslant \mathrm{E}(n + 1)]$
5. $\mathrm{N}(n) \wedge n \leqslant \mathrm{E}(n)$
6. $n < n + 1$ — *T*27, *T*80
7. $\mathrm{E}(n) < \mathrm{E}(n + 1)$ — 5, 6, 2
8. $n < \mathrm{E}(n + 1)$ — 5, 7
9. $\mathrm{I}(n) \wedge \mathrm{I}(\mathrm{E}(n + 1))$ — 5, *DS*8, *A*17, 2
10. $n + 1 \leqslant \mathrm{E}(n + 1)$ — 9, 8, *T*55
11. $\bigwedge n[N(n) \rightarrow n \leqslant \mathrm{E}(n)]$ — 3, 4, *TS*121

According to *TS*142, if two sequences converge, then so do any two corresponding infinite sub-sequences.

*TS*142 1. ~~Show~~ $[\mathrm{C}(n) \underset{n}{\Leftrightarrow} \mathrm{D}(n)] \wedge$
$\bigwedge n \bigwedge m[\mathrm{N}(n) \wedge \mathrm{N}(m) \wedge n < m \rightarrow \mathrm{E}(n) < \mathrm{E}(m)] \wedge$
$\bigwedge n[\mathrm{N}(n) \rightarrow \mathrm{N}(\mathrm{E}(n))] \rightarrow [\mathrm{C}(\mathrm{E}(n)) \underset{n}{\Leftrightarrow} \mathrm{D}(\mathrm{E}(n))]$

| | | |
|---|---|---|
| 2. | $[\mathrm{C}(n) \underset{n}{\Leftrightarrow} \mathrm{D}(n)] \wedge$ $\bigwedge n \bigwedge m[\mathrm{N}(n) \wedge \mathrm{N}(m) \wedge n < m \rightarrow \mathrm{E}(n) < \mathrm{E}(m)] \wedge$ $\bigwedge n[\mathrm{N}(n) \rightarrow \mathrm{N}(\mathrm{E}(n))]$ | |
| 3. | ~~*Show*~~ $\bigwedge z(0 < z \rightarrow \bigvee k \bigwedge n[\mathrm{N}(n) \wedge k < n \rightarrow |\mathrm{C}(\mathrm{E}(n)) - \mathrm{D}(\mathrm{E}(n))| < z])$ | |
| 4. | $0 < z$ | |
| 5. | $\bigwedge n[\mathrm{N}(n) \wedge k_0 < n \rightarrow |\mathrm{C}(n) - \mathrm{D}(n)| < z]$ | 2 (1st conjunct), *DS*10, 4, EI |
| 6. | ~~*Show*~~ $\bigwedge n[\mathrm{N}(n) \wedge k_0 < n \rightarrow |\mathrm{C}(\mathrm{E}(n)) - \mathrm{D}(\mathrm{E}(n))| < z]$ | |
| 7. | $\mathrm{N}(n) \wedge k_0 < n$ | |
| 8. | $n \leqslant \mathrm{E}(n)$ | 2, *TS*141, 7 |
| 9. | $k_0 < \mathrm{E}(n)$ | 7, 8 |
| 10. | $\mathrm{N}(\mathrm{E}(n))$ | 2, 7 |
| 11. | $|\mathrm{C}(\mathrm{E}(n)) - \mathrm{D}(\mathrm{E}(n))| < z$ | 5, 10, 9 |
| 12. | $\bigvee k \bigwedge n[\mathrm{N}(n) \wedge k < n \rightarrow |\mathrm{C}(\mathrm{E}(n)) - \mathrm{D}(\mathrm{E}(n))| < z]$ | 6, EG |
| 13. | $\mathrm{C}(\mathrm{E}(n)) \underset{n}{\Leftrightarrow} \mathrm{D}(\mathrm{E}(n))$ | 3, *DS*10 |

We now consider the familiar notion of the *convergence of a sequence to a number*, which emerges as a special case of the convergence of two sequences. *TS*143, which is simply an instance of *DS*10, gives a necessary and sufficient condition for a sequence to converge to a number $x$.

*TS*143 $\mathrm{C}(n) \underset{n}{\Leftrightarrow} x \leftrightarrow \bigwedge z(0 < z \rightarrow \bigvee k \bigwedge n[\mathrm{N}(n) \wedge k < n \rightarrow |\mathrm{C}(n) - x| < z])$

Thus the sequence

$$\mathrm{C}(0), \mathrm{C}(1), \ldots, \mathrm{C}(n), \ldots$$

is regarded as converging to the number $x$ if, for any preassigned interval surrounding $x$ (no matter how small), all terms occurring sufficiently late in the sequence lie within that interval.

We state the following theorems as examples of convergence to a number.

*T*144 $\dfrac{n+1}{n} \underset{n}{\Leftrightarrow} 1$

*T*145 $\qquad \frac{1}{n} \underset{n}{\Leftrightarrow} 0$

The infinite sequences involved in these two theorems are respectively

$$\frac{1}{0}, \frac{2}{1}, \frac{3}{2}, \frac{4}{3}, \ldots$$

and

$$\frac{1}{0}, \frac{1}{1}, \frac{1}{2}, \frac{1}{3}, \ldots .$$

The assertions *T*144 and *T*145 are thus intuitively true. (The fact that our axioms do not determine the value of $\frac{1}{0}$ does not affect the truth of *T*144 and *T*145; indeed, questions of convergence are always independent of the initial terms of the sequences involved.) We prove *T*144 and leave *T*145 to the reader.

1. ~~*Show*~~ $\frac{n+1}{n} \underset{n}{\Leftrightarrow} 1$
2. ~~*Show*~~ $\bigwedge z\Big(0 < z \rightarrow \bigvee k \bigwedge n[\mathrm{N}(n) \wedge k < n \rightarrow \left|\frac{n+1}{n} - 1\right| < z\Big)$
3. $0 < z$
4. ~~*Show*~~ $\bigwedge n\left[\mathrm{N}(n) \wedge \frac{1}{z} < n \rightarrow \left|\frac{n+1}{n} - 1\right| < z\right]$
5. $\mathrm{N}(n) \wedge \frac{1}{z} < n$
6. $\frac{n+1}{n} - 1 = \frac{n}{n} + \frac{1}{n} - 1$ — *T*67
7. $\qquad = \frac{1}{n}$ — 6
8. $0 < n$ — 3, 5
9. $0 < \frac{1}{n}$ — 8

10. $\left|\frac{1}{n}\right| = \frac{1}{n}$ — 9

11. $\left|\frac{n+1}{n} - 1\right| = \frac{1}{n}$ — 7, 10

12. $1 < n \cdot z$ — 3, 5

13. $\frac{1}{n} < z$ — 8, 12

14. $\left|\frac{n+1}{n} - 1\right| < z$ — 13, 11

15. $\bigvee k \bigwedge n \left[ \mathrm{N}(n) \wedge k < n \rightarrow \left|\frac{n+1}{n} - 1\right| < z \right]$ — 4, EG

16. $\frac{n+1}{n} \underset{n}{\Leftrightarrow} 1$ — 2, *TS*143

If two numbers converge, then they are identical:

*T*146 $\quad x \underset{n}{\Leftrightarrow} y \rightarrow x = y$

An immediate corollary of *T*146, *TS*137, and *TS*138 is that a sequence can converge to at most one number:

*TS*147 $\quad \mathrm{C}(n) \underset{n}{\Leftrightarrow} x \wedge \mathrm{C}(n) \underset{n}{\Leftrightarrow} y \rightarrow x = y$

Some sequences do not converge to any number; for example:

*T*148 $\quad \sim \bigvee x\, n \underset{n}{\Leftrightarrow} x$

Proofs concerning specific instances of convergence are simplified by *TS*134 – *TS*139; for example:

*T*149 $\quad$ 1. ~~*Show*~~ $\frac{n(n+1)}{n^2} \underset{n}{\Leftrightarrow} 1$

2. $1 \underset{n}{\Leftrightarrow} 1$ — *TS*136

3. $\frac{1}{n} \underset{n}{\Leftrightarrow} 0$ — *T*145

4. $1 + \frac{1}{n} \underset{n}{\Leftrightarrow} 1 + 0$ — 2, 3, *TS*139

5. $1 + \frac{1}{n} \underset{n}{\Leftrightarrow} 1$ — 4, *A*3, LL

6. ~~Show~~ $\bigwedge n(\mathrm{N}(n) \wedge 0 < n \rightarrow \frac{n(n+1)}{n^2} = 1 + \frac{1}{n}$

7. $\mathrm{N}(n) \wedge 0 < n$

8. $\frac{n(n+1)}{n^2} = \frac{n^2}{n^2} + \frac{n}{n^2}$ — *A*9

9. $= 1 + \frac{1}{n}$ — 8, 7

10. $\frac{n(n+1)}{n^2} \underset{n}{\Leftrightarrow} 1$ — 5, 6, *TS*140

The relation of convergence is not a congruence relation under the operation of multiplication. For example,

$$\frac{1}{n} \underset{n}{\Leftrightarrow} 0$$

and

$$n \underset{n}{\Leftrightarrow} n \quad ;$$

yet it is not the case that

$$\frac{1}{n} \cdot n \underset{n}{\Leftrightarrow} 0 \cdot n \quad ,$$

for if it were, we should have

$$1 \underset{n}{\Leftrightarrow} 0 \quad .$$

However, if two sequences converge to numbers, their product converges to the product of those numbers.

*TS*150 1. ~~Show~~ $\mathrm{C}(n) \underset{n}{\Leftrightarrow} x \wedge \mathrm{D}(n) \underset{n}{\Leftrightarrow} y \rightarrow \mathrm{C}(n) \cdot \mathrm{D}(n) \underset{n}{\Leftrightarrow} x \cdot y$

2. $C(n) \underset{n}{\Leftrightarrow} x \wedge \mathrm{D}(n) \underset{n}{\Leftrightarrow} y$

3. ~~Show~~ $\bigwedge z(0 < z \rightarrow \bigvee k \bigwedge n[\mathrm{N}(n) \wedge k < n \rightarrow |(\mathrm{C}(n) \cdot \mathrm{D}(n)) - (x \cdot y)| < z])$

| | | |
|---|---|---|
| 4. | $0 < z$ | |
| 5. | $0 < \frac{z}{2(\lvert y\rvert + 1)}$ | 4 |
| 6. | $\bigwedge n\left[N(n) \wedge k_0 < n \rightarrow \lvert C(n) - x\rvert < \frac{z}{2(\lvert y\rvert + 1)}\right]$ | 2, *DS*10, 5, EI |
| 7. | $0 < \frac{z}{2\lvert x\rvert + z}$ | 4 |
| 8. | $\bigwedge n\left[N(n) \wedge k_1 < n \rightarrow \lvert D(n) - y\rvert < \frac{z}{2\lvert x\rvert + z}\right]$ | 2, *DS*10, 7, EI |
| 9. | ~~Show~~ $k_0 \leqslant k_1 \rightarrow \bigvee k \bigwedge n[N(n) \wedge k < n \rightarrow \lvert(C(n) \cdot D(n)) - (x \cdot y)\rvert < z]$ | |
| 10. | $k_0 \leqslant k_1$ | |
| 11. | ~~Show~~ $\bigwedge n[N(n) \wedge k_1 < n \rightarrow \lvert(C(n) \cdot D(n)) - (x \cdot y)\rvert < z]$ | |
| 12. | $N(n) \wedge k_1 < n$ | |
| 13. | $\lvert C(n) - x\rvert < \frac{z}{2(\lvert y\rvert + 1)}$ | 10, 12, 6 |
| 14. | $\lvert y \cdot (C(n) - x)\rvert \leqslant \lvert y\rvert \cdot \frac{z}{2(\lvert y\rvert + 1)}$ | 13, *T*104, *T*116 |
| 15. | $\lvert D(n) - y\rvert < \frac{z}{2\lvert x\rvert + z}$ | 12, 8 |
| 16. | $\lvert C(n) \cdot (D(n) - y)\rvert \leqslant \lvert C(n)\rvert \cdot \frac{z}{2\lvert x\rvert + z}$ | 15, *T*104, *T*116 |
| 17. | $2 \leqslant 2(\lvert y\rvert + 1)$ | *T*106 |
| 18. | $\frac{z}{2(\lvert y\rvert + 1)} \leqslant \frac{z}{2}$ | 17, 4, *T*95 |

| | | |
|---|---|---|
| 19. | $\lvert x\rvert + \lvert C(n) - x\rvert < \lvert x\rvert + \frac{z}{2}$ | 13, 18 |
| 20. | $C(n) < \lvert x\rvert + \frac{z}{2}$ | 19, *T*112 |
| 21. | $\lvert C(n) \cdot (D(n) - y)\rvert < \left(\lvert x\rvert + \frac{z}{2}\right) \cdot \left(\frac{z}{2\lvert x\rvert + z}\right)$ | 7, 20, 16 |
| 22. | $\lvert C(n) \cdot (D(n) - y)\rvert + \lvert y \cdot (C(n) - x)\rvert < \left[\left(\lvert x\rvert + \frac{z}{2}\right) \cdot \left(\frac{z}{2\lvert x\rvert + z}\right)\right] + \left[\lvert y\rvert \cdot \frac{z}{2(\lvert y\rvert + 1)}\right]$ | 14, 21 |
| 23. | $(C(n) \cdot D(n)) - (x \cdot y) = (C(n) \cdot (D(n) - y)) + (y \cdot (C(n) - x))$ | A9 |
| 24. | $\lvert (C(n) \cdot D(n)) - (x \cdot y)\rvert < \left[\left(\lvert x\rvert + \frac{z}{2}\right) \cdot \left(\frac{z}{2\lvert x\rvert + z}\right)\right] + \left(\lvert y\rvert \cdot \frac{z}{2(\lvert y\rvert + 1)}\right)$ | 23, *T*112, 22 |
| 25. | $\frac{\lvert y\rvert}{\lvert y\rvert + 1} < 1$ | *T*106, *T*27, *T*80, *T*99 |
| 26. | $\lvert y\rvert \cdot \frac{z}{2(\lvert y\rvert + 1)} < \frac{z}{2}$ | 4, 25 |
| 27. | $\lvert x\rvert + \frac{z}{2} = \frac{2\lvert x\rvert + z}{2}$ | *T*67, *T*68 |
| 28. | $\left(\lvert x\rvert + \frac{z}{2}\right) \cdot \left(\frac{z}{2\lvert x\rvert + z}\right) = \frac{z}{2}$ | 4, *T*106, 27 |
| 29. | $\lvert (C(n) \cdot D(n)) - (x \cdot y)\rvert < \frac{z}{2} + \frac{z}{2}$ | 24, 26, 28 |

30. $|(C(n) \cdot D(n)) - (x \cdot y)| < z$ 29, *T*65

31. $\bigvee k \bigwedge n[N(n) \wedge k < n \rightarrow |(C(n) \cdot D(n)) - (x \cdot y)| < z]$ 11, EG

32. ~~Show~~ $k_1 \leqslant k_0 \rightarrow \bigvee k \bigwedge n[N(n) \wedge k < n \rightarrow |(C(n) \cdot D(n)) - (x \cdot y)| < z]$

Similar to derivation of line 9

33. $\bigvee k \bigwedge n[N(n) \wedge k < n \rightarrow |(C(n) \cdot D(n)) - (x \cdot y)| < z]$ 9, 32

34. $C(n) \cdot D(n) \underset{n}{\Leftrightarrow} x \cdot y$ 3, *DS*10

*TS*151 and *TS*152 have straightforward proofs, which are left to the reader. The latter provides a means of expressing the convergence of two sequences in terms of the convergence of a sequence to a number.

*TS*151 $C(n) \underset{n}{\Leftrightarrow} x \wedge D(n) \underset{n}{\Leftrightarrow} y \wedge \bigvee k \bigwedge n[N(n) \wedge k < n \rightarrow C(n) \leqslant D(n)] \rightarrow x \leqslant y$

*TS*152 $C(n) \underset{n}{\Leftrightarrow} D(n) \leftrightarrow C(n) - D(n) \underset{n}{\Leftrightarrow} 0$

The last two principles of this section depend very heavily on the Continuity Schema. *TS*153 asserts that every nondecreasing sequence whose terms have an upper bound converges to a number; *TS*154 is the analogue for nonincreasing sequences whose terms have a lower bound, and can be derived simply from *TS*153.

*TS*153 1. ~~Show~~ $\bigwedge n \bigwedge m[N(n) \wedge N(m) \wedge n \leqslant m \rightarrow C(n) \leqslant C(m)] \wedge \bigvee y \bigwedge n[N(n) \rightarrow C(n) \leqslant y] \rightarrow \bigvee z\, C(n) \underset{n}{\Leftrightarrow} z$

2. $\bigwedge n \bigwedge m[N(n) \wedge N(m) \wedge n \leqslant m \rightarrow C(n) \leqslant C(m)] \wedge \bigvee y \bigwedge n[N(n) \rightarrow C(n) \leqslant y]$

3. $\bigvee m \bigvee n[N(n) \wedge m = C(n)]$ T122

4. ~~Show~~ $\bigvee y \bigwedge m(\bigvee n[N(n) \wedge m = C(n)] \rightarrow m \leqslant y)$

Elementary, using 2nd conjunct of 2

5. $\bigwedge m(\bigvee n[N(n) \wedge m = C(n)] \rightarrow m \leqslant z_0) \wedge \bigwedge y[\bigwedge m(\bigvee n[N(n) \wedge m = C(n)] \rightarrow m \leqslant y) \rightarrow z_0 \leqslant y]$ — 3, 4, *AS*19, EI
6. ~~Show~~ $\bigwedge z[0 < z \rightarrow \bigvee k \bigwedge n(N(n) \wedge k < n \rightarrow |C(n) - z_0| < z)]$
7. $0 < z$
8. $\sim z_0 \leqslant z_0 - z$ — 7
9. $\bigwedge m(\bigvee n[N(n) \wedge m = C(n)] \rightarrow m \leqslant z_0 - z) \rightarrow z_0 \leqslant z_0 - z$ — 5 (2nd conjunct)
10. $\sim(\bigvee n[N(n) \wedge m_0 = C(n)] \rightarrow m_0 \leqslant z_0 - z)$ — 8, 9, EI
11. $N(n_0) \wedge m_0 = C(n_0) \wedge z_0 - z < m_0$ — 10, EI
12. ~~Show~~ $\bigwedge n(N(n) \wedge m_0 < n \rightarrow |C(n) - z_0| < z)$
13. $N(n) \wedge n_0 < n$
14. $C(n_0) \leqslant C(n)$ — 11, 13, 2 (1st conjunct)
15. $z_0 - z < C(n)$ — 11, 14
16. $C(n) \leqslant z_0$ — 13, 5 (1st conjunct)
17. $C(n) < z_0 + z$ — 7, 16
18. $|C(n) - z_0| < z$ — 15, 17, *T*118, *T*115
19. $\bigvee k \bigwedge n(N(n) \wedge k < n \rightarrow |C(n) - z_0| < z)$ — 12, EG
20. $\bigvee z\, C(n) \underset{n}{\Leftrightarrow} z$ — 6, *TS*143, EG

*TS*154 $\bigwedge n \bigwedge m[N(n) \wedge N(m) \wedge n \leqslant m \rightarrow C(m) \leqslant C(n)] \wedge \bigvee y \bigwedge n[N(n) \rightarrow y \leqslant C(n)] \rightarrow \bigvee z\, C(n) \underset{n}{\Leftrightarrow} z$

**3. A sketch of further developments.** The theory T4, with its definitional extensions, can be identified with what is generally called *differential calculus.* We offer, in partial justification of this claim, definitions of four of the most fundamental concepts in this subject.

If the sequence A($n$) converges to a number $x$, this number is called the limit of A($n$) as $n$ approaches infinity. Accordingly, the mathematical notion of a limit at infinity may be introduced by the following definitional schema:

$$DS11 \qquad \lim_{n} A(n) = \mathrm{I}x\, A(n) \underset{n}{\Leftrightarrow} x$$

(A more customary notation for '$\lim_{n} A(n)$' is '$\lim_{n\to\infty} A(n)$'.)

By $T$148 it is clear that the definite description on the right side of $DS$11 is not always proper; but if the sequence represented by 'A($n$)' converges to some number, then by $TS$147 the definite description is proper. The basic theorems concerning limits at infinity have essentially been obtained in the previous section; for example, $T$144, $T$145, and $TS$150 can be respectively expressed as follows:

$$\lim_{n} \frac{n+1}{n} = 1$$

$$\lim_{n} \frac{1}{n} = 0$$

$$\bigvee x\, C(n) \underset{n}{\Leftrightarrow} x \wedge \bigvee y\, D(n) \underset{n}{\Leftrightarrow} y \rightarrow$$

$$\lim_{n}\, [C(n) \cdot D(n)] = [\lim_{n} C(n)] \cdot [\lim_{n} D(n)] \quad .$$

Let 'A($x$)' represent any term of T4, for instance, '($x^2 + 1$)'. Then, for each value of '$x$' among the real numbers, the corresponding value of 'A($x$)' will also be a real number. It may happen that as the values of '$x$' approach a fixed number $l$, the corresponding values of 'A($x$)' will also approach some number. In this case, we denote the latter number by

$$\lim_{x\to l} A(x) \quad ,$$

which is read 'the limit of A($x$) as $x$ approaches $l$'. Thus, for example, choosing $l$ as 2 and 'A($x$)' as the term '($x^2 + 1$)', we have

$$\lim_{x\to 2} (x^2 + 1) = 5 \quad .$$

Let us examine the situation somewhat more carefully. We say that a number $u$ is the limit of A($x$) as $x$ approaches $l$ if the values of 'A($x$)' can be brought as close to $u$ as we wish by bringing the values of '$x$' sufficiently close to $l$ (without actually taking $l$ itself as a value for '$x$'). Thus, within T4, the notion can be characterized as follows:

$$(1) \qquad \lim_{x \to l} A(x) = \text{⅂} u \bigwedge z(0 < z \rightarrow \bigvee y[0 < y \wedge \bigwedge x(0 < |x - l| \wedge |x - l| < y \rightarrow |A(x) - u| < z)])$$

Strictly speaking, the formula (1) is not a possible definitional schema. For simplicity, in section 2 of this chapter we excluded free variables from appearing in definitional schemata, and in (1) the variable '$l$' is free. But the effect of (1) can be achieved by the following definitional schema, which introduces a constant '$\underset{\rightarrow}{\text{lim}}$' of degree $\langle 0, 1, 2, 0 \rangle$:

$$DS12 \qquad \lim_{x \to B(x)} A(x) = \text{⅂} u \bigwedge z(0 < z \rightarrow \bigvee y[0 < y \wedge \bigwedge x(0 < |x - B(0)| \wedge |x - B(0)| < y \rightarrow |A(x) - u| < z)])$$

Then (1) is an instance of *DS*12 and can be used instead of the latter in developing the mathematical theory of limits at a point.

To say that $A(x)$ is continuous at a number $a$ is to say that the number $A(a)$ is the limit of $A(x)$ as $x$ approaches $a$; recalling (1), we see that the condition of continuity can be expressed as follows:

$$(2) \qquad \underset{x}{\text{Cont}}\,[A(x), a] \leftrightarrow \bigwedge z(0 < z \rightarrow \bigvee y[0 < y \wedge \bigwedge x(0 < |x - a| \wedge |x - a| < y \rightarrow |A(x) - A(a)| < z)])$$

Thus we regard 'Cont' as an operator of degree $\langle 1, 1, 2, 0 \rangle$. On the basis of (2) it is a simple matter to show that, for example,

$$\bigwedge a \underset{x}{\text{Cont}}\,[2 \cdot x + 1, a] \quad ,$$

$$\sim \underset{x}{\text{Cont}}\left[\frac{1}{x}, 0\right] \quad ,$$

$$\underset{x}{\text{Cont}}\,[\text{⅂} y[I(x) \wedge y = 1) \vee (\sim I(x) \wedge y = 0)], a] \leftrightarrow \sim I(a) \quad .$$

Let us for the moment identify real numbers with instants of time. If $A(x)$ is understood as the distance at the instant $x$ of a certain object from an initial position, and the object is assumed to be moving in a straight line, then the derivative of $A(x)$ at $a$ (or $\underset{x}{\text{Der}}\,[A(x), a]$) will be defined in such a way as to be the velocity of the object at the instant $a$. To arrive at a definition of the derivative, we consider first the familiar procedure for computing average velocities: the average velocity of an object, between two instants $a$ and $b$, is the distance traveled in the interval (which may be positive or negative depending on the direction of motion), divided by the elapsed time. Thus we can characterize average velocity as follows:

(3) $$\underset{x}{\mathrm{Av}}\,[\mathrm{A}(x), a, b] = \frac{\mathrm{A}(b) - \mathrm{A}(a)}{b - a} \quad ;$$

here 'Av' is to be an operator of degree ⟨0, 1, 3, 0⟩. The velocity at an instant $a$ is simply the limit of the approximations obtained by taking average velocities over smaller and smaller time-intervals surrounding $a$. Thus we have the following characterization of the derivative (which in this case is interpreted as the instantaneous velocity):

(4) $$\underset{x}{\mathrm{Der}}\,[\mathrm{A}(x), a] = \lim_{h \to 0} \underset{x}{\mathrm{Av}}\,[\mathrm{A}(x), a - h, a + h]] \quad ;$$

here 'Der' is an operator of degree ⟨0, 1, 2, 0⟩. There are of course other interpretations of the derivative, depending on the interpretation assigned to 'A($x$)'. Many of them have the common feature that $\underset{x}{\mathrm{Der}}\,[\mathrm{A}(x), a]$ is the *rate of change* of the values of A($x$) at the instant $a$.

The formulas (2) – (4) correspond, like (1), to definitional schemata, which we may call *DS*13–*DS*15. By $T_5$ we shall understand the definitional extension of $T_4$ obtained by adding the instances of *DS*11 – *DS*14, and by $T_6$ the definitional extension of $T_5$ obtained by adding the instances of *DS*15.

The passage from differential to integral calculus is achieved by adding to $T_6$ the general notion of *finite summation*, that is, the operation that associates with any numbers A(0), . . . , A($n$ – 1), the sum of A(0), . . . , A($n$ – 1). This sum is usually denoted by

$$\mathrm{A}(0) + \ldots + \mathrm{A}(n - 1)$$

or

$$\sum_{i=0}^{n-1} \mathrm{A}(i) \quad ;$$

our notation will be officially

(5) $$\sum i\, n\, \mathrm{A}(i) \quad ,$$

and informally

(6) $$\sum_{i}^{n} \mathrm{A}(i) \quad ;$$

We shall regard 'Σ' as a constant of degree ⟨0, 1, 2, 0⟩; thus the expressions (5) and (6) are terms in which the variable '$i$' is bound and the variable '$n$' is free. 'Σ' may most conveniently be introduced by two new axioms, neither of which has definitional form. Accordingly, $T_7$, or *integral calculus*, is to be that extension of $T_6$ which is obtained by adding 'Σ' to the constants of $T_6$ and all instances (within the present theory) of the following schemata to the axioms of $T_6$;

*AS*20 $$\sum_{i}^{0} \mathrm{A}(i) = 0$$

$$AS21 \qquad N(n) \rightarrow \sum_{i}^{n+1} A(i) = \left[\sum_{i}^{n} A(i)\right] + A(n) \quad .$$

*AS*20 and *AS*21 constitute the usual so-called 'recursive definition' of 'Σ', but it should be observed that $T_7$ is not a definitional extension of $T_6$. 'Σ' *could*, however, have been introduced by definition. In other words, there is a definitional extension of $T_6$ having *AS*20 and *AS*21 as theorem schemata (see Montague [3]). But the construction of the necessary definitions is rather involved, and relies on properties of the real numbers deeper than those considered here.

Before turning to the integral, it is convenient to introduce notation for the least upper bound and the greatest lower bound of a set of real numbers; we use

$$\sup_x F(x)$$

and

$$\inf_x F(x) \quad ,$$

after the Latin '*supremum*' and '*infimum*', to denote respectively the least upper bound and the greatest lower bound of the set of all objects having the property F. (The reader will recall from chapter VIII the conditions under which such bounds exist.) We regard 'sup' and 'inf' as operators of degree ⟨0, 1, 0, 1⟩ and introduce them by the following definitional schemata:

$$DS16 \qquad \sup_x F(x) = \iota z[\bigwedge x(F(x) \rightarrow x \leqslant z) \wedge \bigwedge y(\bigwedge x[F(x) \rightarrow x \leqslant y] \rightarrow z \leqslant y)]$$

$$DS17 \qquad \inf_x F(x) = \iota z[\bigwedge x(F(x) \rightarrow z \leqslant x) \wedge \bigwedge y(\bigwedge x[F(x) \rightarrow y \leqslant x] \rightarrow y \leqslant z)]$$

By $T_8$ we shall understand the definitional extension of $T_7$ obtained by adding the instances of *DS*16 and *DS*17.

One purpose of the integral is to assign numerical measures to such areas as that shaded below.

FIGURE I

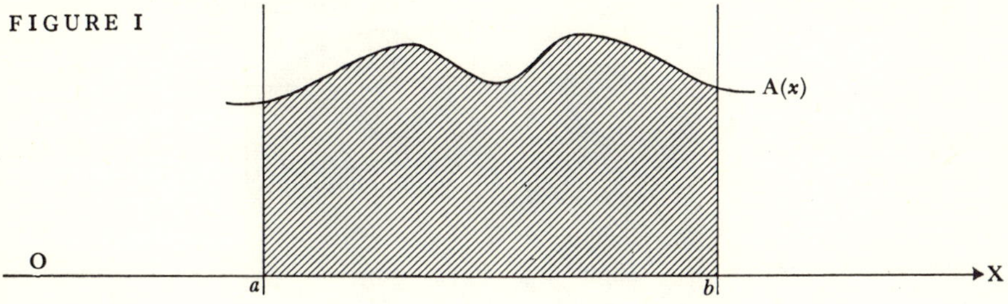

In this figure the line marked 'X' is called the X-*axis*. Each of its points is identified with a real number, in particular, with the distance (in some fixed system of units) of that point from the point marked 'O'. The curve lying above the X-axis is assumed to satisfy the following description: for any point $x$ of the X-axis, the length of the segment joining $x$ to the curve and perpendicular to the X-axis is the number $A(x)$. In this sense we may speak of the curve as given by the term '$A(x)$'. The shaded area may then be described as that enclosed by the X-axis, the perpendiculars to it at the points $a$ and $b$, and the curve given by '$A(x)$'.

An approximation to the shaded area can be computed in the following way: divide the segment of the X-axis between $a$ and $b$ into three equal parts. For each of these parts, construct the largest possible rectangle with that part as base which lies completely beneath the curve. Then compute the total area of the three rectangles so obtained. In this way we arrive at the doubly shaded area in the following figure.

FIGURE II

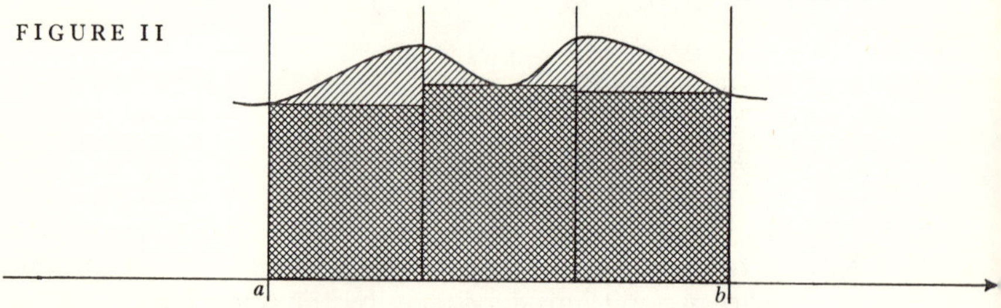


In the light of the familiar definition of area for rectilinear figures, it is seen that the doubly shaded area is given by the number

$$(7)\quad \frac{b-a}{3}\cdot\inf_y \bigvee x\left[A(x) = y \wedge a \leqslant x \wedge x \leqslant a + \frac{b-a}{3}\right] +$$
$$\frac{b-a}{3}\cdot\inf_y \bigvee x\left[A(x) = y \wedge a + \frac{b-a}{3} \leqslant x \wedge x \leqslant a + 2\cdot\frac{b-a}{3}\right] +$$
$$\frac{b-a}{3}\cdot\inf_y \bigvee x\left[A(x) = y \wedge a + 2\cdot\frac{b-a}{3} \leqslant x \wedge x \leqslant a + 3\cdot\frac{b-a}{3}\right] .$$

Now this approximation is not very close, but it can be improved by increasing the number of subdivisions of the interval between $a$ and $b$ (and

FIGURE III

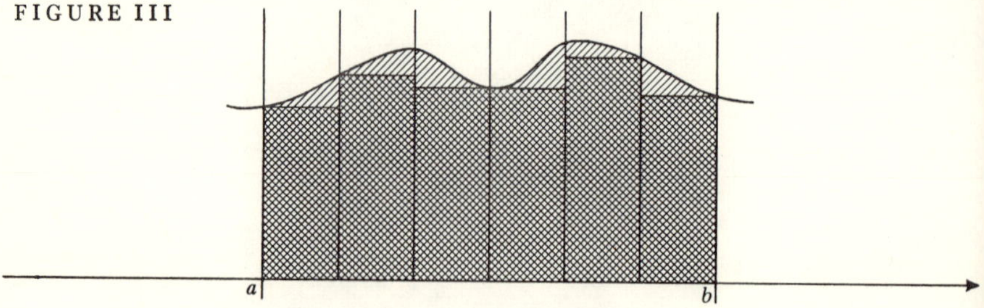

hence of inscribed rectangles). For instance, if we consider six subdivisions, we arrive at the approximation in Figure III.

On the model of (7) above, and with the aid of 'Σ', we can express in general the area of the *inscribed approximation* resulting from $n$ subdivisions; it will be the number

$$\sum_{i}^{n}\left(\frac{b-a}{n}\cdot \inf_{y} \bigvee x\left[A(x)=y \wedge a+i\cdot\frac{b-a}{n}\leqslant x \wedge x\leqslant a+(i+1)\cdot\frac{b-a}{n}\right]\right).$$

In a completely analogous way, we can obtain a family of *circumscribed* rectilinear approximations to the desired area. For instance, in the case of four subdivisions, we obtain the dotted area below.

FIGURE IV

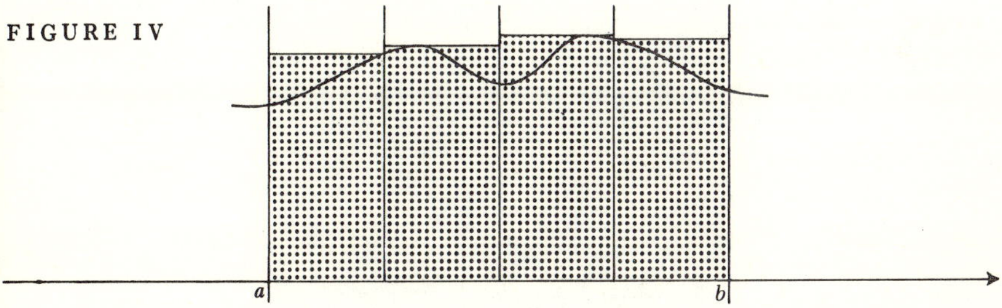


The general expression for the area of the circumscribed approximation resulting from $n$ subdivisions is

$$\sum_{i}^{n}\left(\frac{b-a}{n}\cdot \sup_{y} \bigvee x\left[A(x)=y \wedge a+i\cdot\frac{b-a}{n}\leqslant x \wedge x\leqslant a+(i+1)\cdot\frac{b-a}{n}\right]\right).$$

It is intuitively clear that the shaded area, if it exists at all, is at least equal to each of its inscribed approximations and at most equal to each of its circumscribed approximations. Thus, if $z$ is the desired area, we have:

(8) $$\bigwedge n\left[N(n) \wedge 1\leqslant n \rightarrow \sum_{i}^{n}\left(\frac{b-a}{n}\cdot \inf_{y} \bigvee x\left[A(x)=y \wedge a+i\cdot\frac{b-a}{n}\leqslant x \wedge x\leqslant a+(i+1)\cdot\frac{b-a}{n}\right]\right)\leqslant z \wedge\right.$$

$$z \leqslant \sum_{i}^{n} \left( \frac{b-a}{n} \cdot \sup_{y} \bigvee x \left[ \mathrm{A}(x) = y \wedge a + i \cdot \frac{b-a}{n} \leqslant x \wedge x \leqslant a + (i+1) \cdot \frac{b-a}{n} \right] \right) \Big]$$

Now it turns out that in case A($x$) is continuous at every number between $a$ and $b$ (and even in certain additional cases) there will be exactly one number $z$ satisfying (8). The shaded area can then be defined as this number; accordingly, we arrive at the following characterization of the *integral of* A($x$) *between a and b*, which we identify with this area.

(9) $$\int_a^b \mathrm{A}(x)\,\mathrm{d}x = \daleth z \bigwedge n \left[ \mathrm{N}(n) \wedge 1 \leqslant n \rightarrow \sum_{i}^{n} \left( \frac{b-a}{n} \cdot \inf_{y} \bigvee x \left[ \mathrm{A}(x) = y \wedge a + i \cdot \frac{b-a}{n} \leqslant x \wedge x \leqslant a + (i+1) \cdot \frac{b-a}{n} \right] \right) \leqslant z \wedge z \leqslant \sum_{i}^{n} \left( \frac{b-a}{n} \cdot \sup_{y} \bigvee x \left[ \mathrm{A}(x) = y \wedge a + i \cdot \frac{b-a}{n} \leqslant x \wedge x \leqslant a + (i+1) \cdot \frac{b-a}{n} \right] \right) \right]$$

Here we regard the composite symbol

$$\int \quad \mathrm{d}$$

as an operator of degree $\langle 0, 1, 3, 0 \rangle$; on the left side of (9), '$a$' and '$b$' are free, and '$x$' is bound. As usual, (9) can be replaced by a definitional schema in standard form; '$a$' and '$b$' would be replaced by, say, 'B($x$)' and 'C($x$)'. We should thus arrive at a definitional extension of $T_8$, and hence of integral calculus, in which the usual Riemann integral is available.

4. * **Historical remarks.** The metamathematical notions of this and the last chapter—*definition*, *theory*, *definitional extension*, and the like—constitute an extension of the treatment in Tarski, Mostowski, Robinson [1] beyond the domain of theories with standard formalization, in particular, to theories containing variable-binding operators. The present treatment was developed in collaboration with Professor Dana Scott, and will also appear (along with its model-theoretic rationale) in Montague, Scott, Tarski [1].

For a discussion of fields and ordered fields, see, for instance, van der Waerden [1].

The Continuity Schema is due to Dedekind [1], and the present formulation of the theory of real numbers to Montague [2]. The theory of *real closed fields* (see Tarski [3]) is equivalent to the theory axiomatized by our $A1 - A15$, together with $AS19$. The addition of the predicate 'I' (denoting the set of integers), together with axioms governing it, makes a significant difference: as Tarski has shown, there is an automatic procedure for determining whether a formula is provable in the theory of real closed fields; but this is not true of the theory of real numbers, in view of Church [1]. A theory closely related to the integral calculus of the foregoing section is presented informally, without emphasis on its logical basis, in Landau [1].

It should not be supposed that, because the Principle of Archimedes ($T54$) is provable in the theory of real numbers, every model of this theory is what algebraists call an Archimedean field; indeed, A. Robinson has shown in [1] that the class of Archimedean fields cannot be characterized by any first-order theory (that is, any theory having the logical basis described in this book).

The definition of the integral given in the foregoing section, together with its intuitive justification, was strongly influenced by a conversation with Dr. J. D. Halpern.

## 5. Appendix: list of constants used in chapters VIII and IX.

| LOGICAL CONSTANTS | DEGREE | INTRODUCED ON |
|---|---|---|
| $\sim$ | $\langle 1, 0, 0, 1 \rangle$ | p. 4 |
| $\rightarrow$ | $\langle 1, 0, 0, 2 \rangle$ | p. 4 |
| $\wedge$ | $\langle 1, 0, 0, 2 \rangle$ | p. 39 |
| $\vee$ | $\langle 1, 0, 0, 2 \rangle$ | p. 39 |
| $\leftrightarrow$ | $\langle 1, 0, 0, 2 \rangle$ | p. 39 |
| $\bigwedge$ | $\langle 1, 1, 0, 1 \rangle$ | p. 86 |
| $\bigvee$ | $\langle 1, 1, 0, 1 \rangle$ | p. 86 |
| $=$ | $\langle 1, 0, 2, 0 \rangle$ | p. 213 |
| $\urcorner$ | $\langle 0, 1, 0, 1 \rangle$ | p. 233 |

| OPERATION AND PREDICATE LETTERS | | |
|---|---|---|
| A | $\langle 0, 0, 0, 0 \rangle$ | p. 309 |
| B | $\langle 0, 0, 0, 0 \rangle$ | p. 309 |
| C | $\langle 0, 0, 1, 0 \rangle$ | p. 313 |
| $C_1$ | $\langle 0, 0, 1, 0 \rangle$ | p. 313 |
| D | $\langle 0, 0, 1, 0 \rangle$ | p. 313 |
| $D_1$ | $\langle 0, 0, 1, 0 \rangle$ | p. 313 |
| E | $\langle 0, 0, 1, 0 \rangle$ | p. 313 |
| F | $\langle 1, 0, 1, 0 \rangle$ | p. 287 |

| OPERATION SYMBOLS | DEGREE | INTRODUCED ON |
|---|---|---|
| $+$ | $\langle 0, 0, 2, 0 \rangle$ | p. 279 |
| $-$ | $\langle 0, 0, 1, 0 \rangle$ | p. 279 |
| $0$ | $\langle 0, 0, 0, 0 \rangle$ | p. 279 |
| $\cdot$ | $\langle 0, 0, 2, 0 \rangle$ | p. 279 |
| $^{-1}$ | $\langle 0, 0, 1, 0 \rangle$ | p. 279 |
| $1$ | $\langle 0, 0, 0, 0 \rangle$ | p. 279 |
| $2$ | $\langle 0, 0, 0, 0 \rangle$ | p. 296 |
| $^{2}$ | $\langle 0, 0, 1, 0 \rangle$ | p. 296 |
| $\mid\ \mid$ | $\langle 0, 0, 1, 0 \rangle$ | p. 296 |
| $—$ | $\langle 0, 0, 2, 0 \rangle$ | p. 296 |
| $-$ | $\langle 0, 0, 2, 0 \rangle$ | p. 296 |
| PREDICATES | | |
| $\leqslant$ | $\langle 1, 0, 2, 0 \rangle$ | p. 279 |
| I | $\langle 1, 0, 1, 0 \rangle$ | p. 286 |
| $\neq$ | $\langle 1, 0, 2, 0 \rangle$ | p. 296 |
| $<$ | $\langle 1, 0, 2, 0 \rangle$ | p. 296 |
| N | $\langle 1, 0, 1, 0 \rangle$ | p. 301 |
| R | $\langle 1, 0, 1, 0 \rangle$ | p. 301 |
| VARIABLE-BINDING OPERATORS | | |
| $\Leftrightarrow$ | $\langle 1, 1, 2, 0 \rangle$ | p. 310 |
| lim | $\langle 0, 1, 1, 0 \rangle$ | p. 324 |
| $\lim\limits_{\rightarrow}$ | $\langle 0, 1, 2, 0 \rangle$ | p. 325 |
| Cont | $\langle 1, 1, 2, 0 \rangle$ | p. 325 |
| Av | $\langle 0, 1, 3, 0 \rangle$ | p. 326 |
| Der | $\langle 0, 1, 2, 0 \rangle$ | p. 326 |
| $\Sigma$ | $\langle 0, 1, 2, 0 \rangle$ | p. 326 |
| sup | $\langle 0, 1, 0, 1 \rangle$ | p. 327 |
| inf | $\langle 0, 1, 0, 1 \rangle$ | p. 327 |
| $\int \mathrm{d}$ | $\langle 0, 1, 3, 0 \rangle$ | p. 330 |

## 6. Appendix: list of axioms, definitions, and theorems of chapters VIII and IX.

AXIOMS:

*A1* $x + (y + z) = (x + y) + z$

*A2* $x + y = y + x$

*A3* $x + 0 = x$

*A*4 $x + -x = 0$

*A*5 $x \cdot (y \cdot z) = (x \cdot y) \cdot z$

*A*6 $x \cdot y = y \cdot x$

*A*7 $x \cdot 1 = x$

*A*8 $\sim x = 0 \rightarrow x \cdot x^{-1} = 1$

*A*9 $x \cdot (y + z) = (x \cdot y) + (x \cdot z)$

*A*10 $\sim 0 = 1$

*A*11 $0 \leqslant x \vee 0 \leqslant -x$

*A*12 $\sim x = 0 \rightarrow \sim 0 \leqslant x \vee \sim 0 \leqslant -x$

*A*13 $0 \leqslant x \wedge 0 \leqslant y \rightarrow 0 \leqslant x + y$

*A*14 $0 \leqslant x \wedge 0 \leqslant y \rightarrow 0 \leqslant x \cdot y$

*A*15 $x \leqslant y \leftrightarrow 0 \leqslant y + -x$

*A*16 $\mathrm{I}(0)$

*A*17 $\mathrm{I}(x) \rightarrow \mathrm{I}(x + 1) \wedge \mathrm{I}(x + -1)$

*A*18 $\mathrm{I}(x) \wedge \mathrm{I}(y) \wedge x \leqslant y \wedge y \leqslant x + 1 \rightarrow y = x \vee y = x + 1$

*AS*19 $\bigvee x\mathrm{F}(x) \wedge \bigvee y \bigwedge x(\mathrm{F}(x) \rightarrow x \leqslant y) \rightarrow$
$\bigvee z[\bigwedge x(\mathrm{F}(x) \rightarrow x \leqslant z) \wedge \bigwedge y(\bigwedge x[\mathrm{F}(x) \rightarrow x \leqslant y] \rightarrow z \leqslant y)]$

*AS*20 $\sum_{i}^{0} \mathrm{A}(i) = 0$

*AS*21 $\mathrm{N}(n) \rightarrow \sum_{i}^{n+1} \mathrm{A}(i) = \left[\sum_{i}^{n} \mathrm{A}(i)\right] + \mathrm{A}(n)$

DEFINITIONS (in the form given in chapter IX):

*DS*1 $\mathrm{A} \neq \mathrm{B} \leftrightarrow \sim \mathrm{A} = \mathrm{B}$

*DS*2 $\mathrm{A} < \mathrm{B} \leftrightarrow \mathrm{A} \leqslant \mathrm{B} \wedge \sim \mathrm{A} = \mathrm{B}$

*DS*3 $2 = 1 + 1$

*DS*4 $\mathrm{A}^2 = \mathrm{A} \cdot \mathrm{A}$

*DS*5 $|\mathrm{A}| = \daleth z[(0 \leqslant \mathrm{A} \wedge z = \mathrm{A}) \vee (\sim 0 \leqslant \mathrm{A} \wedge z = -\mathrm{A})]$

*DS*6 $\mathrm{A} - \mathrm{B} = \mathrm{A} + -\mathrm{B}$

*DS*7 $\frac{\mathrm{A}}{\mathrm{B}} = \mathrm{A} \cdot \mathrm{B}^{-1}$

*DS*8 $\mathrm{N}(\mathrm{A}) \leftrightarrow \mathrm{I}(\mathrm{A}) \wedge 0 \leqslant \mathrm{A}$

*DS*9 $\mathrm{R}(\mathrm{A}) \leftrightarrow \bigvee y \bigvee z[\mathrm{I}(y) \wedge \mathrm{I}(z) \wedge 0 \leqslant z \wedge \sim 0 = z \wedge \mathrm{A} = y \cdot z^{-1}]$

*DS*10 $$A(n) \underset{n}{\Leftrightarrow} B(n) \leftrightarrow \bigwedge z(0 < z \rightarrow \bigvee k \bigwedge n[N(n) \wedge k < n \rightarrow |A(n) - B(n)| < z])$$

*DS*11 $$\lim_{n} A(n) = \iota x\, A(n) \underset{n}{\Leftrightarrow} x$$

*DS*12 $$\lim_{x \to B(x)} A(x) = \iota u \bigwedge z(0 < z \rightarrow \bigvee y[0 < y \wedge \bigwedge x(0 < |x - B(x)| \wedge |x - B(x)| < y \rightarrow |A(x) - u| < z)])$$

*DS*13 $$\underset{x}{\mathrm{Cont}}\, [A(x), B(x)] \leftrightarrow \bigwedge z(0 < z \rightarrow \bigvee y[0 < y \wedge \bigwedge x(0 < |x - B(x)| \wedge |x - B(x)| < y \rightarrow |A(x) - A(B(x))| < z)])$$

*DS*14 $$\underset{x}{\mathrm{Av}}\, [A(x), B(x), C(x)] = \frac{A(C(0)) - A(B(0))}{C(0) - B(0)}$$

*DS*15 $$\underset{x}{\mathrm{Der}}\, [A(x), B(x)] = \lim_{h \to 0} [\underset{x}{\mathrm{Av}}\, [A(x), B(x) - h, B(x) + h]]$$

*DS*16 $$\sup_{x} F(x) = \iota z[\bigwedge x(F(x) \rightarrow x \leqslant z) \wedge \bigwedge y(\bigwedge x[F(x) \rightarrow x \leqslant y] \rightarrow z \leqslant y)]$$

*DS*17 $$\inf_{x} F(x) = \iota z[\bigwedge x(F(x) \rightarrow z \leqslant x) \wedge \bigwedge y(\bigwedge x[F(x) \rightarrow y \leqslant x] \rightarrow y \leqslant z)]$$

*DS*18 $$\int_{B(x)}^{C(x)} A(x)\, dx = \iota z \bigwedge n\Big[N(n) \wedge 1 \leqslant n \rightarrow \sum_{i}^{n} \Big(\frac{C(0) - B(0)}{n} \cdot \inf_{y} \bigvee x\Big[A(x) = y \wedge B(0) + i \cdot \frac{C(0) - B(0)}{n} \leqslant x \wedge x \leqslant B(0) + (i + 1) \cdot \frac{C(0) - B(0)}{n}\Big]\Big) \leqslant z \wedge z \leqslant \sum_{i}^{n} \Big(\frac{C(0) - B(0)}{n} \cdot \sup_{y} \bigvee x\Big[A(x) = y \wedge B(0) + i \cdot \frac{C(0) - B(0)}{n} \leqslant x \wedge x \leqslant B(0) + (i + 1) \cdot \frac{C(0) - B(0)}{n}\Big]\Big)\Big]$$

THEOREMS:

*T*1 $x + z = y + z \rightarrow x = y$

*T*2 $z + x = z + y \rightarrow x = y$

*T*3 $\sim z = 0 \wedge x \cdot z = y \cdot z \rightarrow x = y$

*T*4 $\sim z = 0 \wedge z \cdot x = z \cdot y \rightarrow x = y$

*T*5 $-0 = 0$

*T*6 $x \cdot 0 = 0$

*T*7 $x + -y = 0 \leftrightarrow x = y$

*T*8 $x = -y \leftrightarrow y = -x$

*T*9 $x = 0 \leftrightarrow -x = 0$

*T*10 $--x = x$

*T*11 $-(x + y) = -x + -y$

*T*12 $-x \cdot y = -(x \cdot y)$

*T*13 $x \cdot -y = -(x \cdot y)$

*T*14 $-x \cdot -y = x \cdot y$

*T*15 $\sim x = 0 \rightarrow (-x)^{-1} = -(x^{-1})$

*T*16 $x \cdot y = 0 \leftrightarrow x = 0 \vee y = 0$

*T*17 $\sim x \cdot y = 0 \rightarrow (x \cdot y)^{-1} = x^{-1} \cdot y^{-1}$

*T*18 $x \leqslant x$

*T*19 $x \leqslant y \wedge y \leqslant x \rightarrow x = y$

*T*20 $x \leqslant y \wedge y \leqslant z \rightarrow x \leqslant z$

*T*21 $x \leqslant y \vee y \leqslant x$

*T*22 $0 \leqslant x \cdot x$

*T*23 $0 \leqslant 1$

*T*24 $x \leqslant y \rightarrow x + z \leqslant y + z$

*T*25 $x + z \leqslant y + z \rightarrow x \leqslant y$

*T*26 $x \leqslant y \wedge z \leqslant w \rightarrow x + z \leqslant y + w$

*T*27 $\sim y + 1 \leqslant y$

*T*28 $\sim y \leqslant y + -1$

*T*29 $x \leqslant y \rightarrow -y \leqslant -x$

*T*30 $-y \leqslant -x \rightarrow x \leqslant y$

*T*31 $x \leqslant 0 \leftrightarrow 0 \leqslant -x$

*T*32 $0 \leqslant x \wedge y \leqslant 0 \rightarrow x \cdot y \leqslant 0$

*T*33 $x \leqslant y \wedge 0 \leqslant z \rightarrow x \cdot z \leqslant y \cdot z$

*T*34 $x \leqslant y \wedge z \leqslant 0 \rightarrow y \cdot z \leqslant x \cdot z$

*T*35 $0 \leqslant z \wedge \sim z = 0 \rightarrow 0 \leqslant z^{-1}$

*T*36 $x \cdot z \leqslant y \cdot z \wedge 0 \leqslant z \wedge \sim z = 0 \rightarrow x \leqslant y$

*T*37 $y \cdot z \leqslant x \cdot z \wedge z \leqslant 0 \wedge \sim z = 0 \rightarrow x \leqslant y$

*T*38 $x = y \rightarrow x \leqslant y$

*T*39 $\mathrm{I}(y) \wedge 0 \leqslant y \wedge \sim y = 0 \rightarrow 1 \leqslant y$

*T*40 $I(y) \wedge y \leqslant 0 \wedge \sim y = 0 \rightarrow y \leqslant -1$

*TS*41 $\bigvee x F(x) \rightarrow \bigvee x F(-x)$

*TS*42 $\bigvee x F(x) \wedge \bigvee y \bigwedge x(F(x) \rightarrow y \leqslant x) \rightarrow \bigvee z[\bigwedge x(F(x) \rightarrow z \leqslant x) \wedge \bigwedge y(\bigwedge x[F(x) \rightarrow y \leqslant x] \rightarrow y \leqslant z)]$

*TS*43 $\bigwedge x[F(x) \rightarrow I(x)] \wedge \bigwedge x(F(x) \rightarrow y \leqslant x) \wedge \bigwedge z(\bigwedge x[F(x) \rightarrow z \leqslant x] \rightarrow z \leqslant y) \rightarrow F(y)$

*TS*44 $\bigwedge x[F(x) \rightarrow I(x)] \wedge \bigwedge x(F(x) \rightarrow x \leqslant y) \wedge \bigwedge z(\bigwedge x[F(x) \rightarrow x \leqslant z] \rightarrow y \leqslant z) \rightarrow F(y)$

*TS*45 $F(0) \wedge \bigwedge x[I(x) \wedge F(x) \rightarrow F(x+1) \wedge F(x+-1)] \rightarrow \bigwedge x[I(x) \rightarrow F(x)]$

*T*46 $I(x) \rightarrow I(-x)$

*T*47 $I(x) \wedge I(y) \rightarrow I(x+y)$

*T*48 $I(x) \wedge I(y) \rightarrow I(x \cdot y)$

*TS*49 $\bigvee x[I(x) \wedge F(x)] \wedge \bigwedge x[I(x) \wedge F(x) \rightarrow F(x+1) \wedge F(x+-1)] \rightarrow \bigwedge x[I(x) \rightarrow F(x)]$

*T*50 $I(x) \rightarrow \bigvee y(I(y) \wedge x = y + y) \vee \bigvee y(I(y) \wedge x = y + y + 1)$

*T*51 $I(x) \rightarrow \sim \bigvee y(I(y) \wedge x + x = y + y + 1)$

*T*52 $I(x) \rightarrow [\bigvee y(I(y) \wedge x = y + y) \leftrightarrow \sim \bigvee y(I(y) \wedge x = y + y + 1)]$

*T*53 $0 \leqslant x \wedge \sim x = 0 \rightarrow \bigvee n(I(n) \wedge y \leqslant n \cdot x)$

*T*54 $\sim x = 0 \rightarrow \bigvee n(I(n) \wedge y \leqslant n \cdot x)$

*T*55 $I(n) \wedge I(m) \wedge n < m \rightarrow n + 1 \leqslant m$

*T*56 $x - x = 0$

*T*57 $-(x - y) = y - x$

*T*58 $x - y = (z - y) - (z - x)$

*T*59 $(x - y) + (z - w) = (x + z) - (y + w)$

*T*60 $(x + y)^2 = x^2 + 2xy + y^2$

*T*61 $1^2 = 1$

*T*62 $(x - y)^2 = x^2 - 2xy + y^2$

*T*63 $(x \cdot y)^2 = x^2 \cdot y^2$

*T*64 $y \neq 0 \rightarrow \left(\frac{x}{y}\right)^2 = \frac{x^2}{y^2}$

*T*65 $\frac{x}{2} + \frac{x}{2} = x$

*T*66 $\frac{0}{x} = 0$

*T*67 $\frac{x + y}{z} = \frac{x}{z} + \frac{y}{z}$

*T*68 $xz \neq 0 \rightarrow \frac{xy}{xz} = \frac{y}{z}$

*T*69 $y \neq 0 \rightarrow -\left(\frac{x}{y}\right) = \frac{-x}{y} \wedge -\left(\frac{x}{y}\right) = \frac{x}{-y}$

*T*70 $0 \leqslant x \wedge x \leqslant 1 \rightarrow x^2 \leqslant x$

*T*71 $0 \leqslant x^2$

*T*72 $x \neq 0 \rightarrow 0 < x^2$

*T*73 $x \leqslant x + y^2$

*T*74 $0 < 1$

*T*75 $1 < 2$

*T*76 $\sim x < x$

*T*77 $x < y \wedge y < z \rightarrow x < z$

*T*78 $x \leqslant y \wedge y < z \rightarrow x < z$

*T*79 $x < y \wedge y \leqslant z \rightarrow x < z$

*T*80 $x \leqslant y \leftrightarrow \sim y < x$

*T*81 $x \leqslant y \leftrightarrow x < y \vee x = y$

*T*82 $x < y \leftrightarrow x + z < y + z$

*T*83 $x < y \wedge z < w \rightarrow x + z < y + w$

*T*84 $0 < x \rightarrow x < 2x$

*T*85 $x < y \leftrightarrow -y < -x$

*T*86 $x < 0 \leftrightarrow 0 < -x$

*T*87 $0 < z \rightarrow 0 < z^{-1}$

*T*88 $z < 0 \rightarrow z^{-1} < 0$

*T*89 $0 < z \rightarrow [x < y \leftrightarrow xz < yz]$

*T*90 $z < 0 \rightarrow [x < y \leftrightarrow yz < xz]$

*T*91 $0 < x \wedge 0 < y \rightarrow [x \leqslant y \leftrightarrow y^{-1} \leqslant x^{-1}]$

*T*92 $x < 0 \wedge y < 0 \rightarrow [x \leqslant y \leftrightarrow y^{-1} \leqslant x^{-1}]$

*T*93 $0 < x \wedge 0 < y \rightarrow [x < y \leftrightarrow y^{-1} < x^{-1}]$

*T*94 $x < 0 \wedge y < 0 \rightarrow [x < y \leftrightarrow y^{-1} < x^{-1}]$

| | |
|---|---|
| *T*95 | $0 < x \wedge x \leqslant y \wedge 0 \leqslant z \to \frac{z}{y} \leqslant \frac{z}{x}$ |
| *T*96 | $0 < x \wedge x < y \wedge 0 < z \to \frac{z}{y} < \frac{z}{x}$ |
| *T*97 | $x \leqslant y \wedge 0 < z \to \frac{x}{z} \leqslant \frac{y}{z}$ |
| *T*98 | $x < y \wedge 0 < z \to \frac{x}{z} < \frac{y}{z}$ |
| *T*99 | $0 < x \wedge x < y \to 0 < \frac{x}{y} \wedge \frac{x}{y} < 1$ |
| *T*100 | $0 \leqslant y \wedge x^2 \leqslant y^2 \to x \leqslant y$ |
| *T*101 | $0 \leqslant x \wedge 0 < y \to 0 \leqslant \frac{x}{y}$ |
| *T*102 | $0 < x \wedge 0 < y \to 0 < \frac{x}{y}$ |
| *T*103 | $\bigvee t \bigwedge z[(0 \leqslant x \wedge z = x) \vee (\sim 0 \leqslant x \wedge z = -x) \leftrightarrow z = t]$ |
| *T*104 | $0 \leqslant x \to \lvert x\rvert = x$ |
| *T*105 | $x < 0 \to \lvert x\rvert = -x$ |
| *T*106 | $0 \leqslant \lvert x\rvert$ |
| *T*107 | $\lvert x\rvert = 0 \leftrightarrow x = 0$ |
| *T*108 | $x \leqslant \lvert x\rvert$ |
| *T*109 | $-x \leqslant \lvert x\rvert$ |
| *T*110 | $\lvert -x\rvert = \lvert x\rvert$ |
| *T*111 | $\lvert x\rvert \leqslant y \leftrightarrow x \leqslant y \wedge -x \leqslant y$ |
| *T*112 | $\lvert x + y\rvert \leqslant \lvert x\rvert + \lvert y\rvert$ |
| *T*113 | $\lvert x - y\rvert \leqslant \lvert x - z\rvert + \lvert z - y\rvert$ |
| *T*114 | $\lvert x - y\rvert \leqslant \lvert x\rvert + \lvert y\rvert$ |
| *T*115 | $\lvert x - y\rvert = \lvert y - x\rvert$ |
| *T*116 | $\lvert x \cdot y\rvert = \lvert x\rvert \cdot \lvert y\rvert$ |
| *T*117 | $y \neq 0 \to \left\lvert\frac{x}{y}\right\rvert = \frac{\lvert x\rvert}{\lvert y\rvert}$ |
| *T*118 | $\lvert x - y\rvert < z \leftrightarrow x - z < y \wedge y < x + z$ |
| *TS*119 | $\bigvee x\mathrm{F}(x) \wedge \bigwedge x[\mathrm{F}(x) \to \mathrm{N}(x)] \to \bigvee x[\mathrm{F}(x) \wedge \bigwedge y(\mathrm{F}(y) \to x \leqslant y])$ |

*TS*120 $\bigwedge x[N(x) \wedge \bigwedge y(N(y) \wedge y < x \rightarrow F(y)) \rightarrow F(x)] \rightarrow \bigwedge x[N(x) \rightarrow F(x)]$

*TS*121 $F(0) \wedge \bigwedge x[N(x) \wedge F(x) \rightarrow F(x+1)] \rightarrow \bigwedge x[N(x) \rightarrow F(x)]$

*T*122 $N(0)$

*T*123 $N(x) \rightarrow N(x+1)$

*T*124 $N(x) \wedge N(y) \wedge x + 1 = y + 1 \rightarrow x = y$

*T*125 $N(x) \rightarrow x + 1 \neq 0$

*T*126 $N(x) \wedge N(y) \rightarrow N(x+y) \wedge N(x \cdot y)$

*T*127 $I(x) \wedge \bigvee y(I(y)) \wedge x^2 = 2y) \rightarrow \bigvee y(I(y) \wedge x = 2y)$

*T*128 $R(x) \rightarrow x^2 \neq 2$

*T*129 $0 < x \rightarrow \bigvee y\, x = y^2$

*T*130 $0 \leqslant x \leftrightarrow \bigvee y\, x = y^2$

*T*131 $x \leqslant y \leftrightarrow \bigvee z\, x + z^2 = y$

*T*132 $\dfrac{n^2+1}{n} \underset{n}{\Leftrightarrow} n$

*T*133 $\sim n + 1 \underset{n}{\Leftrightarrow} n$

*TS*134 $C(n) \underset{n}{\Leftrightarrow} D(n) \leftrightarrow C(m) \underset{m}{\Leftrightarrow} D(m)$

*TS*135 $\bigwedge n\, C(n) = C_1(n) \wedge \bigwedge n\, D(n) = D_1(n) \rightarrow [C(n) \underset{n}{\Leftrightarrow} D(n) \leftrightarrow C_1(n) \underset{n}{\Leftrightarrow} D_1(n)]$

*TS*136 $C(n) \underset{n}{\Leftrightarrow} C(n)$

*TS*137 $C(n) \underset{n}{\Leftrightarrow} D(n) \rightarrow D(n) \underset{n}{\Leftrightarrow} C(n)$

*TS*138 $C(n) \underset{n}{\Leftrightarrow} D(n) \wedge D(n) \underset{n}{\Leftrightarrow} E(n) \rightarrow C(n) \underset{n}{\Leftrightarrow} E(n)$

*TS*139 $C(n) \underset{n}{\Leftrightarrow} C_1(n) \wedge D(n) \underset{n}{\Leftrightarrow} D_1(n) \rightarrow C(n) + D(n) \underset{n}{\Leftrightarrow} C_1(n) + D_1(n)$

*TS*140 $\bigvee k \bigwedge n[N(n) \wedge k < n \rightarrow C(n) = D(n)] \rightarrow C(n) \underset{n}{\Leftrightarrow} D(n)$

*TS*141 $\bigwedge n \bigwedge m[N(n) \wedge N(m) \wedge n < m \rightarrow E(n) < E(m)] \wedge \bigwedge n[N(n) \rightarrow N(E(n))] \rightarrow \bigwedge n[N(n) \rightarrow n \leqslant E(n)]$

*TS*142 $[C(n) \underset{n}{\Leftrightarrow} D(n)] \wedge \bigwedge n \bigwedge m[N(n) \wedge N(m) \wedge n < m \rightarrow E(n) < E(m)] \wedge \bigwedge n[N(n) \rightarrow N(E(n))] \rightarrow [C(E(n)) \underset{n}{\Leftrightarrow} D(E(n))]$

*TS*143 $C(n) \underset{n}{\Leftrightarrow} x \leftrightarrow \bigwedge z(0 < z \rightarrow \bigvee k \bigwedge n[N(n) \wedge k < n \rightarrow |C(n) - x| < z])$

*T*144 $\frac{n+1}{n} \underset{n}{\Leftrightarrow} 1$

*T*145 $\frac{1}{n} \underset{n}{\Leftrightarrow} 0$

*T*146 $x \underset{n}{\Leftrightarrow} y \rightarrow x = y$

*TS*147 $C(n) \underset{n}{\Leftrightarrow} x \wedge C(n) \underset{n}{\Leftrightarrow} y \rightarrow x = y$

*T*148 $\sim \bigvee x \; n \underset{n}{\Leftrightarrow} x$

*T*149 $\frac{n(n+1)}{n^2} \underset{n}{\Leftrightarrow} 1$

*TS*150 $C(n) \underset{n}{\Leftrightarrow} x \wedge D(n) \underset{n}{\Leftrightarrow} y \rightarrow C(n) \cdot D(n) \underset{n}{\Leftrightarrow} x \cdot y$

*TS*151 $C(n) \underset{n}{\Leftrightarrow} x \wedge D(n) \underset{n}{\Leftrightarrow} y \wedge \bigvee k \bigwedge n[N(n) \wedge k < n \rightarrow C(n) \leqslant D(n)] \rightarrow x \leqslant y$

*TS*152 $C(n) \underset{n}{\Leftrightarrow} D(n) \leftrightarrow C(n) - D(n) \underset{n}{\Leftrightarrow} 0$

*TS*153 $\bigwedge n \bigwedge m[N(n) \wedge N(m) \wedge n \leqslant m \rightarrow C(n) \leqslant C(m)] \wedge \bigvee y \bigwedge n[N(n) \rightarrow C(n) \leqslant y] \rightarrow \bigvee z \; C(n) \underset{n}{\Leftrightarrow} z$

*TS*154 $\bigwedge n \bigwedge m[N(n) \wedge N(m) \wedge n \leqslant m \rightarrow C(m) \leqslant C(n)] \wedge \bigvee y \bigwedge n[N(n) \rightarrow y \leqslant C(n)] \rightarrow \bigvee z \; C(n) \underset{n}{\Leftrightarrow} z$

# BIBLIOGRAPHY


ACKERMANN, W.

[1] *Solvable cases of the decision problem*, Amsterdam, 1954.

BEHMANN, H.

[1] Beiträge zur Algebra der Logik, insbesondere zum Entscheidungsproblem, *Mathematische Annalen*, vol. 86 (1922), pp. 163–229.

BERNAYS, P., and SCHÖNFINKEL, M.

[1] Zum Entscheidungsproblem der mathematischen Logik, *Mathematische Annalen*, vol. 99 (1928), pp. 342–72.

BOOLE, G.

[1] *The mathematical analysis of logic*, London and Cambridge, 1847.

[2] *An investigation of the laws of thought*, London, 1854.

CARNAP, R.

[1] *Logische Syntax der Sprache*, Vienna, 1934. English translation: New York, 1937.

[2] *Meaning and necessity*, Chicago, 1947.

CHURCH, A.

[1] An unsolvable problem of elementary number theory, *American Journal of Mathematics*, vol. 58 (1936), pp. 345–63.

[2] A note on the *Entscheidungsproblem*, *Journal of Symbolic Logic*, vol. 1 (1936), pp. 40–41; Correction, *ibid.*, pp. 101–02.

[3] *Introduction to mathematical logic*, vol. 1, Princeton, 1956.

COPI, I. M.

[1] *Symbolic logic*, New York, 1954.

COUTURAT, L.

[1] *La Logique de Leibniz*, Paris, 1901.

[2] *Opuscules et Fragments inédits de Leibniz*, Paris, 1903.

[3] *Les Principes des Mathématiques*, Paris, 1905.

DEDEKIND, R.

[1] *Stetigkeit und irrationale Zahlen*, Braunschweig, 1872. English translation in *Essays on the theory of numbers*, La Salle, Illinois, 1901.

DE MORGAN, A.

[1] *Formal logic*, London, 1847.

EUCLID

[1] Book X, *The thirteen books of Euclid's elements*, translated by Sir T. L. Heath, Cambridge, 1908; second edition, 1926.

FREGE, G.

[1] *Begriffsschrift*, Halle, 1879.

[2] Über Sinn und Bedeutung, *Zeitschrift für Philosophie und Kritik*, vol. 100 (1892), pp. 25 – 50.

[3] *Grundegesetze der Arithmetik*, vol. 1, Jena, 1893.

GENTZEN, G.

[1] Untersuchungen über das logische Schliessen, *Mathematische Zeitschrift*, vol. 39 (1934 – 35), pp. 176 – 210 and 405 – 31.

GÖDEL, K.

[1] Die Vollständigkeit der Axiome des logischen Funktionenkalküls, *Monatshefte für Mathematik und Physik*, vol. 37 (1930), pp. 349 – 60.

[2] Über formal unentscheidbare Sätze der *Principia Mathematica* und verwandter Systeme I, *Monatshefte für Mathematik und Physik*, vol. 38, (1931), pp. 173 - 98.

HERBRAND, J.

[1] Sur la théorie de la démonstration, *Comptes Rendus des Séances de l'Académie des Sciences*, vol. 186 (Paris, 1928), pp. 1274 – 76.

[2] *Recherches sur la théorie de la démonstration* (Travaux de la Société des Sciences et des Lettres de Varsovie, Classe III, No. 33, 1930, 128 pp.).

HILBERT, D., and ACKERMANN, W.

[1] *Grundzüge der theoretischen Logik*, Berlin, 1928.

[2] *Grundzüge der theoretischen Logik*, second edition, Berlin, 1938.

[3] *Grundzüge der theoretischen Logik*, third edition, Berlin, 1949.

[4] *Grundzüge der theoretischen Logik*, fourth edition, Berlin, 1959.

HILBERT, D., and BERNAYS, P.

[1] *Grundlagen der Mathematik*, vol. 1, Berlin, 1934.

JAŚKOWSKI, S.

[1] On the rules of suppositions in formal logic, *Studia Logica*, no. 1 (Warsaw, 1934).

KALMÁR, L.

[1] Über die Axiomatisierbarkeit des Aussagenkalküls, *Acta Scientiarum Mathematicarum*, vol. 7, 1934 – 35, pp. 222 – 43.

KLEENE, S. C.

[1] *Introduction to metamathematics*, Princeton, 1952.

LANDAU, E.

[1] *Differential and integral calculus*, New York, 1951.

LEIBNIZ, G. W.

[1] Scientia Generalis. Characteristica XIX and XX, in Gerhardt's *Die Philosophischen Schriften von G. W. Leibniz*, vol. 7 (1890), pp. 228 – 35. English translation in *A survey of symbolic logic*, by C. I. Lewis, Berkeley, 1918, pp. 373 – 79 and New York, 1960, pp. 291 – 97.

ŁUKASIEWICZ, J.

[1] *Elementy logiki matematycznej* (Elements of Mathematical Logic), Warsaw, 1929.

[2] Zur Geschichte der Aussagenlogik, *Erkenntnis*, vol. 5 (1935 – 36), pp. 111 – 31.

LUKASIEWICZ, J., and TARSKI, A.

[1] Untersuchungen über den Aussagenkalkül, *Comptes Rendus des Séances de la Société des Sciences et des Lettres de Varsovie*, Classe III, vol. 23 (1930), pp. 30 – 50. English translation: Tarski [4], Article IV.

MACCOLL, H.

[1] The calculus of equivalent statements and integration limits, *Proceedings of the London Mathematical Society*, vol. 9 (1877 – 78), pp. 9 – 20 and 177 – 86, vol. 10 (1878 – 79), pp. 16 – 28, vol. 11 (1879 – 80), pp. 113 – 21.

MATES, B.

[1] *Stoic logic* (University of California Publication in Philosophy, vol. 26), Berkeley and Los Angeles, 1953 and 1961.

MONTAGUE, R.

[1] On the paradox of grounded classes, *Journal of Symbolic Logic*, vol. 20 (1955), p. 140.

[2] Semantical closure and non-finite axiomatizability I, in *Infinitistic Methods*, Proceedings of the Symposium on Foundations of Mathematics, Warsaw, 1959, pp. 45 – 69.

[3] Deterministic theories, in *Decisions, values, and groups*, vol. 2, Oxford, 1963.

MONTAGUE, R., and KALISH, D.

[1] Remarks on descriptions and natural deduction, *Archiv für mathematische Logik und Grundlagenforschung*, vol. 3 (1957), pp. 50 – 64; vol. 3 (1957), pp. 65 – 73.

MONTAGUE, R., SCOTT, D., TARSKI, A.

[1] *An axiomatic approach to set theory*, Amsterdam, forthcoming.

PAGER, D.

[1] An emendation of the axiom system of Hilbert and Ackermann for the restricted calculus of predicates, *Journal of Symbolic Logic*, vol. 27 (1962), pp. 131 – 38.

PEANO, G.

[1] Sul concetto di numero, *Revista di matematica*, vol. 1 (1891), pp. 87 – 102, 256 – 67.

PEIRCE, C. S.

[1] *Collected papers of Charles Sanders Peirce*, edited by Hartshorne and Weiss, Cambridge, Mass., 1933; see various papers from 1870 to 1903 in vol. 3.

[2] On the algebra of logic: a contribution to the philosophy of notation, *American Journal of Mathematics*, vol. 7 (1885), pp. 180 – 202.

[3] The logic of relatives, *The Monist*, vol. 7 (1897), pp. 161 – 217.

POST, E. L.

[1] Introduction to a general theory of elementary propositions, *American Journal of Mathematics*, vol. 43 (1921), pp. 163 – 85.

QUINE, W. V.

[1] *A system of logistic*, Cambridge, Mass., 1934.

[2] *Mathematical logic*, New York, 1940; revised edition, Cambridge, Mass. (Harvard University Press), 1951.

[3] *Methods of logic*, New York, 1950; revised edition, 1959.

[4] *From a logical point of view*, Cambridge, Mass. (Harvard University Press), 1953.

[5] A proof procedure for quantification theory, *Journal of Symbolic Logic*, vol. 20 (1955), pp. 141 – 49.

ROBINSON, A.

[1] *On the metamathematics of algebra*, Amsterdam, 1951.

ROSSER, B.

[1] *Logic for mathematicians*, New York, 1953.

RUSSELL, B.

[1] On denoting, *Mind*, vol. 14 (1905), pp. 479 – 93.

[2] The theory of implication, *American Journal of Mathematics*, vol. 28 (1906), pp. 159 – 202.

[3] *Introduction to mathematical philosophy*, London, 1919.

SCHOLZ, H.

[1] *Metaphysik als strenge Wissenschaft*, Cologne, 1941.

SCHRÖDER, E.

[1] *Algebra der Logik*, vol. 1, Leipzig, 1890.

SUPPES, P.

[1] *Introduction to logic*, Princeton, 1957.

TARSKI, A.

[1] Über einige fundamentale Begriffe der Metamathematik, *Comptes Rendus des Séances de la Société des Sciences et des Lettres de Varsovie*, Classe III, vol. 23 (1930), pp. 22 – 29. English translation: Tarski [4], Article III.

[2] *Projęcie prawdy w językach nauk dedukcyjnych* (The concept of truth in the languages of the deductive sciences) (Travaux de la Société des Sciences et des Lettres de Varsovie, Classe III, no. 34, 1933, vii, 116 pp.). German translation in *Studia Philosophica*, vol. 1 (1936), pp. 261 – 405. English translation: Tarski [4], Article VIII.

[3] *A decision method for elementary algebra and geometry*, second edition, Berkeley and Los Angeles, 1951.

[4] *Logic, semantics, metamathematics*, Oxford, 1956.

TARSKI, A., MOSTOWSKI, A., ROBINSON, R. M.

[1] *Undecidable theories*, Amsterdam, 1953.

VAN DER WAERDEN, B. L.

[1] *Modern algebra*, New York, 1949.

WHITEHEAD, A. N., and RUSSELL, B.

[1] *Principia Mathematica*, vol. 1, London, 1910; 2nd edition, 1925.

# INDEX OF PROPER NAMES

# INDEX OF SUBJECTS

C D E 7 F 8 G 9 H 0 I 1 J 2